# Advanced Treatment Techniques for Industrial Wastewater

Athar Hussain
*Chaudhary Brahm Prakash Government Engineering College, India*

Sirajuddin Ahmed
*Jamia Millia Islamia, India*



A volume in the Advances in Environmental Engineering and Green Technologies (AEEGT) Book Series

IGI Global
DISSEMINATOR OF KNOWLEDGE

Published in the United States of America by
IGI Global
Engineering Science Reference (an imprint of IGI Global)
701 E. Chocolate Avenue
Hershey PA, USA 17033
Tel: 717-533-8845
Fax: 717-533-8661
E-mail: cust@igi-global.com
Web site: http://www.igi-global.com



Library of Congress Cataloging-in-Publication Data

Names: Hussain, Athar, 1976- editor. | Ahmed, Sirajuddin, 1967- editor.
Title: Advanced treatment techniques for industrial wastewater / Athar Hussain and Sirajuddin Ahmed, editors.
Description: Hershey PA : Engineering Science Reference (an imprint of IGI Global), [2019] | Includes bibliographical references.
Identifiers: LCCN 2017055219| ISBN 9781522557548 (hardcover) | ISBN 9781522557555 (ebook)
Subjects: LCSH: Sewage disposal. | Factory and trade waste.
Classification: LCC TD741 .A5 2019 | DDC 628.3--dc23 LC record available at https://lccn.loc.gov/2017055219

This book is published in the IGI Global book series Advances in Environmental Engineering and Green Technologies (AEEGT) (ISSN: 2326-9162; eISSN: 2326-9170)

British Cataloguing in Publication Data
A Cataloguing in Publication record for this book is available from the British Library.



For electronic access to this publication, please contact: eresources@igi-global.com.

# Advances in Environmental Engineering and Green Technologies (AEEGT) Book Series

Sang-Bing Tsai
University of Electronic Science and Technology of China
Zhongshan Institute, China
Ming-Lang Tseng
Lunghwa University of Science and Technology, Taiwan
Yuchi Wang
University of Electronic Science and Technology of China
Zhongshan Institute, China

ISSN:2326-9162
EISSN:2326-9170

## Mission

Growing awareness and an increased focus on environmental issues such as climate change, energy use, and loss of non-renewable resources have brought about a greater need for research that provides potential solutions to these problems. Research in environmental science and engineering continues to play a vital role in uncovering new opportunities for a "green" future.

The **Advances in Environmental Engineering and Green Technologies (AEEGT)** book series is a mouthpiece for research in all aspects of environmental science, earth science, and green initiatives. This series supports the ongoing research in this field through publishing books that discuss topics within environmental engineering or that deal with the interdisciplinary field of green technologies.

## Coverage

- Cleantech
- Green Transportation
- Contaminated Site Remediation
- Air Quality
- Industrial Waste Management and Minimization
- Water Supply and Treatment
- Waste Management
- Alternative Power Sources
- Biofilters and Biofiltration
- Pollution Management

IGI Global is currently accepting manuscripts for publication within this series. To submit a proposal for a volume in this series, please contact our Acquisition Editors at Acquisitions@igi-global.com or visit: http://www.igi-global.com/publish/.

The Advances in Environmental Engineering and Green Technologies (AEEGT) Book Series (ISSN 2326-9162) is published by IGI Global, 701 E. Chocolate Avenue, Hershey, PA 17033-1240, USA, www.igi-global.com. This series is composed of titles available for purchase individually; each title is edited to be contextually exclusive from any other title within the series. For pricing and ordering information please visit http://www.igi-global.com/book-series/advances-environmental-engineering-green-technologies/73679. Postmaster: Send all address changes to above address. 

## Titles in this Series

*For a list of additional titles in this series, please visit: www.igi-global.com/book-series*

*Food Systems Sustainability and Environmental Policies in Modern Economies*
Abiodun Elijah Obayelu (Federal University of Agriculture – Abeokuta (FUNAAB), Nigeria)
Engineering Science Reference • copyright 2018 • 371pp • H/C (ISBN: 9781522536314) • US $195.00 (our price)

*Microbial Biotechnology in Environmental Monitoring and Cleanup*
Pankaj (G. B. Pant University of Agriculture and Technology, India) and Anita Sharma (G. B. Pant University of Agriculture and Technology, India)
Engineering Science Reference • copyright 2018 • 427pp • H/C (ISBN: 9781522531265) • US $235.00 (our price)

*Handbook of Research on Renewable Energy and Electric Resources for Sustainable Rural Development*
Valeriy Kharchenko (Federal Scientific Agroengineering Center VIM, Russia) and Pandian Vasant (Universiti Teknologi PETRONAS, Malaysia)
Engineering Science Reference • copyright 2018 • 672pp • H/C (ISBN: 9781522538677) • US $345.00 (our price)

*Green Production Strategies for Sustainability*
Sang-Binge Tsai (University of Electronic Science and Technology of China (Zhongshan Institute), China & Civil Aviation University of China, China) Bin Liu (Shanghai Maritime University, China) and Yongian Li (Nankai University, China)
Engineering Science Reference • copyright 2018 • 325pp • H/C (ISBN: 9781522535379) • US $235.00 (our price)

*Innovative Strategies and Frameworks in Climate Change Adaptation Emerging Research and Opportunities*
Alexander G. Flor (University of the Philippines Open University, Philippines) and Benjamina Gonzalez Flor (University of the Philippines, Philippines)
Engineering Science Reference • copyright 2018 • 165pp • H/C (ISBN: 9781522527671) • US $165.00 (our price)

*Utilizing Innovative Technologies to Address the Public Health Impact of Climate Change Emerging Research and Opportunities*
Debra Weiss-Randall (Florida Atlantic University, USA)
Engineering Science Reference • copyright 2018 • 295pp • H/C (ISBN: 9781522534143) • US $185.00 (our price)

*Economical and Technical Considerations for Solar Tracking Methodologies and Opportunities for Energy Management*
S. Soulayman (Higher Institute for Applied Sciences and Technology, Syria)
Engineering Science Reference • copyright 2018 • 647pp • H/C (ISBN: 9781522529507) • US $245.00 (our price)

IGI Global
DISSEMINATOR OF KNOWLEDGE

701 East Chocolate Avenue, Hershey, PA 17033, USA
Tel: 717-533-8845 x100 • Fax: 717-533-8661
E-Mail: cust@igi-global.com • www.igi-global.com

*Dedicated to my beloved parents who are the backbone of what I am today.*

# Table of Contents

# Detailed Table of Contents



The chapter gives an idea about water as a life-sustaining medium and the sources of its pollution along with the deteriorating effects of over burdening of natural resources and the effect of various heavy metal ions discharged into the water bodies on human health and wellbeing. Several human diseases and disorders that are caused due to intake of water polluted by toxic heavy metal ions are also listed. The need and urgency in determination and removal of heavy metal ions from the water sources in order to release load on aquifers by making water safe for reuse is emphasized. The procedure and advantages of composite cation-exchangers along with the work carried out in the field to develop various lead, cadmium, chromium, and mercury selective cation-exchangers are also included. The utility of organic-inorganic composite material for the detection of heavy metals, which render portable water unsafe for use and pose a threat to the wellbeing of man, is also discussed.



Biochars have been extensively applied in soil remediation, carbon sequestration, and in climate change mitigation. However, in recent years, there has been a significant increase in biochar research in water treatment due to their stupendous adsorptive properties for various contaminants. This is attributed to their large surface areas, pore structures, chemical compositions, and low capital costs involved making them suitable candidates for replacing activated carbons. This chapter discusses the preparation methods and properties of biochars and their removal efficacy for organic contaminants and microbial control. Factors affecting adsorption and the mechanisms of adsorption of organic pollutants on biochars are also concisely discussed. Biochars present environmentally benign and low-cost adsorbents for removal of both organic pollutants and microbial control for wastewater purification systems.



Soil pollution is rising rapidly due to the existence of pollutants or natural alterations in the soil. It makes the drinking water ineffective and unusable by the human beings. The major cause of the soil contamination is agricultural activities, industrial activities, and inadmissible disposal of waste in the soil. The most common pollutants to accumulate in the soil are petroleum hydrocarbons, solvents, pesticides, lead, and other heavy metals. The important technology to remediate the pollutants or contaminants in the soil is bioremediation. The utilization of bioremediation in the contaminated soil is increasing rapidly due to the presence of toxic pollutants. It is the most advanced technologies which make use of organisms to deteriorate the harmful compounds in order to prevent the soil pollution. The aim of the chapter is to describe the available bioremediation technologies and their application in removing the pollutants exist in the soil.



High-quality water is one of the most important challenges around the world. Conventional techniques of wastewater treatment need to be developed. Therefore, finding sustainable, environmentally friendly, and efficient treatment techniques is required. In this regard, due to the extraordinary potential of nanotechnology resulted from nanoscale size characteristics, recently nanomaterials have been the subject of novel research and development worldwide. In this chapter, the authors review recent development of the direct applications of nanomaterial as an adsorbent in adsorption systems for integrating nanoparticles into conventional treatment technologies for wastewater treatment, especially a wide range of candidate nanomaterials and its properties. In addition, advantages and limitations as compared to existing processes are discussed.



Plastic, one of the most preferred materials in today's industrial world, is posing a serious threat to the environment and consumer health in many direct and indirect ways. The global plastic production increased over years due to the vast applications of plastics in many sectors. More than 50% of the plastic waste generated in the country is recycled and used in the manufacture of various plastic products. The

In many regions in the world, groundwater represents an important source of fresh water. It is now established that several contaminants enter groundwater from a number of sources and pathways. These sources are both natural and anthropogenic. Contamination of groundwater resources by a variety of anthropogenic pollutants from both point and nonpoint sources represents a key global environmental problem. The most frequently identified contaminant sources are industrial manufacturing, agricultural activities, municipal landfills, and wastes. Frequently detected contaminants included nitrates, volatile organic compounds, arsenic, and fluorides. Other contaminant species include solvents, fuel hydrocarbons, heavy metals, pesticides, disinfectants, detergents, and radionuclides. In this chapter, the main sources and pathways for contaminants in groundwater are reviewed. It identifies challenges that need to be met to minimize risk to drinking water and ecosystems. Particular attention is paid to the occurrence of known and potential endocrine disrupting substances in groundwater.



The objective of this work was to investigate the occurrence and fate of five heavy metals in water, sludge, and sediments from a conventional municipal wastewater treatment facility in Kisumu City, Kenya. The effluent quality was compared with the effluent quality parameters stipulated by the National Environmental Management Authority (NEMA) to assess the efficiency of the plant and potential effect of the discharged effluent on the recipient river. The levels of the heavy metals recorded in the sludge samples were significantly higher than those in the corresponding water samples. The order of the metal percentage removal efficiency (%R) from the treatment plant was Mg>Cu>Mn>Fe>Zn. It is concluded that the plant is a point source for Zn loading into the recipient waters which poses potential risk to end users downstream. The heavy metal-laden sludge was within permissible limits for utilization in agricultural lands.



A lot of solid waste is generated in every country of the world. Among various disposal methods used for management of solid waste, open dumping is mostly used in many countries, especially developing countries. Due to open dumping, the environment becomes polluted. It also creates aesthetic problems at the site. Industrial and agricultural wastes are simple to handle being of specific characteristics. This chapter describes in detail the composition of municipal solid waste, various problems arising due to solid waste, approaches used for solid waste management, and environmental recycling system (ERS). ERS, its components, and methods for segregation of waste have also been discussed in detail in this chapter.

# Preface

Environmental consciousness is nowadays playing a growing role in production and logistics. Planning, developing, and controlling of manufacturing processes and technologies should not only support the goal of high productivity but should also respond to the need of resource and energy conservation and pollution prevention. Environmental awareness is driven mainly by the scarcity of natural resources and by more strict legal regulations. The modern enterprise policy should look at the relations between economic actions and ecological consequences. Many small communities, even in industrialized countries, do not have such resources to meet the challenges. For long-term sustainability, incorporation of the most advanced technologies may not be feasible for small communities in developed countries and for most communities in developing countries. To respond to this crucial need, appropriate technologies are discussed in this book. A unique feature of the book is the way in which chapters interact. Cross- references between theory and application foster overall integration of subject matter.

This volume is aimed at assisting process engineers, plant managers, environmental consultants, water treatment plant operators, and students. This book is intended to cover wastewater treatment technologies, air pollution controls, solid waste management and highlights various other range of topics such as agriculture pollution, hazardous waste management, and sewage farming.

Some of the sustainability strategies that need to be examined in detail are: (a) reduction in chemical and energy use in water treatment, (b) production of water that contains less pathogens and disinfection byproducts compared to the use of surface water, (c) focus on water utilities and communities (e.g., water treatment plants) to improve source water quality to reduce further treatment of the filtrate, (d) utilization of alternate low-cost wastewater treatment. If the watersheds are protected and the source water is of high quality, treatment technologies can be less costly and thus sustainable

Specifically, Chapter 1 first gives an overview on the fundamentals in the field of Advance water treatment systems and their applications.

Chapters 2 aims to push forward the practical use of biochar by recognizing the gap between biochar research and application and pinpointing the barriers to field uses of biochar. It presents the whole picture of biochar in its production, characterization, application, and development, deliberating the scientific findings, uncertainties, and barriers to practice of biochar amendment for sustaining soil fertility, improving crop production, promoting animal performance, remediating water and land, and mitigating greenhouse gas emissions. With emphasis on the mechanisms and processes of biochar amendment for achieving various agricultural and environmental benefits, the book elucidates the chemical composition and quality characteristics of biochar as influenced by feedstock, production conditions, and post-handling; the interactions of biochar with contaminants and soil constituents; and the prospective achievements of biochar amendment in improving soil physical, chemical, and biological quality and animal health; reducing soil greenhouse gas emissions; and decontaminating storm water and mine sites.

Chapter 3 explains several different remediation strategies used around the world to treat soil contaminated with toxic metals and/or organic chemicals. These strategies either individually or in combination with each other have been routinely implemented by the remediation industry to successfully treat contaminated soil. Biological remediation technologies requires knowledge of interdisciplinary sciences, involving microbiology, chemistry, hydrogeology, engineering, soil and plant sciences, geology, and ecology. Biological processes are typically implemented at a relatively low cost, and biological remediation methods have been successfully used to treat polluted soils, oily sludges, and groundwater contaminated by petroleum hydrocarbons, solvents, pesticides, and other chemicals

Chapter 4 address the fundamentals in the controlled synthesis of nanomaterials toward fulfilling specific functions, and is intended for the scientists and engineers making efforts in the fields related to nanotechnology provide basic knowledge relative to nanomaterial synthesis and allow the readers to have a clear picture for making desired nanoproducts via direct synthesis or seeding growth.

Chapter 5 attempts to review the current practices prevalent in India to deal with this plastic waste and problems associated with it. Plastic, one of the most preferred materials in today's industrial world is posing a serious threat to the environment and consumer's health in many direct and indirect ways. The global plastic production increased over years due to the vast applications of plastics in many sectors. In this chapter, author examines the environmental and health effects and harm caused by the burning of plastics in detail and focuses on the current status of plastic waste management in India and industries working under the extended producer responsibility.

Chapters 6 to 8 is intended to introduce the practice of industrial wastewater treatment to senior undergraduate and postgraduate environmental engineering students. Practitioners of the field may also find it useful as a quick overview of the subject. This chapter focuses on systems that incorporate a physical, chemical biological treatment process within the treatment train. It does not delve into the details of theory or the "mathematics" of design, but instead discusses the issues concerning industrial wastewater treatment in an accessible manner. Some prior knowledge of the theory behind the unit processes discussed and the manner in which they are supposed to work is assumed. This chapter approaches the development of suit-able treatment strategies by first identifying and addressing important wastewater characteristics. The chapters also covers different treatment technologies for the removal of emerging contaminants and includes membrane bioreactor, membrane filtration, ultrafiltration and nanofiltration and advanced sorbent materials together with more conventional natural systems. The MBR is an emerging technology based on the use of membranes in combination with traditional biological treatment. It is considered as a promising technology able to achieve more efficient removal of micro-pollutants in comparison to conventional wastewater treatment plants.

Chapter 9 provides a concise, yet comprehensive, overview of the environmental pollution caused by agriculture, offering a unique combination of clear scientific and technical understanding of the most important sources of agricultural pollution, with an understanding of: the general nature of agricultural production systems and their different relationships with the environment; the principles and practice of pollution control on farms, new technologies; statutory control and regulation; financial incentives within agri-environmental policy; and the emergence of `alternative' agricultural systems. This chapter signifies biotic and abiotic side effects of cultivating practices that outcome in infection and deterioration of the environment and encompassing biological communities, as well as causing health effects to humans. Thus there is a need for securing agriculture sector for crop productivity and handling country's economy.

Chapter 10 discusses the dramatic rise in pesticide usage throughout the globe along with the ever-increasing human population and crop production. The increasing grave misuse of pesticides is causing critical damage to the environment. The status of pesticide pollution was analyzed on a global scale to protect human health and endangered plant and animal species. During this phase, pesticide development and usage were also considered and reviewed. The global pesticide consumption pattern has changed significantly over the last 50 years. Compared to pesticides, herbicide consumption has rapidly escalated, while the proportionate consumption of insecticides and fungicides/bactericides has declined.

Chapter 11 provides a robust, practical introduction to groundwater quality, and a succinct summary of concentration of heavy metals in the polluted groundwater. It is hoped that this chapter will contribute to improving groundwater quality: firstly, through enhanced efficiencies in research investment, by avoiding needless overlap/ competition between approaches; secondly, by achieving a clearer recognition of situations for which a particular approach is best suited, and of the specialist capabilities essential to the informed delivery of each; and, finally, by achieving recognition for the value and maturity of the whole field of groundwater remediation.

Chapter 12 attempts to juxtapose both field and experimental results, so as to judge the extent of toxicity of the heavy metal, and to determine whether its toxicity can be reduced or not. This chapter investigates the occurrence and fate of five heavy metals in water, sludge, and sediments from a conventional municipal wastewater treatment facility in Kisumu City-Kenya. The effluent quality was compared with the effluent quality parameters stipulated by the National Environmental Management Authority (NEMA) to assess the efficiency of the plant and potential effect of the discharged effluent on the recipient river. .

Chapter 13 explains municipal waste management system which includes the collection, diversion (reuse, recycling, recovery/composting), and disposal of the municipal solid waste.

Experts in the area of environmental microbiology, wastewater treatment, solidwaste management, and bioremediation, from diverse institutions worldwide, have contributed to this book. We would like to express our sincere appreciation to each contributor for his/her work and for their patience and attention to detail during the entire production process. We sincerely hope these eminent contributors will encourage us in the future as well, in the greatest interest of academia. The book will be of interest to a broad audience of analytical chemists, environmental chemists, water management operators and technologists working in the field of wastewater treatment, or newcomers who want to learn more about the topic.

Globally, e-waste is a big issue not only for developing countries but also for developed countries. Chapter 14 explains that how informal sectors is creating a lots of problems for human health and environment. This chapter shows the current global scenario and its more focus is on Indian scenario and companies which are working on e-waste recycling and what's new in e-waste management rules 2016 and harmful effects of electronic waste.

# Acknowledgment

*Coming together is a beginning;*
*Keeping together is progress;*
*Working together is a success*

The *Advanced Treatment Techniques for Industrial Wastewater* has been the culmination of ideation and brainstorming with a lot of people. It is only natural that we should gratefully acknowledge their valuable contributions sincerely. First, I would like to thank each one of the authors for their contributions. I wish to express my gratitude to my co-editor Prof. Sirajuddin Ahmed, Civil Engineering Department, Faculty of Engineering and Technology, Jamia Millia Islamia, New Delhi for initiating the preparation of this book and my encouragement.

I am extremely thankful and indebt of Mr. Ayushman Bhattacharya who did everything that was foisted on him and left no stone unturned in pursuit of excellence—your pivotal contributions are gratefully acknowledged.

Special thanks to my wife Mrs. Bushra Jamal who willingly took over all my household responsibilities as I began sprinting towards the completion of the book. Thank you for being the support system.

# Introduction

Environment and Development are two sides of same coin. The need for development through socio-economic activities, especially with the growing population and a desire for higher standards of living with industrial production is obvious. Practically these activities contribute to environmental degradation including soil, water, air, etc. The topics covered in this book are fairly wide and would meet the training requirement for both applied science as well as engineering disciplines including chemical, civil, biochemical etc. This volume is intended to serve as a general introduction to environmental engineering for a senior-level student or graduate student. The reader is expected to have a background in basic engineering concepts and design; only topics specific to the discipline will be discussed herein. It is not intended to be a comprehensive authority on each subject, but rather to serve as a reference and concept review for the upper-level reader. Many facts are presented in charts, graphs, and figures to help illustrate the scope of environmental issues, the text's main focus is on identifying major issues and giving appropriate examples to illustrate the complex interactions that are characteristic of all environmental problems. The authors have endeavored to present a balanced view of issues, diligently avoiding personal biases and fashionable philosophies. It is not the purpose of this textbook to tell you what to think. Rather, our goal is to provide access to information and the conceptual framework needed to understand complex issues so that you can comprehend the nature of environmental problems and formulate your own views. Water treatment is not only a very important subject, but it is extremely interesting. Its importance is simply one of environmental protection and public safety, because after all, water is one of the basic natural elements we rely upon for survival.

The extensive use of this book by students, faculty and practicing engineers will bring satisfaction to the authors who have put in a lot of effort in writing this book.

# Chapter 1
# Advanced Water Treatment Systems and Their Applications

**Tauseef Ahmad Rangreez**
*National Institute of Technology Srinagar, India*

**Rizwana Mobin**
*Government College for Women, Cluster University Srinagar, India*

**Hamida-Tun-Nisa Chisti**
*National Institute of Technology Srinagar, India*

**Rafia Bashir**
*National Institute of Technology Srinagar, India*

**Tabassum Ara**
*National Institute of Technology Srinagar, India*

## ABSTRACT

*The chapter gives an idea about water as a life-sustaining medium and the sources of its pollution along with the deteriorating effects of over burdening of natural resources and the effect of various heavy metal ions discharged into the water bodies on human health and wellbeing. Several human diseases and disorders that are caused due to intake of water polluted by toxic heavy metal ions are also listed. The need and urgency in determination and removal of heavy metal ions from the water sources in order to release load on aquifers by making water safe for reuse is emphasized. The procedure and advantages of composite cation-exchangers along with the work carried out in the field to develop various lead, cadmium, chromium, and mercury selective cation-exchangers are also included. The utility of organic-inorganic composite material for the detection of heavy metals, which render portable water unsafe for use and pose a threat to the wellbeing of man, is also discussed.*

DOI: 10.4018/978-1-5225-5754-8.ch001

## 1. INTRODUCTION

The rapid, unplanned and ambitious industrial development has lead to environmental problems throughout the world. The natural sources are being exploited in the name of the industrial growth and development. The natural resources are currently being over-burdened, so the ecological balance is disturbed, which threatens the very survival of life on the planet (Deblonde, Leguille, & Hartemann, 2011). The industrial processes (mining, extraction, processing) and domestic discharges pollute air, water and soil. The environmentalists are facing a serious task to provide safe, clean and favorable environment for life. Pollution is defined as any adverse changes in the physical, chemical or biological properties of air, land and water that may adversely affect human life or other desired species or industrial production processes, living conditions and cultural assets (Obasi, 1999). These pollutants are a threat not only to humans but other life forms also. Pollutants can be solid, liquid or gaseous substances and may have chemical or energy form. Pollutants are even natural, but in this context these don't lead to serious environmental problems.

A collective problem now-a-days is the water pollution, more importantly, means and measures that should be adapted to control and monitor this threat. The new trends in research should focus not only on the effective handling of pollution, but to prevent and minimize the production of toxic and hazardous substances in the first place.

## 2. WATER EMINENCE

It is not possible to even think about life in absence of water. About 70% of the earth's surface is covered with water. About 97% of the water is present in the ocean, 2% as ice and only 1% of the water as present in lakes and rivers maintains the life on the earth (Dessouky & Ettouney, 2002).

Water is essential for food, health, economy, energy (Moore, Gould, & Keary, 2003) and plays a leading role in photosynthesis and breathing, which are the processes of life. It is used as solvent in a number of metabolic processes for plants and animals. Water constitutes about 69% of the body weight, 70% of the brain and 80% of the blood by volume. The Royal Society of Chemistry has placed water at first place in the list of green solvents. Annual utilization of fresh water is about 3240 cubic kilometres, agriculture (69%), industry (23%), only 8% of the water is available for domestic use (Sethi, 2001).

## 3. POLLUTION OF WATER

Any toxic chemical (organic and inorganic), biological or radioactive substance that enters water and changes the physical, chemical and biological properties of water is known as water pollutant. The industrial revolution and uncontrolled use of natural water resources, population growth, urbanization and changing ways of life are a major cause of serious water pollution (Shannon et al., 2008). Water quality deterioration is mainly due to industrial wastes and human activities. Water quality deterioration affects aquatic ecosystems, groundwater and human life. Annually about 10-20 billion deaths are reported due to water borne diseases, (Leonard, 2003) alone diarrhoea causes about 6,000 children to die every day (Ashbolt, 2004). It is estimated that about 0.78 million people throughout the world have no access to safe drinking water (WHO, UNICEF, 2013).

## 4. WATER POLLUTION SOURCES

The wastewater discharge, runoff from urban and rural areas, the increased use of chemicals (fertilizers, pesticides and surfactants), animal matter into surface and ground waters, municipal emissions into water bodies constitute the main pollution sources (Fawell & Nieuwenhuijsen, 2003; Mozaz, Alda, & Barcelo, 2004). The water bodies are constantly under pressure due to the growing demand for food and energy (Godfray, 2010). The sources of water pollution are generally divided into two types:

1. Point source and
2. Non-point source

Point sources are confined sources of pollution, including the discharge from industries and these are also called localized sources whereas non-point sources are non-localized diffusible sources of water pollution (Figure 1). Depending on the source and quality of the waste, these are subject to regulation by various environmental protection agencies. The non-point sources are more troublesome in comparison to point sources, because they contain different type of pollutants.

Water pollutants are generally classified into the following four types: 1) organic contaminants 2) inorganic contaminants, 3) suspended solids and 4) radioactive materials. Organic pollutants are carcinogenic, while heavy metal ions are not biodegradable and prolonged exposure to these ions can cause health risks and even death.

*Figure 1. Point and non-point sources of water pollution*

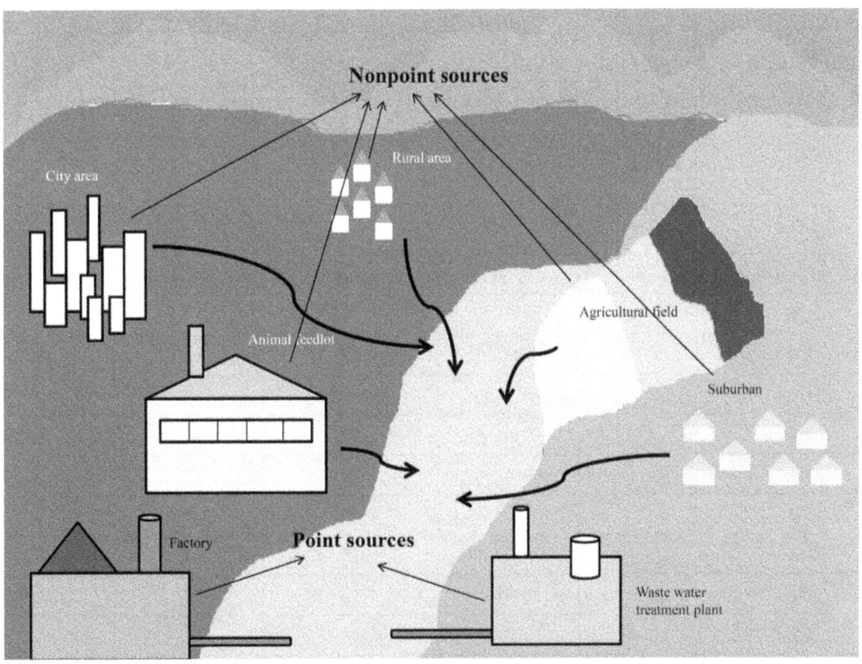

## 5. HEAVY METAL ION POLLUTION

Heavy metal pollution is a result of unabated discharge from the industries, geological weathering, volcanic eruptions, forest fires, treatment of ore, metal coating, the use of chemical substances, fertilizers in agriculture and so on. Metals are generally used in manufacturing processes, batteries, dyes, as catalysts in the oil industry, protective metal coatings, synthetic chemistry and metal processing plants (Low & Lee, 1991). The heavy metals from such sources and other human activities make their entry into different environmental spheres and are thus a serious threat to life. The main problem arises from the non-biodegradable and bio-accumulative nature of these materials (Argun, Dursun, Ozdemir, & Karatas, 2007). Some metals (copper, zinc and iron) must be in the human body at very low concentrations where as some heavy metals / metalloids have very high toxicity, such as mercury, cadmium, arsenic and lead. Mercury, lead and cadmium are also known as "big three" toxic heavy metals (Naja, & Volesky, 2009). Because of their high toxicity and detrimental effect on human health, several heavy metals are placed in the priority list of various environmental agencies. Table 1 shows the maximum limits of some heavy metals and metalloids in drinking water (United States Public Health, USPH, 2012; WHO, 2011; EWQS, 2013).

## 6. EFFECT ON HUMAN HEALTH DUE TO HEAVY METAL/METALLOIDS

The heavy metal ions get accumulated and biomagnified on their entry into the food chain. The heavy metals are toxic and hazardous to humans and several other life forms. These toxic metal ions lead to discomfort and cause damage to several vital organs. Figure 2 represents harmful effects of heavy toxic metals on human health.

Some of these metals/ metalloids are as follows.

*Table 1. Maximum permissible limits ($\mu g\ L^{-1}$) of heavy metals/metalloids in drinking water*

| Heavy Metal Ion/Metalloids | WHO | European Union | United States | Canada | ISI |
|---|---|---|---|---|---|
| Cd | 3 | 5 | 5 | 5 | 3 |
| Cr | 50 | 50 | 100 | 50 | 50 |
| Cu | - | 2000 | - | 1000 | 1500 |
| Pb | - | 10 | 15 | 10 | 300 |
| Hg | 6 | 1 | 2 | 1 | 1 |
| Ni | - | 20 | - | - | 20 |
| As | 10 | 10 | 10 | 10 | 50 |
| Se | 40 | 10 | 50 | 10 | 10 |
| Co | 50 | - | - | - | - |

*Figure 2. Human health and heavy toxic metals ions*

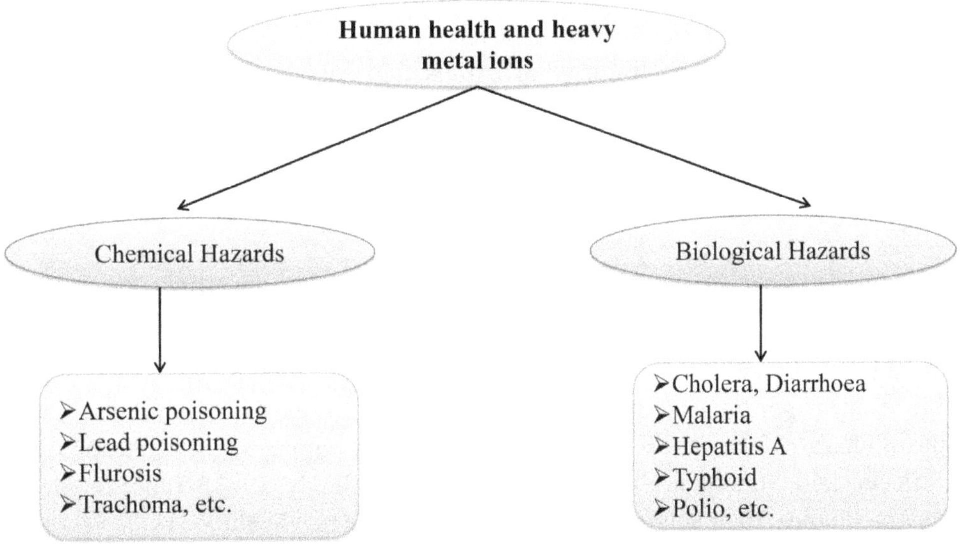


## 6.1 Cadmium

Cadmium is usually found in industrial workplaces and possesses extremely toxic properties. It enters the water bodies by the industrial emissions from the nickel-cadmium batteries, phosphate fertilizers, fuels and alloys (Low & Lee, 1991) and causes gastrointestinal tract erosion, pulmonary, renal damages, stomach pain, vomiting, damage to the kidneys and destruction of blood red blood cells (Alizadeh, Ganjali, Nourozi, Zare, & Hoseini, 2011; Nriagu & Pacyna, 1988; Borsari, 1994).

## 6.2 Lead

This is one of the most toxic heavy metal. It is released into the environment through natural and man-made activities. The main sources include volcanoes, sea aerosols, manufacturing lead batteries, metal processing plants, paints, lead pipes and solders. Lead poisoning is also known as plumbism. Nervous system is most badly affected by lead poisoning. Its exposure causes irritability, insomnia, damage to the kidneys, brain, reproductive and central nervous system (Connell, Birkinshaw, & Dwyer, 2008; Kadirvelu, Thamaraiselvi, & Namasivayam, 2001; Pearce, 2007).

## 6.3 Arsenic

Arsenic exposure is associated with the occurrence of hypertension and peripheral vascular disease (Abhyankar, Jones, Guallar, & Acien, 2012; Chen et al., 2007). The human body/skin is most affected by the arsenic exposure. Squamous cell carcinoma (SCC) and Bowen's disease (BD) are also reported for chronic arsenic exposure (Debendra et al., 1998; Ghosh et al., 2007). It is also associated with cancer of the several body parts such as urinary bladder, lung and pancreas (Catterina, 2013). It is released into the environment by natural and human activities (Anawar, Akai, & Sakugawa, 2004).

### 6.4 Chromium

Chromium may exist as a hexavalent or trivalent form in aqueous solution. Chromium (VI) is highly toxic and carcinogenic than Cr (III). It is used in textiles, thermal power plants, hard chromium plating, dyes and sugar mills (Deepali & Gangwar, 2010; Javed, 2013). Cr (VI) is a powerful oxidising agent causes gastric ulcers, respiratory diseases, liver and kidney damages.

### 6.5 Mercury

Mercury is a major environmental toxic pollutant. High reactivity and relative solubility of mercury in water and other tissues makes mercury a potent toxic heavy metal (James, David, Gary, & Anton, 2002). It is used in the production of batteries, camcorders, cathodic tubes, calculators, dental amalgam fillings and mercury vapor lamps (Khan & Inamuddin, 2006). It easily accumulates in the food chain and causes health hazards like nausea, abdominal pain, kidney damages, gastrointestinal tract, hearing, teratogenic effects (Jain, Sondhi, & Sharma, 2000; Eyssen & Ruedy, 1983). Table 2 shows some of the heavy metals / metalloids, their sources and their impact on human health.

## 7. REMOVAL OF HEAVY METALS

The water bodies around the world are depleting at a very fast rate due to various factors. Advanced sewage treatment technology can release stress on water bodies, human health, well being and climate change. The same can be done through the treatment and reuse of waste water, which in effect involves the development of effective, novel, reliable and cost-effective materials and methods of detection and determination (Amin, Alazba, & Manzoor, 2014). Some methods that have been used for the processes are:

*Table 2. Sources and effects of metals/metalloids on human health*

| Heavy Metal/metalloid | Sources | Effect on Human Health |
|---|---|---|
| **Chromium** | Cooling tower, dyes, electroplating, ink, anodizing, paints, tanning etc. | Cancer |
| **Cadmium** | Coal combustion, metal plating, water pipe, pigments, phosphate fertilizers etc. | Cardiovascular disease, reproductive failure, kidney damage and hypertension |
| **Lead** | Metal processing plants batteries, paints etc. | Affects nervous and renal systems, headache, constipation, cancer etc. |
| **Nickel** | Diesel oil, coal, steel & non-ferrous alloys, tobacco smoke etc. | Lung cancer, respiratory Symptoms and haemorrhage |
| **Zinc** | Galvanizing, alloys rayon, paper etc. | Cancer |
| **Mercury** | Coal combustion, electrical batteries, Chlor-alkali industry | Nerves damage, death, kidney and brain damage |
| **Cobalt** | Alloys, steel, electroplating glass, enamel etc. | Cancer |
| **Manganese** | Metal alloys, power plants, gasoline etc. | Nervous system damage |

- Chemical and electrochemical precipitation
- Evaporation
- Membrane filtration
- Reverse osmosis
- Adsorption
- Electrochemical sensors
- Ion-exchange process

Biological treatment as activated sludge and filters cannot eliminate a wide range of contaminants as most of the compounds are soluble (Servos, 2005; Urase & Kikuta, 2005). Likewise, physicochemical treatments such as flocculation and lime softening are ineffective in the removal of various interfering endocrine disrupting compounds (EDC) and drugs (Servos, 2005; Urase & Kikuta, 2005). Chlorinated water provides protection against the re-emergence of bacteria and pathogens (Szewzyk, Szewzyk, Manz, & Schleifer, 2000; Zhang, Giano, 2002) but creates an unwanted taste and odour (Suffet et al., 1995) along with the establishment of various disinfection products (disinfection by-products) in portable water, whereas ozonization is not an effective choice due to its high cost. Ultraviolet and ion exchange are considered to be advanced treatment methods.

## 8. ION-EXCHANGE PROCESS

Ion-exchange chromatography allows the separation of ions/polar molecules on the basis of their charge. It is used for both positive and negative charged molecules. The ion exchange process occurs between the mobile phase and the ion exchange groups bound on the supporting material. In ancient Egypt and Greece, people used some soils, sand and natural zeolites as materials for desalting. H. Thompson and J.T. Way treated various clays with ammonium sulphate/ carbonate in solution to extract ammonium and remove calcium (Thompson, 1850; Way & Roy, 1850). The synthesis and technical applications of inorganic cationic exchangers was developed by Gans (Gans, 1905). The natural cation exchangers were known but not used until 1930, when the first organic ion-exchanger was synthesized (Adams, & Holmes 1935).

The ion-exchange process retains analyte molecules on the column. The functional groups on stationary phase interact with analyte ions of opposite charge. Based on the type of ion (cation or anion) exchanged, it is referred to as cation/anion-exchange chromatography. Cation-exchange chromatography retains positively charged cations because the stationary phase displays a negatively charged functional group on the other hand anion-exchange chromatography retains anions using positively charged functional group. The ion-exchange processes are widely used for removal of heavy metals from wastewaters. These ion-exchangers have several properties as high ion-exchange capacity, removal efficiency and fast kinetics (Kang, Lee, Moon, & Kim, 2004). Ion-exchange resins have the specific ability to exchange cations with the metal ions in the wastewater. The most common cation exchangers are strongly acidic resins with sulfonic acid groups ($-SO_3H$) and weakly acid resins with carboxylic acid groups (-COOH). Hydrogen ions in the sulfonic group or carboxylic group of the resin serve as exchangeable ions with metal cations. The solution containing heavy metal ions passes through the ion-exchange column where metal ions are exchanged for the hydrogen ions on the resin with the following ion-exchange process:

*Table 3. Some important heavy metal ion selective composite ion-exchangers*

| S. No. | Composite Material | Application | References |
|---|---|---|---|
| 1 | Polypyrrole Th(IV) phosphate | Pb(II) selective | (Khan, Inamuddin, & Alam, 2005) |
| 2 | Poly-*o*-toluidine Ce(IV) phosphate | Cd(II) selective | (Khan & Akhtar, 2011) |
| 3 | Polyaniline Sn(IV) phosphate | Pb(II) selective | (Khan & Inamuddin, 2006) |
| 4 | Poly(methylmethacrylate) Zr(IV) phosphate | Pb(II) selective | (Siddiqui, Khan, & Inamuddin, 2007) |
| 5 | Polyacrylamide Th(IV) phosphate | Pb(II) selective | (Islam & Patel, 2008) |
| 6 | Acrylamide aluminumtungstate | Pb(II) selective | (Nabi, Ganai, & Shalla, 2008) |
| 7 | Poly-*o*-anisidine Sn(IV) arsenophosphate | Pb(II) selective | (Khan, Habiba, & Khan, 2009) |
| 8 | Acrylonitrile Sn(IV) tungstate | Pb(II) selective | (Nabi, Naushad, & Bushra, 2009) |
| 9 | Polyacrylamide apatite | Pb(II) selective | (Ulusoy & Akkaya, 2009) |
| 10 | Macrocyclic diamide sulphur | Cd(II) selective | (Shamsipur *et al.*, 2009) |
| 11 | $B_2O_3/TiO_2$ nano-composite | Cd(II) selective | (Kalfa, Yalcinkaya, & Turker) |
| 12 | Ethylenediamine tetra acetic acid Sn(IV) iodate | Pb(II) selective | (Nabi & Shalla, 2009) |
| 13 | Acrylamide Zr(IV) arsenate | Pb(II) selectve | (Nabi & Shalla, 2009) |
| 14 | Acrylamide stannic silicomolybdate | Pb(II) selective | (Khan, Ganai, & Nabi, 2009) |
| 15 | Poly(methylmethacrylate) Ce(IV) molybdate | Pb(II) selective | (Bushra *et al.*, 2014) |
| 16 | Polyaniline Ce(IV) molybdate | Cd(II) selective | (Alam & Inamuddin, 2010) |
| 17 | Poly-*o*-methoxyaniline Zr(IV) molybdate | Cd(II) selective | (Inamuddin & Ismail, 2010) |
| 18 | Poly-*o*-toluidine stannic molybdate | Pb(II) selective | (Nabi, Bushra, Naushad, & Khan, 2010) |
| 19 | Poly-*o*-anisidine Sn(IV) phosphate | Cd(II) selective | (Khan & Khan, 2010) |
| 20 | Polyaniline Sn(IV) molybdate | Pb(II) selective | (Othman, Naushad, & Nilchi, 2011) |
| 21 | Polyaniline Zirconium titanium phosphate | Hg(II) selective | (Khan, Paquiza, 2011) |
| 22 | Polyaniline Humic acid | Hg(II) selective | (Zhang, Li, Sun, Tang, & Zhai, 2010) |
| 23 | Nylon-6,6 Sn(IV) phosphate | Hg(II) selective | (Khan & Akhtar, 2009) |
| 24 | Poly-o-toluidine Zr(IV) phosphate | Hg(II) selective | (Khan & Akhtar, 2008) |
| 25 | Poly-o-toluidine Th(IV) phosphate | Hg(II) selective | (Khan, Khan, & Inamuddin, 2007) |
| 26 | Polyaniline Sn(IV) phosphate | Hg(II) selective | (Khan & Inamuddin, 2006) |
| 27 | Poly-*o*-anisidine Sn(IV) phosphate | Sb(II) selective | (Khan & Khan, 2009) |
| 28 | cellulose acetate-Zr(IV) molybdophosphate | Cr(III) selective | (Nabi & Naushad, 2008) |
| 29 | Polyaniline-Sn(IV)tungstophosphate | Pb(II) selective | (Nabi, Akhtar, Khan, & Khan, 2014) |
| 30 | Polyaniline Sn(IV)tungstoarsenate | Cd(II) & Hg(II) selective | (Alam, Othman, & Naushad, 2013) |
| 31 | Polypyrrole polyantimonic acid | Cd(II) & Hg(II) selective | (Alam, Othman, & Naushad, 2013) |
| 32 | Polyaniline Sn(IV) silicomolybdate | Pb(II) selective | (Nabi, Ganai, & Khan, 2011) |
| 33 | Polyaniline zirconium titanium phosphate | Pb(II) selective | (Khan & Paquiza, 2011) |
| 34 | Acrylamide Sn(IV) molybdate | Cd(II) selective | (Shahadat, Shalla, & Raeissi, 2012) |
| 35 | Polyaniline Zr(IV) arsenate | Pb(II) selective | (Nabi, Bushra, & Shahadat, 2012) |
| 36 | Poly-*o*-toluidine Zr(IV) tungstate | Pb(II) selective | (Bushra, Shahadat, Raeissi, & Nabi, 2012) |
| 37 | Polyaniline Ti(IV) arsenate | Pb(II) selective | (Shahadat *et al.*, 2012) |

*continued on following page*

*Table 3. Continued*

| S. No. | Composite Material | Application | References |
|---|---|---|---|
| 38 | Polyaniline Ti(IV) phosphate | Pb(II) selective | (Khan & Baig, 2012) |
| 39 | Polyaniline Sn(IV) tungstoarsenate | Cd(II) selective | (Alam, Othman, & Naushad, 2013) |
| 40 | Polyaniline Sn(IV) silicate | Cd(II) selective | (Naushad, Othman, & Islam, 2013) |
| 41 | Polyvinyl alcohol zinc oxide | Cd(II) & Pb(II) selective | (Khan, Rafiuddin, & Inamuddin, 2013) |
| 42 | Poly-*o*-toluidine Sn(IV) tungstate | Pb(II) selective | (Khan & Shaheen, 2013) |
| 43 | Polyaniline Sn(IV) tungstomolybdate | Pb(II) selective | (Bushra, Naushad, Adnan, Ibrahim, & Rafatullah, 2014) |
| 44 | Polyaniline Ti(IV) arsenophosphate | Cd(II) selective | (Bushra *et al.*, 2014) |
| 45 | Polyaniline Sn(IV) silicophosphate | Pb(II) selective | (Khan, Ahmad, Umar, & Nabi, 2015) |
| 46 | Poly(3,4-ethylenedioxythiophene)-poly(styrenesulfonate) Zr(IV) phosphate | Cd(II) selective | (Mohammad, Inamuddin, & Hussain, 2014) |
| 47 | Polyaniline tungstophosphate | Pb(II) selective | (Khan *et al.*, 2013) |
| 48 | Carboxymethyl cellulose Sn(IV) phosphate | Pb(II) selective | (Inamuddin, Naushad, Rangreez, & Othman, 2015) |
| 49 | Polyaniline Zr(IV) sulfosalicylate | Pb(II) selective | (Shahadat & Bushra, 2015) |
| 50 | Poly(3,4-ethylenedioxythiophene)-poly(styrenesulfonate) Zr(IV) monothiophosphate | Pb(II) selective | (Rangreez, Inamuddin, Naushad, & Ali, 2015) |
| 51 | Cellulose acetate Zr(IV) molybdophosphate | Cd(II) & Pb(II) selective | (Khan, Inamuddin, & Naushad, 2015) |
| 52 | Single walled carbon nanotubes Ce(IV) phosphate | Cd(II) selective | (Inamuddin, Rangreez, & Khan, 2015) |
| 53 | Hydrous ferric oxide-201 composite | Cr(IV) selective | (Hua *et al.*, 2017) |
| 54 | Alginate-gelatin binary bio-composite | Cr(IV) selective | (Vishwanathan, 2015) |
| 55 | Polyaniline-magnetic mesoporous silica composite | Cr(IV) selective | (Tang *et al.*, 2014) |
| 56 | Chitosan-graft-polyacrylamide magnetic composite | Hg(II) selective | (Li *et al.*, 2015) |
| 57 | Polypyrrole Zirconium titanium phosphate | Th(IV) selective | (Khan, Paquiza, & Khan, 2010) |
| 58 | Parachlorophenol Tin antimonate | Bi(III) and Cu(II) Selective | (Chithra, Raveendran, & Beena, 2008) |

$$nR\text{-}SO_3 + M^{n+} \rightarrow (R\text{-}SO_3\text{-})_n M^{n+} + nH^+ \quad (1)$$

$$nR\text{-}COOH + M^{n+} \rightarrow (R\text{-}COO^-)_n M^{n+} + nH^+ \quad (2)$$

The uptake of heavy metal ions by ion-exchange resins is rather affected by certain variables such as pH, temperature, initial metal concentration and contact time (Gode & Pehlivan, 2006). For multivalent ions such as $Cr^{3+}$, $Ni^{2+}$, $Co^{2+}$, $Ce^{4+}$, $Fe^{3+}$ and $Pb^{2+}$ it was found that the adsorption capacity of these ions decreases with decreasing ionic charge (Gode & Pehlivan, 2006; Farha, Aal, Ashourb, & Garamon, 2009).

The organic-inorganic composite materials represent materials with unusual properties often modified in order to optimize their performance for specific use. These are developed by the incorporation of organic conducting polymers into inorganic precipitate of polyvalent metal acid salts. These materials have great deal of attention because of their special mechanical, chemical and electrochemical properties. Table 3 represents various cation-exchangers along with their respective selective heavy metal ion.

## CONCLUSION

The water bodies are under constant threat mostly due to anthropogenic activities. In order to relieve pressure from aquifers, advanced technologies are a must to tackle water scarcity and health concerns. The novel methods of purification and reuse have to be reliable, efficient and cost effective.

The composite cation-exchangers are developed by the incorporation of organic polymers into inorganic precipitates of polyvalent metal acid salts. The composite ion-exchangers listed here have been found to be selective for heavy metal ions viz, Pb (II), Cd(II), Cr(III) and Hg(II). These ions are very toxic and hence their detection and removal is of utmost importance. A lot of work has been done in the field of ion-exchangers, but the field is so vast and there is so much scope for the improvement, particularly the nano-composite exchangers and is yet a simple, reliable and easy method for the detection and removal of heavy metal ions.

## REFERENCES

Abhyankar, L. N., Jones, M. R., Guallar, E., & Acien, A. N. (2012). Arsenic exposure and hypertension: A systematic review. *Environmental Health Perspectives*, *120*(4), 494–500. doi:10.1289/ehp.1103988 PMID:22138666

Abo-Farha, S. A., Abdel-Aal, A. Y., Ashourb, I. A., & Garamon, S. E. (2009). Removal of some heavy metal cations by synthetic resin purolite. C100. *Journal of Hazardous Materials*, *169*(1-3), 190–194. doi:10.1016/j.jhazmat.2009.03.086 PMID:19403237

Adamsand, B. A., & Holmes, E. L. (1935). Adsorptive properties of synthetic resins. *Journal of the Chemical Society*, *54*, 1–6.

Al Othman, Z. A., Naushad, M., & Nilchi, A. (2011). Development, characterization and ion exchange thermodynamics for a new crystalline composite cation exchange material: Application for the removal of $Pb^{2+}$ ion from a standard sample (rompin hematite). *Journal of Inorganic and Organometallic Polymers and Materials*, *21*(3), 547–559. doi:10.100710904-011-9491-9

Alam, M. M., Alothman, Z. A., & Naushad, M. (2013). Analytical and environmental applications of polyaniline Sn(IV) tungstoarsenate and polypyrrolepolyantimonic acid composite cation-exchangers. *Journal of Industrial and Engineering Chemistry*, *19*(6), 1973–1980. doi:10.1016/j.jiec.2013.03.006

Alam, M. M., Alothman, Z. A., & Naushad, M. (2013). Analytical and environmental applications of polyaniline Sn(IV) tungstoarsenate and polypyrrolepolyantimonic acid composite cation-exchangers. *Journal of Industrial and Engineering Chemistry*, *19*(6), 1973–1980. doi:10.1016/j.jiec.2013.03.006

Alam, Z., Inamuddin, & Nabi, S. A. (2010). Synthesis and characterization of a thermally stable strongly acidic Cd(II) ion selective composite cation-exchanger: PolyanilineCe(IV) molybdate. *Desalination, 250*(2), 515–522. doi:10.1016/j.desal.2008.09.008

Alizadeh, T., Ganjali, M. R., Nourozi, P., Zare, M., & Hoseini, M. (2011). A carbon paste electrode impregnated with $Cd^{2+}$ imprinted polymer as a new and high selective electrochemical sensor for determination of ultra-trace $Cd^{2+}$ in water samples. *Journal of Electroanalytical Chemistry, 657*(1-2), 98–106. doi:10.1016/j.jelechem.2011.03.029

Amin, M. T., Alazba, A. A., & Manzoor, U. (2014). A review of removal of pollutants from water/wastewater using different types of nanomaterials. *Advances in Materials Science and Engineering*.

Anawar, H. M., Akai, J., & Sakugawa, H. (2004). Mobilization of arsenic from subsurface sediments by effect of bicarbonate ions in groundwater. *Chemosphere, 54*(6), 753–762. doi:10.1016/j.chemosphere.2003.08.030 PMID:14602108

Argun, M. E., Dursun, S., Ozdemir, C., & Karatas, M. (2007). Heavy metal adsorption by modified oak sawdust: Thermodynamics and kinetics. *Journal of Hazardous Materials, 141*(1), 77–85. doi:10.1016/j.jhazmat.2006.06.095 PMID:16879919

Ashbolt, N. J. (2004). Microbial contamination of drinking water and disease outcomes in developing regions. *Toxicology, 198*(1-3), 229–238. doi:10.1016/j.tox.2004.01.030 PMID:15138046

Borsari, M. (1994). *Encyclopedia of Inorganic Chemistry* (1st ed.). New York: Wiley.

Bushra, R., Shahadat, M., Ahmad, A., Nabi, S. A., Umar, K., Oves, M., ... Muneer, M. (2014). Synthesis, characterization, antimicrobial activity and applications of polyanilineTi (IV) arsenophosphate adsorbent for the analysis of organic and inorganic pollutants. *Journal of Hazardous Materials, 264*, 481–489. doi:10.1016/j.jhazmat.2013.09.044 PMID:24238807

Bushra, R., Shahadat, M., Khan, M. A., Inamuddin, Adnan, R., & Rafatullah, M. (2014). Optimization of polyaniline supported Ti(IV) arsenophosphate composite cation exchanger based ion-selective membrane electrode for the determination of lead. *Industrial & Engineering Chemistry Research, 53*(50), 19387–19391. doi:10.1021/ie5034655

Bushra, R., Shahadat, M., Raeissi, A. S., & Nabi, S. A. (2012). Development of nano-composite adsorbent for removal of heavy metals from industrial effluent and synthetic mixtures; its conducting behaviour. *Desalination, 289*, 1–11. doi:10.1016/j.desal.2011.12.013

Catterina, F., Yan, Y., Jacqueline, C., Hugo, B., Roxana, P. L., Johanna, A., ... Craig, S. (2013). Arsenic, tobacco smoke, and occupation: Associations of multiple agents with lung and bladder cancer. *Epidemiology (Cambridge, Mass.), 24*(6), 898–905. doi:10.1097/EDE.0b013e31829e3e03 PMID:24036609

Chen, Y., Litvak, P. F., Howe, G. R., Graziano, J. H., Rauf, P. B., & Parvez, F. (2007). Arsenic exposure from drinking water, dietary intakes of B vitamins and folate, and risk of high blood pressure in Bangladesh: A population-based, cross-sectional study. *American Journal of Epidemiology, 165*(5), 541–552. doi:10.1093/aje/kwk037 PMID:17164464

Chithra, P. G., Raveendran, R., & Beena, B. (2008). Parachlorophenol anchored tin antimonite-An inorgano-organic ion-exchanger selective towards heavy metals like Bi(III) and Cu(II). *Desalination*, *232*(1-3), 20–25. doi:10.1016/j.desal.2008.01.006

Connell, D. W. O., Birkinshaw, C., & Dwyer, T. F. O. (2008). Heavy metal adsorbents prepared from the modification of cellulose: A review. *Bioresource Technology*, *99*(15), 6709–6724. doi:10.1016/j.biortech.2008.01.036 PMID:18334292

Debendra, N., Mazumder, G., Haque, R., Ghosh, N., Binay, K. D., Santra, A., ... Smith, A. H. (1998). Arsenic levels in drinking water and the prevalence of skin lesionsin West Bengal, India. *International Journal of Epidemiology*, *27*(5), 871–877. doi:10.1093/ije/27.5.871 PMID:9839746

Deblonde, T., Leguille, C. C., & Hartemann, P. (2011). Emerging pollutants in wastewater: A review of the literature. *International Journal of Hygiene and Environmental Health*, *214*(6), 442–448. doi:10.1016/j.ijheh.2011.08.002 PMID:21885335

Deepali, & Gangwar, K.K. (2010). Metals concentration in textile and tannery effluents, Distribution of estrogens, 17beta-estradiol andestrone, in Canadian municipal wastewater treatment plants. *Science Total Environment, 336*, 155-170.

El-Dessouky, H., & Ettouney, H. (2002). Teaching desalination-a multidiscipline engineering science. *Heat Transfer Engineering*, *23*(5), 1–3. doi:10.1080/01457630290090590

Eyssen, G. E. M., & Ruedy, J. (1983). Methyl mercury exposure in northern quebec I. neurologic findings in adults. *American Journal of Epidemiology*, *118*(4), 461–469. doi:10.1093/oxfordjournals.aje.a113651 PMID:6637973

Fawell, J., & Nieuwenhuijsen, M. J. (2003). Contaminants in drinking water. *British Medical Bulletin*, *68*(1), 199–208. doi:10.1093/bmb/ldg027 PMID:14757718

Gans, R. (1905). Jahrb Preuss Geol. *Landesanstalt Berlin*, *26*, 179.

Ghosh, P., Banerjee, M., Chaudhuri, S. D., Das, J. K., Sarma, N., Basu, A., & Giri, A. K. (2007). Increased chromosome aberration frequencies in the Bowen's patients compared to non-cancerous skin lesions individuals exposed to arsenic. *Mutation Research/Genetic Toxicology and Environmental Mutagenesis*, *632*(1-2), 104–110. doi:10.1016/j.mrgentox.2007.05.005 PMID:17600756

Gode, F., & Pehlivan, E. (2006). Removal of chromium (III) from aqueous solutions using Lewatit S 100: The effect of pH, time, metal concentration and temperature. *Journal of Hazardous Materials*, *136*(2), 330–337. doi:10.1016/j.jhazmat.2005.12.021 PMID:16439060

Godfray, H. C. J., Beddington, J. R., Crute, I. R., Haddad, L., Lawrence, D., Muir, J. F., ... Toulmin, C. (2010). Food Security: The Challenge of Feeding 9 Billion People. *Science*, *327*(5967), 812–818. doi:10.1126cience.1185383 PMID:20110467

Gopalakannan, V., & Viswanathan, N. (2016). One pot synthesis of metal ion anchored alginate–gelatin binary biocomposite for efficient Cr(VI) removal. *International Journal of Biological Macromolecules*, *83*, 450–459. doi:10.1016/j.ijbiomac.2015.10.010 PMID:26456290

Hua, M., Yang, B., Shan, C., Zhang, W., He, S., Lv, L., & Pan, B. (2017). Simultaneous removal of As(V) and Cr(VI) from water by macroporous anion exchanger supported nanoscale hydrous ferric oxide composite. *Chemosphere*, *171*, 126–133. doi:10.1016/j.chemosphere.2016.12.051 PMID:28012384

Inamuddin, N., Naushad, M., Rangreez, T. A., & ALOthman, Z. A. (2015). Ion-selective potentiometric determination of Pb(II) ions using PVC-based carboxymethyl cellulose Sn(IV) phosphate composite membrane electrode. *Desalination and Water Treatment*, *56*(3), 806–813. doi:10.1080/19443994.2014.941307

Inamuddin, R.T.A., & Khan, A. (2015). Synthesis of single-walled carbon nanotubes cerium(IV) phosphate composite cation exchanger: Ion exchange studies and its application as ion-selective membrane electrode for determination of Cd(II) ions. *Polymer Composites*. doi:10.1002/pc

Inamuddin, & Ismail, Y.A. (2010). Synthesis and characterization of electrically conducting poly-o-methoxyanilineZr(1V) molybdate Cd(II) selective composite cationexchanger. *Desalination, 250*, 523-529.

Islam, M., & Patel, R. (2008). Polyacrylamide thorium (IV) phosphate as an importantlead selective fibrousion exchanger: Synthesis, characterization and removal study. *Journal of Hazardous Materials*, *156*(1-3), 509–520. doi:10.1016/j.jhazmat.2007.12.046 PMID:18242841

Jain, A. K., Sondhi, S. M., & Sharma, V. K. (2000). Synthesis, characterization and Hg(II) ion selectivity of 1-(2-Nitro-4-methyl phenyl)-6-methyl-6-methoxy-1,4,5,6-tetrahydropyrimidine- 2-(3H) thione (TPT). *Electroanalysis*, *12*(4), 301–305. doi:10.1002/(SICI)1521-4109(20000301)12:4<301::AID-ELAN3013.0.CO;2-2

James, G. W., David, P. K., Gary, H. H., & Anton, M. S. (2002). *Handbook of Ecotoxicology* (2nd ed.). CRC Press.

Javed, M., & Usmani, N. (2013). Assessment of heavy metal (Cu, Ni, Fe, Co, Mn, Cr, Zn) pollution in effluent dominated rivulet water and their effect on glycogen metabolism and histology of Mastacembelusarmatus. *SpringerPlus*, *2*(1), 390–402. doi:10.1186/2193-1801-2-390 PMID:24133639

Kadirvelu, K., Thamaraiselvi, K., & Namasivayam, C. (2001). Removal of heavy metals from industrial wastewaters by adsorption onto activated carbon prepared from an agricultural solid waste. *Bioresource Technology*, *76*(1), 63–65. doi:10.1016/S0960-8524(00)00072-9 PMID:11315812

Kalfa, O. M., Yalcinkaya, O., & Turker, A. R. (2009). Synthesis of nano $B_2O_3/TiO_2$ composite material and its application to preconcentration and separation of cadmium. *Journal of Hazardous Materials*, *166*, 455–461. doi:10.1016/j.jhazmat.2008.11.112 PMID:19135793

Kang, S. Y., Lee, J. U., Moon, S. H., & Kim, K. W. (2004). Competitive adsorption characteristics of $Co^{2+}$, $Ni^{2+}$, and $Cr^{3+}$ by IRN 77 cation exchange resin in synthesized, wastewater. *Chemosphere*, *56*(2), 141–147. doi:10.1016/j.chemosphere.2004.02.004 PMID:15120560

Khan, A., Asiri, A.M., Rub, M.A., Azum, N., Khan, A.A.P., Khan, S.B., Rahman, M.M., & Khan, I. (2013). Synthesis, characterization of silver nanoparticle embedded polyaniline tungstophosphate-nanocomposite cation exchanger and its application for heavy metal selective membrane. *Composite B Engineering, 45*, 1486-1492.

Khan, A.A., & Inamuddin. (2006). Preparation, physico-chemical characterization, analytical applications and electrical conductivity measurement studies of an 'organic– inorganic' composite cation-exchanger: PolyanilineSn(IV) phosphate. *Reactive and Functional Polymer, 66*, 1649-1663.

Khan, A. A., & Akhtar, T. (2008). Preparation, physico-chemical characterization and electrical conductivity measurement studies of an organic–inorganic nanocomposite cation-exchanger: Poly-o-toluidine Zr(IV) phosphate. *Electrochimica Acta, 53*(17), 5540–5548. doi:10.1016/j.electacta.2008.03.002

Khan, A. A., & Akhtar, T. (2009). Synthesis, characterization and ion-exchange properties of a fibrous type 'polymeric-inorganic' composite cation-exchanger Nylon-6,6 Sn(IV) phosphate: Its application in making Hg(II) selective membrane electrode. *Electrochimica Acta, 54*(12), 3320–3329. doi:10.1016/j.electacta.2008.12.038

Khan, A. A., & Akhtar, T. (2011). Synthesis, characterization and analytical application of nano-composite cation-exchange material, poly-o-toluidine Ce(IV) phosphate: Its application in making Cd(II) ion selective membrane electrode. *Solid State Sciences, 13*(3), 559–568. doi:10.1016/j.solidstatesciences.2010.12.026

Khan, A. A., & Baig, U. (2012). Electrically conductive membrane of polyaniline–titanium(IV)phosphate cation exchange nanocomposite: Applicable for detection of Pb(II) using its ion-selective electrode. *Journal of Industrial and Engineering Chemistry, 18*(6), 1937–1944. doi:10.1016/j.jiec.2012.05.008

Khan, A. A., Habiba, U., & Khan, A. (2009). Synthesis and characterization of organicinorganic nano-composite poly-o-anisidinesn(IV) arsenophosphate: Its analytical applications as Pb(II) ion-selective membrane electrode. *International Journal of Analytical Chemistry. Article, ID659215*, 10.

Khan, A. A., & Inamuddin. (2006). Applications of Hg(II) sensitive polyanilineSn(IV) phosphate composite cation-exchange material in determination of $Hg^{2+}$ from aqueous solutions and in making ion-selective membrane electrode. *Sensors and Actuators. B, Chemical, 120*(1), 10–18. doi:10.1016/j.snb.2006.01.033

Khan, A. A., & Inamuddin. (2006). Preparation, physico-chemical characterization, analytical applications and electrical conductivity measurement studies of an 'organic–inorganic' composite cation-exchanger: Polyaniline Sn(IV) phosphate. *Reactive & Functional Polymers, 66*(12), 1649–1663. doi:10.1016/j.reactfunctpolym.2006.06.007

Khan, A. A., Inamuddin, & Alam, M. M. (2005). Preparation, characterization and analytical applications of a new and novel electrically conducting fibrous type polymeric–inorganic composite material: Polypyrrole Th(IV) phosphate used as a cation-exchanger and Pb(II) ion-selective membrane electrode. *Materials Research Bulletin, 40*(2), 289–305. doi:10.1016/j.materresbull.2004.10.014

Khan, A. A., & Khan, A. (2009). Synthesis, characterization and electrical conductivity measurement studies of poly-o-anisidine Sn(IV) phosphate nano-composite cation-exchange material. *Materials Science and Engineering B, 158*(1-3), 92–97. doi:10.1016/j.mseb.2009.01.024

Khan, A. A., & Khan, A. (2010). Ion-exchange studies on poly-o-anisidineSn(IV) phosphate nano composite and its application as Cd(II) ion-selective membrane electrode. *Central European Journal of Chemistry, 8*, 396–408.

Khan, A. A., Khan, A., & Inamuddin. (2007). Preparation and characterization of a new organic–inorganic nano-composite poly-o-toluidine Th(IV) phosphate: Its analytical applications as cation-exchanger and in making ion-selective electrode. *Talanta*, *72*(2), 699–710. doi:10.1016/j.talanta.2006.11.044 PMID:19071675

Khan, A. A., & Paquiza, L. (2011). Characterization and ion-exchange behavior of thermally stable nano-composite polyaniline zirconium titanium phosphate: Its analytical application in separation of toxic metals. *Desalination*, *265*(1-3), 242–254. doi:10.1016/j.desal.2010.07.058

Khan, A. A., Paquiza, L., & Khan, A. (2010). An advanced nano-composite cation-exchanger polypyrrole zirconium titanium phosphate as a Th(IV)-selective potentiometric sensor: Preparation, characterization and its analytical application. *Journal of Materials Science*, *45*(13), 3610–3625. doi:10.100710853-010-4407-6

Khan, A. A., & Shaheen, S. (2013). Synthesis and characterization of a novel hybrid nano composite cation exchanger poly-o-toluidine Sn(IV) tungstate: Its analytical applications as ion-selective electrode. *Solid State Sciences*, *16*, 158–167. doi:10.1016/j.solidstatesciences.2012.10.024

Khan, A. M., Ganai, S. A., & Nabi, S. A. (2009). Synthesis of a crystalline organic-inorganic composite exchanger, acrylamide stannic silicomolybdate: Binary and quantitative separation of metal ions. *Colloids and Surfaces. A, Physicochemical and Engineering Aspects*, *337*(1-3), 141–145. doi:10.1016/j.colsurfa.2008.12.012

Khan, M. A., Ahmad, A., Umar, K., & Nabi, S. A. (2015). Synthesis, characterization, and biological applications of nanocomposites for the removal of heavy metals and dyes. *Industrial & Engineering Chemistry Research*, *54*(1), 76–82. doi:10.1021/ie504148k

Khan, M. M. A., Rafiuddin, & Inamuddin. (2013). PVC based polyvinyl alcohol zinc oxide composite membrane: Synthesis and electrochemical characterization for heavy metal ions. *Journal of Industrial and Engineering Chemistry*, *19*(4), 1365–1370. doi:10.1016/j.jiec.2012.12.041

Khan, T. A. (2015). Heavy metal ion-exchange kinetic studies over cellulose acetate Zr(IV) molybdophosphate composite cation-exchanger. *Desalination and Water Treatment*, *53*(6), 1675–1682. doi:10.1080/19443994.2013.855676

Leonard, P., Hearty, S., Brennan, J., Dunne, L., Quinn, J., Chakraborty, T., & Kennedy, R. O. (2003). Advances in biosensors for detection of pathogens in food and water. *Enzyme and Microbial Technology*, *32*(1), 3–13. doi:10.1016/S0141-0229(02)00232-6

Li, K., Wang, Y. W., Huang, M., Yan, H., Yang, H., Xiao, S. J., & Li, A. M. (2015). Preparation of chitosan-graft-polyacrylamide magnetic composite microspheres for enhanced selective removal of mercury ions from water. *Journal of Colloid and Interface Science*, *455*, 261–270. doi:10.1016/j.jcis.2015.05.043 PMID:26073848

Low, K. S., & Lee, C. K. (1991). Cadmium uptake by the Moss, Calymperesdelessertii, Besch. *Bioresource Technology*, *38*(1), 1–6. doi:10.1016/0960-8524(91)90214-5

Mohammad, A., Inamuddin, & Hussain, S. (2014). Poly (3,4-ethylenedioxythiophene): polystyrene sulfonate (PEDOT:PSS) Zr(IV) phosphate composite cation exchanger: sol-gel synthesis and physico-chemical characterization. *Ionics*, *21*(4), 1063–1071. doi:10.100711581-014-1247-4

Moore, M., Gould, P., & Keary, B. S. (2003). Global urbanization and impact on health. *International Journal of Hygiene and Environmental Health*, *206*(4-5), 269–278. doi:10.1078/1438-4639-00223 PMID:12971682

Mozaz, S. R., de Alda, M. J. L., & Barcelo, D. (2004). Monitoring of estrogens, pesticides and bisphenol A in natural waters and drinking water treatment plants by solid phase extraction-liquid chromatography-mass spectrometry. *Journal of Chromatography. A*, *1045*(1-2), 85–92. doi:10.1016/j.chroma.2004.06.040 PMID:15378882

Nabi, S. A., Akhtar, A., Khan, M. D. A., & Khan, M. A. (2014). Synthesis, characterization and electrical conductivity of Polyaniline-Sn(IV)tung- stophosphate hybrid cation exchanger: Analytical application for removal of heavy metal ions from wastewater. *Desalination*, *340*, 73–83. doi:10.1016/j.desal.2014.02.020

Nabi, S. A., Bushra, R., Naushad, M., & Khan, A. M. (2010). Synthesis, characterization and analytical applications of a new composite cation exchange material poly-otoluidine stannic molybdate for the separation of toxic metal ions. *Chemical Engineering Journal*, *165*(2), 529–536. doi:10.1016/j.cej.2010.09.064

Nabi, S. A., Bushra, R., & Shahadat, M. (2012). Removal of toxic metal ions by using compositecation-exchange material. *Journal of Applied Polymer Science*, *125*(5), 3438–3446. doi:10.1002/app.36325

Nabi, S. A., Ganai, A. S., & Shalla, A. H. (2008). New organic-inorganic type acrylamide aluminumtungstate: Preparation, characterization and analytical applications as a cation exchange material. *Separation Science and Technology*, *43*(14), 3695–3711. doi:10.1080/01496390802212674

Nabi, S. A., Ganai, S. A., & Khan, A. M. (2011). Synthesis, characterization and ion exchange behavior of polyaniline stannic silicomolybdate, an organic–inorganic composite material: Quantitative separation of $Pb^{2+}$ ions from industrial effluents. *Journal of Inorganic and Organometallic Polymers*, *21*(1), 25–35. doi:10.100710904-010-9423-0

Nabi, S. A., & Naushad, M. (2008). Synthesis, characterization and analytical applications of a new composite cation exchanger cellulose acetate-Zr (IV) molybdophosphate. *Colloids and Surfaces. A, Physicochemical and Engineering Aspects*, *316*(1-3), 217–225. doi:10.1016/j.colsurfa.2007.09.005

Nabi, S. A., Naushad, M., & Bushra, R. (2009). Synthesis and characterization of a new organic–inorganic $Pb^{2+}$ selective composite cation exchanger acrylonitrile stannic(IV) tungstate and its analytical applications. *Chemical Engineering Journal*, *152*(1), 80–87. doi:10.1016/j.cej.2009.03.033

Nabi, S. A., & Shalla, A. H. (2009). EDTA-stannic (IV) iodate: Preparation, characterization and its analytical applications for metal content determination in real and synthetic samples. *Journal of Porous Materials*, *16*(5), 587–597. doi:10.100710934-008-9236-5

Nabi, S. A., & Shalla, A. H. (2009). Synthesis, characterization and analytical application of hybrid; Acrylamide zirconium (IV) arsenate a cation exchanger, effect of dielectric constant on distribution coefficient of metal ions. *Journal of Porous Materials*, *163*, 657–664. PMID:18722710

Naja, G. M., & Volesky, B. (2009). Treatment of metal-bearing effluents: removal and recovery. In Handbook on Heavy Metals in the Environment. Taylor and Francis.

Naushad, M., AL-Othman, Z. A., & Islam, M. (2013). Adsorption of cadmium ion using a new composite cation-exchanger polyaniline Sn(IV) silicate: Kinetics, thermodynamic and isotherm studies. *International Journal of Environmental Science and Technology*, *10*(3), 567–578. doi:10.100713762-013-0189-0

Nriagu, J. O., & Pacyna, J. M. (1988). Quantitative assessment of worldwide contamination of air, water and soils by trace metals. *Nature*, *333*(6169), 134–139. doi:10.1038/333134a0 PMID:3285219

Obasi, M. N. (1999). Environmental pollution with particular reference to chemical pollutant in Nigeria. White and White Publishers Ltd.

Pearce, J. M. (2007). Burton's line in lead poisoning. *European Neurology*, *57*(2), 118–119. doi:10.1159/000098100 PMID:17179719

Rangreez, T. A., Inamuddin, Naushad, M., & Ali, H. (2015). Synthesis and characterisation of poly(3,4-ethylenedioxythiophene)-poly(styrenesulfonate) (PEDOT:PSS) Zr(IV) monothiophosphate composite cation exchanger: analytical application in the selective separation of lead metal ions. *International Journal of Environmental Analytical Chemistry*, *95*(6), 556–568. doi:10.1080/03067319.2015.1036863

Sethi, A. C. (2001). Key note address. *International Conference on Fresh Water*, Bonn, Germany.

Shahadat, M., Bushra, R. (2015). Synthesis, characterization and significant applications of PANI-Zr(IV) sulphosalicylate nanocomposite. *Advances in Nanotechnology, 6*, 181-206.

Shahadat, M., Nabi, S. A., Bushra, R., Raeissi, A. S., Umar, K., & Ansari, M. O. (2012). Synthesis, characterization, photolytic degradation, electrical conductivity and applications of a nanocomposite adsorbent for the treatment of pollutants. *RSC Advances*, *2*(18), 7207–7220. doi:10.1039/c2ra20589k

Shahadat, M., Shalla, A. H., & Raeissi, A. S. (2012). Synthesis, characterization, and sorption behavior of a novel composite cation exchange adsorbent. *Industrial & Engineering Chemistry Research*, *51*(47), 15525–15529. doi:10.1021/ie3014555

Shamsipur, M., Dezaki, A. S., Akhond, M., Sharghi, H., Paziraee, Z., & Alizadeh, K. (2009). Novel PVC-membrane potentiometric sensors based on a recently synthesized sulfur-containing macrocyclicdiamide for $Cd^{2+}$ ion. Application to flow-injection Potentiometry. *Journal of Hazardous Materials*, *172*(2-3), 566–573. doi:10.1016/j.jhazmat.2009.07.003 PMID:19640638

Shannon, M. A., Bohn, P. W., Elimelech, M., Georgiadis, J. G., Marias, B. J., & Mayes, A. M. (2008). Science and technology for water purification in the coming decades. *Nature*, *452*(7185), 301–310. doi:10.1038/nature06599 PMID:18354474

Siddiqui, W.A., Khan, S.A., & Inamuddin. (2007). Synthesis, characterization and ionexchange properties of a new and novel 'organic–inorganic' hybrid cation- exchanger: Poly(methyl methacrylate) Zr(IV) phosphate. *Colloids and Surfaces A: Physicochemical and Engineering Aspect, 295*, 193-199.

Suffet, I. H., Ho, J., Chou, D., Khiari, D., & Mallevialle, J. (1995). Taste-and- odor problems observed during drinking water treatment. In I. H. Suffet, J. Mallevialle, & E. Kawczynski (Eds.), *Advances in Taste-and-odor Treatment and Control*. American Water Works Association.

Szewzyk, U., Szewzyk, R., Manz, W., & Schleifer, K. H. (2000). Microbiological Safety of Drinking Water. *Annual Review of Microbiology*, *54*(1), 81–127. doi:10.1146/annurev.micro.54.1.81 PMID:11018125

Tang, L., Fang, Y., Pang, Y., Zeng, G., Wang, J., Zhou, Y., ... Chen, J. (2014). Synergistic adsorption and reduction of hexavalent chromium using highly uniform polyaniline–magnetic mesoporous silica composite. *Chemical Engineering Journal*, *254*, 302–312. doi:10.1016/j.cej.2014.05.119

Thompson, H. S. (1850). On the absorbent power of soils. *Journal of the Royal Agricultural Society of England*, *11*, 68–74.

Ulusoy, U., & Akkaya, R. (2009). Adsorptive features of polyacrylamide-apatite composite for $Pb^{2+}$, $UO_2^{2+}$ and $Th^{4+}$. *Journal of Hazardous Materials*, *163*(1), 98–108. doi:10.1016/j.jhazmat.2008.06.064 PMID:18656311

United States Public Health. (2012). *Drinking water standards*. Author.

Urase, T., & Kikuta, T. (2005). Separate estimation of adsorption and degradation of pharmaceutical substances and estrogens in the activated sludge process. *Water Research*, *39*(7), 1289–1300. doi:10.1016/j.watres.2005.01.015 PMID:15862328

Way, J. T. (1850). On the power of soils to absorb manure. *Journal of the Royal Agricultural Society of England*, *11*, 313–379.

WHO and UNICEF. (2013). *Progress on Sanitation and Drinking Water*. Geneva: World Health Organization.

World Health Organisation. (2011). *Guidelines for Drinking Water Quality* (4th ed.).

Zhang, W., & Di Giano, F. A. (2002). Comparison of bacterial regrowth in distribution systems using free chlorine and chloramine: A statistical study of causative factors. *Water Research*, *36*(6), 1469–1482. doi:10.1016/S0043-1354(01)00361-X PMID:11996337

Zhang, Y., Li, Q., Sun, L., Tang, R., & Zhai, J. (2010). High efficient removal of mercury from aqueous solution by polyaniline/humic acid nanocomposite. *Journal of Hazardous Materials*, *175*(1-3), 404–409. doi:10.1016/j.jhazmat.2009.10.019 PMID:19896766

# Chapter 2
# Preparation and Application of Biochars for Organic and Microbial Control in Wastewater Treatment Regimes

**Victor Odhiambo Shikuku**
*Kaimosi Friends University College, Kenya*

**Wilfrida N. Nyairo**
*Maseno University, Kenya*

**Chrispin O. Kowenje**
*Maseno University, Kenya*

## ABSTRACT

*Biochars have been extensively applied in soil remediation, carbon sequestration, and in climate change mitigation. However, in recent years, there has been a significant increase in biochar research in water treatment due to their stupendous adsorptive properties for various contaminants. This is attributed to their large surface areas, pore structures, chemical compositions, and low capital costs involved making them suitable candidates for replacing activated carbons. This chapter discusses the preparation methods and properties of biochars and their removal efficacy for organic contaminants and microbial control. Factors affecting adsorption and the mechanisms of adsorption of organic pollutants on biochars are also concisely discussed. Biochars present environmentally benign and low-cost adsorbents for removal of both organic pollutants and microbial control for wastewater purification systems.*

## INTRODUCTION

Biochar, according to the International Biochar Initiative (IBI), is a solid material obtained by carbonization of biomass (IBI, 2013). Biochar is therefore a carbon-rich solid material derived by heating biomass, under limited or no oxygen (pyrolysis or carbonization) conditions (Sohi, 2012), with possible application

DOI: 10.4018/978-1-5225-5754-8.ch002

in soil for carbon sequestration and as an adsorbent. The varied conditions of carbonization of different parent feedstock materials result in infinite brands of biochars with significantly varied properties primarily intended for agricultural use such as soil remediation and the aforementioned carbon sequestration (Lehmann, 2007; Lehmann et al., 2011). Since 2005 to the present there has been a steady increase in the number of multi-disciplinary scientific publications with the word "biochar" as a key word. This is a testament on the increased biochar research within the scientific community. However, while a large portion of the publications relate to application of biochars as soil amendments, carbon storage agents and in mitigation of climate change, a significant fraction, around one-tenth of the publications between 2005 and 2013, relate to the use of biochars as environmentally benign adsorbents for removal of contaminants from water (Tan et al., 2015). This is attributed to their high surface area, presence of numerous functional groups, porous network and their mineral components which imbues the materials with high adsorptive properties. These attributes make biochar a potentially low-cost and effective adsorbent and a suitable alternative to activated carbons which are widely applied in conventional wastewater treatment regimes worldwide. In terms of capital costs involved, activated carbons require high preparation temperatures including activation steps whereas biochars require relatively lower temperatures hence reduced energy and infrastructural requirements besides being a renewable resource (Ahmad et al., 2012; Lu et al., 2012). Agricultural biomass and other solid wastes such as manures, sawdust, waste tyres, water hyacinth etc which form the precursor materials for development of biochar can be readily obtained in abundance. Furthermore, biochar synthesis adds economic value to these otherwise waste materials and would potentially eliminate their negative environmental impacts. Hence there is a synergistic "win-win" relation between environmental management and environmental protection (Dong et al., 2013). This chapter discusses the preparation methods and properties of the resulting biochars, the adsorptive potential of biochars for organic pollutants, factors affecting adsorption processes and the adsorption mechanisms of organic pollutants onto biochars. Application of biochars for microbial control is also discussed. The chapter demonstrates that biochars present an environmentally benign technology for water and wastewater treatment.

## PREPARATION METHODS AND PHYSICOCHEMICAL PROPERTIES OF BIOCHARS

Biochar, as earlier defined, is a porous material produced through thermochemical conversion of biomass in a minimally oxygenated high temperature environment (IBI, 2013). The production utilizes organic materials such as wood (Schimmelpfennig & Glaser, 2012; Liu et al., 2010), animal manure (Park et al., 2011) and agricultural waste (Demirbas, 2004). Most of these starting materials are readily available and quite often the choice of biomass will depend on the most abundant one in a given area.

Due to the diversity of these organic materials, field and laboratory studies have shown that the yielding biochars have different properties such as composition, porosity, BET surface area and total pore volume (Ozcimen and Ersoy-Mericboyu, 2010). Apart from the choice of initial feedstock biomass, the physicochemical properties of the final biochar product are also determined by preparation conditions such as heating temperature and reaction duration and activation methods applied (Onay, 2007). These properties play a vital role in determining their performance and application in the environment. The preparation methods used for biochar production include; pyrolysis, gasification and hydrothermal carbonization (HTC)

## Preparation Methods

### Pyrolysis

This is the most common method used for biochar production. It is a thermochemical decomposition of biomass through heating at elevated temperature in the absence of oxygen. The process results to the formation of three main products, which are; a solid product (biochar) which is carbon rich, a liquid phase (bio-oil) and 'non-condensable' gases like CO and $CH_4$ (Mohan et al., 2006). Pyrolysis is categorized as either fast or slow pyrolysis depending on the heating rate, temperature and residence time (Vamvuka, 2011). The biomass in slow pyrolysis is heated is at a temperature range of (300-650°C) at low heating rates (10-30°C/min) and relatively long residence time (a few minutes to a couple of hours). On the other hand, fast pyrolysis produces biochar at a high heating rates (above10-30°C/min) and short residence time (less than 10 s). Slow pyrolysis produces a higher solid biochar yield (25-35%) compared to fast pyrolysis.

### Gasification

This is partial combustion of biomass at high temperature (above 700°C) for a short residence time (10-20s) in the presence of a controlled amount of oxidizing agent. The primary product is a gaseous mixture (syngas containing CO, $H_2$, $CO_2$, $CH_4$, and small amounts of hydrocarbons). The oxidizing agent used in gasification is air, oxygen, steam or a mixture of these gases. The typical biochar yield of gasification averages about 10 wt% of biomass (Brewer et al, 2009; Qian et al, 2013)

### Hydrothermal Carbonization (HTC)

This is a thermochemical process that converts organic feedstock into a carbon rich solid product. HTC takes place in water at elevated temperature (180-260°C) and elevated reaction pressure (more than 1 atm) for 5-240 min (Hu et al, 2010). Unlike pyrolysis, HTC is not affected by the moisture content of the feedstock since it is carried out in water. This eliminates the financial load that accompanies pre-treatment of wet feedstock (Benavente at al., 2014). The HTC process results to the formation three main products which are; solid (hydrochar) (40-70%), liquid (bio-oil mixed with water) and fractions of gases (mainly $CO_2$)

## Physicochemical Properties of Biochars

Different thermochemical treatments have different operational parameters resulting to a final product with different physical and chemical characteristics. Surface morphology such as pore distribution, surface area and surface functionality are the properties of particular importance in the application of chars as adsorbents.

The Brunauer-Emmet-Teller (BET) surface area of the char particles formed via slow pyrolysis and HTC, particularly, has been reported to be affected mainly by high heating rate (HHR) and high reaction temperature (HRT). This is because HHR and HRT cause the melting of the cell structure of biomass thus reducing its surface area (Downie et al., 2009). An experiment on biochar obtained from wood species by Schimmelpfennig & Glaser (2012) showed that while an increase in temperature between (200-700°C) resulted to increase in BET surface area 10-500 $m^2/g$): a further increase in temperature

(800°C), however significantly reduced the BET surface area. On the other hand, during the pyrolysis of sunflower seeds by Angin (2013), at varying reaction temperatures of 400-600°C the surface area of the biochar was reported to reduce with increase in heating rates (10-50°C). Similar results have also been reported by other researchers (Downie et al., 2009).

In contrast, hydrochars have relatively low surface areas compared to biochars. In an experiment carried out on pinewood samples, a BET-surface area of 21 $m^2/g$ was reported for samples prepared through HTC (at 300°C for 20 min at the heating rate of 10°C/min) while with the same starting material under pyrolysis (at 700°C for 2 h at the heating rate of 10°C/min) (Liu et al., 2010).

## Biochars for Removal of Organic Contaminants

In this section, application of biochars in water purification is briefly discussed under two subsections, namely (a) removal of dyes (b) removal of antibiotics. A summary of selected biochars derived from various feedstocks is given in Table 1. The adsorption mechanisms and the factors influencing the adsorption process are also concisely discussed.

### Removal of Dyes

Dye contaminated industrial effluents present a serious challenge due to the complex composition of the wastewater such as toxic compounds, acids and/or bases, oil and greese, dissolved solids, besides color. Of all the contaminants, coloration stands out because of its visibility. Unfortunately, conventional wastewater treatment strategies are not well designed to remove color and textile dyes from water. This is due to the recalcitrance of the dyes to aerobic digestion and general stability towards light and oxidizing agents.

Hameed and El-khaiary (2008) prepared biochar by slow pyrolysis from rice husk biomass and examined its ability to sorb malachite green from synthetic wastewater. The biochar exhibited high adsorptive potential posting over 95% removal efficiency for initial pollutant concentration of 25 mg/L. The adsorption reaction was strongly pH dependant with maximum adsorption reported at pH 5 at 303 K. The theoretical Langmuir maximum adsorption capacity was 149 mg/g.

Hibiscus cannabinus char (HCC) was synthesized by at 1000°C and chemically activated through acid treatment and applied for sequestration of methylene blue (MB) (Mahmoud et al., 2012). Acid activation increased the surface area of the chars with higher oxygen content than the unmodified biochar. Consequently, the acid treated biochar possessed a higher adsorption capacity for MB than HCC.

In a separate study, bamboo wastes were pyrolyzed at 400–900°C to yield biochars. The amount of char produced was inversely proportional to the pyrolysis temperature possibly due to partial gasification. On the other hand, the surface area of the biochars increased with increase in pyrolysis temperatures. The bamboo derived biochar had adsorption potential for methylene blue, Acid blue 25 and acid yellow 117 with the highest adsorption capacity reported for methylene blue (Mui et al., 2010). In a related study, bamboo biochar was also shown to have the capacity for adsorbing metal complex dye acid black 172 from water (Yang et al., 2013). These results indicate that the performances of the biochars not only depend on the experimental conditions, type of dye, the biomass used as the precursor but also on the preparation conditions of the biochar.

## Removal of Antibiotics

Pharmaceuticals and personal care products (PPCPs) are a group of emerging contaminants and their use is continuously increasing around the world. They are extensively used in human and veterinary medicine to prevent illness and promote growth in fish and livestock farming as well as in agriculture. After administration, these chemicals can transform into more polar and soluble forms as metabolites or as conjugates of glucuronic and sulphuric acid (Heberer et al., 2002; Nikolaou et al., 2007). Pharmaceuticals and their metabolites are readily excreted with urine and faeces and enter into urban wastewater treatment plants (WWTPs). These compounds find their way into the aquatic environment through discharge of treated wastewater, seepage from landfills, septic systems, sewer lines and disposal of expired drugs into water systems among other pathways (González-Naranjo et al., 2013). Most of these compounds are found in hospital effluents and municipal sewage in high concentrations (Michael et al., 2013). Due to their toxic effects, their removal from water is of utmost importance. Unfortunately, conventional wastewater treatment strategies are insuffient for complete removal of PPCPs (Kimosop et al., 2016). Several biochars have been reported as possible efficacious and low-cost alternatives to commercial activated carbons and only a selected few are mentioned hereunder.

Water hycinth derived biochar prepared under slow pyrolysis at 350°C for 60 min was found to adsorb about 60% caffeine, a stimulant and constituent of analgesics, and ciprofloxacin from water within 5 min with 5mg/L initial concentration at 25°C (Ng'eno et al., 2016). The isotherm data was described by the Langmuir model while the kinetic data best fitted the pseudo-second order kinetic model. Thermodynamically, the adsorption processes were spontaneous and exothermic in nature. Beyond the $pK_a$ of ciprofloxacin, the adsorption extent was greatly influence by electrostatic repulsive forces. The adsorption was generally attributed to hydrophobic interactions. Baggasse biochar imbued with magnetic properties induced by impegnation with haematite ($\alpha$-$Fe_2O_3$) particles was demonstrated to have the capacity to adsorb carbamazepine, a pharmaceutical ingredient used for control of epileptic seizures and emergent water contaminant (Kimosop et al., 2017). The experimental data was described by the Freundlich model implying multilayer adsorption sequences. The adsorption capacity was strongly pH dependent with maximum adsorption occuring at pH 6. Noteworthy, the exhausted biochar was recoverable *via* an external magnetic field. Magnetization is an emerging technique for recovery of contaminant-laden adsorbent. Table 1 lists selected biochars for removal of different organic contaminants.

*Table 1. Various biochars for selected organic contaminants*

| Biochar precusor | Pyrolysis temperature C | Contaminant | Reference |
|---|---|---|---|
| Swine manure | 400 | Paraquat | (Tsai and Chen 2013) |
| Eucalyptus | 400 | Methylene Blue | (Sun et al., 2013) |
| Grass | 400 | Fluorinated herbicides | (Sun et al., 2011) |
| Orange peel | 300, 700 | Hydrophobic organic compounds | (Chen 2012a) |
| Pine wood shavings | 150-700 | Naphthalene | (Chen 2012b) |
| Brazilian pepper wood | 450, 600 | Sulfamethoxazole | (Yao et al., 2012) |

It is also conceivable that the presence of the mineral ions in the biochar network influenced the adsorption mechanism. Therefore, the characterization of the adsorption mechanisms is an important part of evaluation of the efficiency of an adsorbent for removal of certain contaminants. Adsorption mechanims are not only controlled by the reaction conditions and nature of the adsorbents but also by the properties of the contaminants. That is, inorganic contaminants such as heavy metals and organic molecules exhibit different reaction mechanisms. The following section discusses the adsorption mechanisms of organic compounds onto biochars. Generally, pore-filling, coulombic interactions, hydrogen bonding and hydrophobic interactions represent the main mechanisms of organic molecules adsorption onto biochars. As aforementioned, though the specific interactions vary between different organic molecules, they can adequately be correlated with the chemical and surface properties of the biochars. In terms of the surface properties, the difference in mechanistic processes involved can be attributed to the presence of carbonized and non-carbonized phases of the biochar which interact in varied ways with the adsorbates.

## Pore Filling

Biochars are porous materials characterized by presence of micropores (less than 2 nm), mesopores (between 2–50 nm), and macropores (above 50 nm). However, it is argued that micropores and small mesopores (2–20 nm) constitute the bulk to the biochar surface area and therefore play a significant role in the sequestration of organic compounds (Pignatello et al., 2006). It follows therefore that pore-filling as an adsorption mechanism of organic molecules onto biochars becomes predominant and is controlled by the total micropore and mesopore volumes of the adsorbents especially low adsorbate concentrations (Hao et al., 2013; Kasozi et al., 2010). To evaluate the pore-filling effects and rate determining steps, the sorption kinetics data is fitted to various kinetic models and the linearity of the plots examined. Among the widely used models, the intra-particle diffusion model is the most common amongst researchers. In pore-filling mechanisms, the kinetic diameter of the organic adsorbates relative to the pore diameters of the biochars play a significant role in promoting or inhibiting the adsorption mechanism (Nguyen et al., 2007).

## Hydrophobic Interactions

Adsorption of organic compounds by the hydrophobic interaction mechanism comprises both adsorptions of the molecules onto the organic matter network found in the non-carbonized fraction of the biochar, also called partitioning and hydrophobic adsorption (Murphy et al., 1994). Freshly prepared biochars with relatively minimum surface oxidation are generally hydrophobic in nature, and can therefore adsorb hydrophobic or neutral organic moeties by both partitioning adsorption and hydrophobic interaction mechanisms. For example, the adsorption of ciprofloxacin onto water hyacinth derived biochar was reported to be dominated by hydrophobic interactions (Ng'eno et al., 2016). Compared to partitioning however, hydrophobic adsorption occurs on sorbent surfaces with low hydration energy, as a consequence of competitive effects between the sorbate apolar molecules and water molecules (Murphy et al., 1994; Zhu et al., 2005). Therefore, in a binary or multi-sorbate adsorption system, maximum uptake of the competing molecules is a function of their octanol–water partition coefficients ($K_{ow}$), a measure of hydrophobicity (Murphy et al., 1994).

## Electrostatic Interaction

Coulombic attractions constitute one of the main mechanisms for the uptake of ionic forms of organic compounds in solution (Zheng et al., 2013). In principle, unlike charges attract and like charges repel. Therefore, positively charged organic species are adsorbed onto negatively charged biochar surfaces. On the other hand, mineral-laden biochars tend to have positively charged binding sites which would attract anionic organic compounds (Ahmad et al., 2014). The net surface charge of biochar is a function of the solution pH, and the pH at which the surface charge is electrically neutral is known as the point of zero charge ($pH_{pzc}$). Below the $pH_{pzc}$, the surface charge of the biochar is positively charged, whereas at pH above $pH_{pzc}$, the surface is becomes negative (Mukherjee et al., 2011). The solution pH also affects the ionic form in which the organic compound exists depending on its $pK_a$. Generally, at pH below the $pK_a$ value, the compound is protonated and exists in its cationic form. Therefore, if the solution pH is below both the $pH_{pzc}$ and $pK_a$ of the biochar and organic adsorbate, respectively, both the adsorbent surface and the adsorbate will possess a net positive charge and electrostatic repulsion is expected. The same occurs when the solution pH is above both the biochar $pH_{pzc}$ and the ionizable compound's $pK_a$ value. Maximum adsorption is therefore achieved when the adsorbent surface and the sorbate molecules posses opposite charges. These effects were demonstrated for the adsorption of carbamazepine onto biochar prepared from baggasse (Kimosop et al., 2017). To overcome the difficulty of electrostatic repulsion, the ionic strength of the solution should be increased to promote sorption (Inyang et al., 2014). However, increase in the ionic strength of the bulk solutions decreases adsorption capacity when the coulombic interactions are attractive and the adsorbates are cationic (e.g $CaCl_2$) due to competition between the cations from the solution used to control ionic strength and the cationic organic molecules.

## Hydrogen Bonding

Hydrogen bonding (H-bonding) is a plausible mechanism for the sorption of polar organic compounds on biochars (Sun et al., 2011; Chen et al., 2015). The abundant polar groups, such as phenols, on biochar matrices promotes adsorption of water molecules and consequently enhances H-bonding between biochars or bound water molecules and the electronegative elements, such as oxygen atoms, on the organic molecules (Chen et al., 2015).

It is important to note that the adsorption process is often multi-mechanistic in nature involving several mechanisms operating simultaneously. For example, Qiu et al. (2009) proposed π–π interactions between dye molecules and graphene layers of biochar, coulombic attractive and repulsive forces and hydrogen bonding as the adsorption mechanisms adsorption of dye onto a straw-derived biochar. In a separate study, Zhang et al. (2013) prepared pig manure-based biochars and proposed a multi-mechanistic adsorption process for removal of atrazine and carbaryl involving hydrophobic interactions, pore-filling and π–π electron donor–acceptor interactions. Figure 1 portrays a summary of the adsorption mechanisms.

*Figure 1. Summary of adsorption mechanisms of organic pollutants onto biochars*

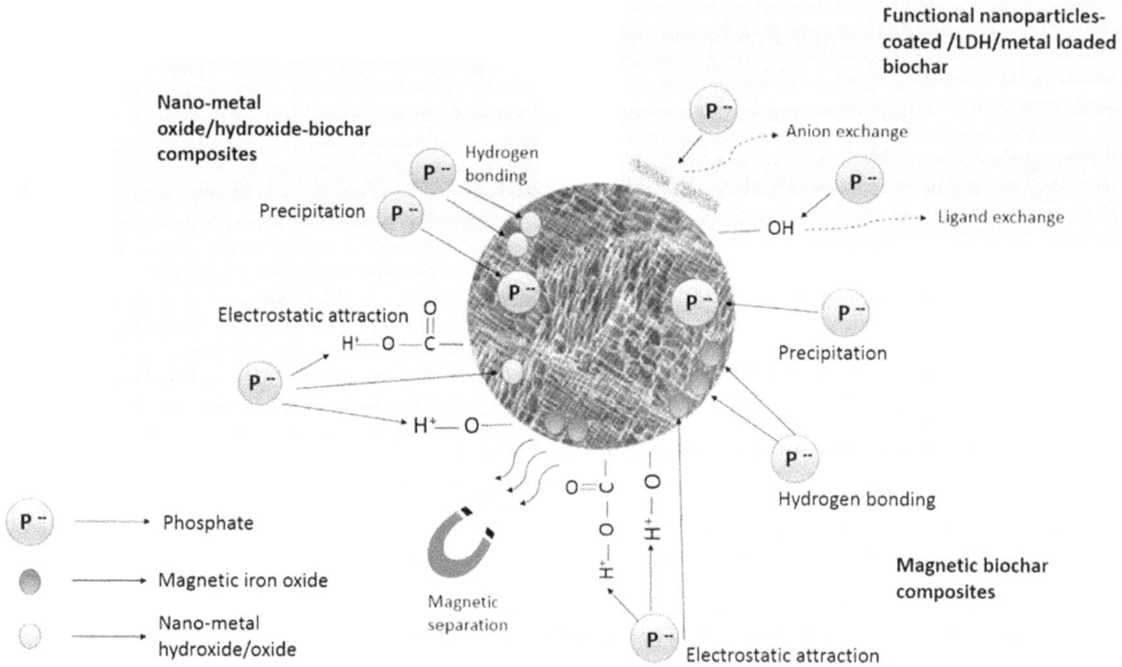


## FACTORS INFLUENCING ADSORPTION OF ORGANIC CONTAMINANTS ONTO BIOCHARS

The performance of biochars for removal of organic contaminants is a function of the experimental conditions such as temperature, pH, solute-adsorbent ratio, and presence of competing ions and the properties of the biochars including the pre-treatment effects.

### Effect of Biochar Properties

As earlier mentioned, the preparation conditions of the biochar, such as temperature, residence time, source of biomass and the thermochemical technology used, significantly influences the properties of the biochar which consequently affects its adsorptive capability. The pyrolitic conditions have been shown to directly affect the properties of the generated biochar. These properties in turn have a corresponding effect on the adsorption capacity of the adsorbent for specific contaminants. Among the key parameters is the pyrolysis temperature that has been shown to have a direct impact on the structural features of the biochar. Specifically, increase in temperature resulted to biochars with higher specific surface areas and degree in carbonization resulting to increased adsorption capability (Chen et al., 2008). These findings are corroborated by findings by Ahman and co-workers (2013) whereby biochars generated under high pyrolysis temperature exhibited higher adsorption capacities for trichloroethylene. Under similar preparation conditions, the properties of the chars are influenced by the type of biomass used due to variance in their mineral compositions (Sun et al., 2014). Similar findings, of varied adsorption capacities for biochars derived from different feedstocks, were obtained by Xu et al. (2011) when biochars from four different feedstocks namely canola, peanut, soybean and rice hull straws prepared under similar conditions were evaluated for their ability to adsorb methyl violet. Finally, the biochar properties are not only influenced by the pyrolysis parameters such as temperature and residence time and the type of biomass used but also properties are also impacted by the type of pyrolysis process employed. Biochars produced by hydrothermal carbonization vary in properties from those produced by pyrolysis and consequently their adsorption capacities would vary depending on the type of adsorbate (Kumar et al., 2011).

## Effect of pH

Solution pH is a fundamental factor that influences the efficiency of adsorption processes. The solution pH affects the adsorbent surface charge and the speciation and degree of ionization of the adsorbate thereby influencing both the efficiency of removal and mechanisms controlling the adsorption process. Biochars, like other adsorbent materials, possess various surface functional groups whose activities are influenced by the solution pH. The functional groups include carboxylates, carbonyls and hydroxyl groups among others. As earlier discussed, for solution pH below the so-called point of zero charge ($pH_{pzc}$) of the biochar, the net surface charge of the biochar is positive which would only promote adsorption of anionic adsorbates while inhibiting the uptake of cationic ones (Ng'eno et al., 2016). Besides electrostatic repulsion to cationic adsorbates for a positively charged adsorbent surface at low solution pH, the decreased adsorption is also attributed to competition between the adsorbate cations and $H^+$ and $H_3O^+$ ions for the adsorption sites. This phenomenon is well portrayed in study on the adsorption of a cationic dye onto a food waste derived biochar (Parshetti et al., 2014). In the study, the adsorption capacity of the biochar was significantly high at alkaline pH where the adsorbent surface charge was negative and the adsorption evidently attributed to electrostatic attraction.

## Effect of Solute-Sorbent Ratio

Increase in adsorbent dosages generally result in increase in extent of adsorption due to increased number of active sites (Tsai and Chen, 2013). However, other studies have reported a decrease in adsorption efficiencies with increase in adsorbent dosage (Chen et al., 2011). This can be attributed to overlapping of the binding sites with increase in adsorbent particles. Therefore, to achieve synergistic combination of capital costs involved and maximum removal efficiency, empirical determination of optimum adsorbent dosage is paramount. Theoretical models, such as the Langmuir isotherm model, can be used for prediction of the adsorbent dosages required for large scale real-life water purification systems (Shikuku et al., 2017).

## Effect of Temperature

Temperature changes have direct impact on adsorption processes, namely: under pre-equilibrium conditions, temperature changes alter the rate of adsorption while under post-equilibrium conditions; temperature changes shift the equilibrium position of the adsorbent for a specific adsorbate (Shikuku et al., 2017). Generally, adsorption processes involving biochars tend to be endothermic in nature, denoted by positive change in enthalpy values ($\Delta H$). As such, the amount of adsorbate migrating to the solid phase increases with increase in temperature (Chen et al., 2011; Zhang et al., 2013). This can be explained in terms of increase in diffusion rate of the adsorbate molecules with rise in temperature due to increased thermal energy. The change in free energy ($\Delta G$) reflects the thermodynamic favorability of the adsorption process where negative and positive $\Delta G$ values imply a spontaneous and non-spontaneous process, respectively. Noteworthy, increase in magnitudes of $\Delta G$, for a spontaneous process, with rise in temperatures denotes increased favorability at high temperatures (Parshetti et al., 2014). It is important to note that the thermodynamic parameters, $\Delta H$ and $\Delta G$, also give insights into the adsorption mechanisms involved. In literature, $\Delta H$ values below 40 kJ/mol are associated with physisorption mechanisms (Shikuku et al., 2017).

## DISINFECTION AND MICROBIAL CONTROL

Chlorination and ozonation are the most widely used techniques for disinfection and microbial control attributed to their demonstrable effectiveness (Li et al., 2008). However, the reactions between the natural organic matter (NOM) in water and the applied disinfectants, chlorine or ozone, result in formation of toxic disinfection by-products (DBPs). This has called for search for alternative effective and safer disinfection strategies. In recent years, studies have shown that certain transition metal oxide nanoparticles such as zinc oxide and titanium dioxide, have stupendous inherent capacity to significantly reduce the toxicity of bacterial pathogens (e.g E. coli and Staphylococcus aureus) in wastewater down to zero concentration levels (Adams et al., 2006; Ren et al., 2009). For instance, silver nanoparticles are known for their antimicrobial and toxic effects on bacteria, such as Pseudomonas aeruginosa and E. coli, largely present in wastewaters (Li et al., 2008). In a separate study, biochar pyrolytically derived from bamboo biomass and chemically modified with magnetic zero-valent iron (ZVI) particles restricted the growth of E. Coli by nearly 50% relative to the blank control samples (Zhou et al., 2014). Zhou and co-workers (2014) also reported that the ZVI-modified bamboo biochars augmented with Ag ions (AgNPs-ZVI) exhibited 100% inhibition to the growth of E. coli. These results indicate that biochars impregnated with such like nanoparticles are potential cost effective and environmentally benign alternative strategies for inhibition and control of microbial growth with concomitant reduction in the formation of toxic disinfection by-products during water treatment.

## POTENTIAL LIMITATION TO BIOCHAR APPLICATION

Despite the self-evident adsorption efficacy and low capital investments associated with biochars, possible negative effects of biochars must be evaluated before practical use. Among the most serious potential limitations of biochars is their chemical composition. Leaching of toxic chemical constituents of biochars, such as extractable heavy metals and polycyclic aromatic hydrocarbons (PAHs) compounds, into the aqueous system during treatment presents a concern requiring monitoring and control. It may suffice to say that, generally, empirical data indicates that toxic heavy metals associated with charred biomass is minimal and fall within tolerable limits (Freddo et al., 2012; Agrafioti et al., 2013). However, this does not negate the need for monitoring of leachate quality. Pyrolysis has been documented to be appropriate technique for biochar development that inhibits possible release of toxic heavy metals into the aqueous media (Zhang et al., 2013). The variation of the toxic constituents with type of biomass and pyrolysis temperature and technology used has been demonstrated. Nevertheless, there still remains a paucity of information on the interplay and influence of preparation conditions with the chemical composition of the biochars generated. This calls for further research before extrapolation of laboratory scale results into large scale field application.

## REFERENCES

Adams, L., Lyon, D., & Alvarez, P. (2006). Comparative eco-toxicity of nanoscale TiO2, SiO2, and ZnO water suspensions. *Water Resources*, *40*(19), 3527–3532. PMID:17011015

Agrafioti, E., Kalderis, D., & Diamadopoulos, E. (2013). Ca and Fe modified biochars as adsorbents of arsenic and chromium in aqueous solutions. *Journal of Environmental Management*, *146*, 444–450. doi:10.1016/j.jenvman.2014.07.029 PMID:25199600

Ahmad, M., Lee, S. S., Dou, X., Mohan, D., Sung, J. K., Yang, J. E., & Ok, Y. S. (2012). Effects of pyrolysis temperature on soybean stover-and peanut shell-derived biochar properties and TCE adsorption in water. *Bioresource Technology*, *118*, 536–544. doi:10.1016/j.biortech.2012.05.042 PMID:22721877

Ahmad, M., Lee, S. S., Rajapaksha, A. U., Vithanage, M., Zhang, M., Cho, J. S., ... Ok, Y. S. (2013). Trichloroethylene adsorption by pine needle biochars produced at various pyrolysis temperatures. *Bioresource Technology*, *143*, 615–622. doi:10.1016/j.biortech.2013.06.033 PMID:23838320

Ahmad, M., Rajapaksha, A., Lim, J., Zhang, M., Bolan, N., Mohan, D., ... Ok, Y. (2014). Biochar as a sorbent for contaminant management in soil and water: A review. *Chemosphere*, *99*, 19–33. doi:10.1016/j.chemosphere.2013.10.071 PMID:24289982

Angin, D. (2013). Effect of pyrolysis temperature and heating rate on biochar obtained from pyrolysis of safflower seeds press cake. *Bioresource Technology*, *128*, 593–597. doi:10.1016/j.biortech.2012.10.150 PMID:23211485

Benavente, V., Calabuig, E., & Fullanan, A. (2014). (in press). Upgrading of moist agro-industrial wastes by hydrothermal carbonization. *Journal of Analytical and Applied Pyrolysis*.

Brewer, C. E., Schmidt-Rohr, K., Satrio, J. A., & Brown, R. C. (2009). Characterization of biochar from fast pyrolysis and gasification systems. *Environmental Progress & Sustainable Energy*, *28*(3), 386–396. doi:10.1002/ep.10378

Chen, B., Zhou, D., & Zhu, L. (2008). Transitional adsorption and partition of nonpolar and polar aromatic contaminants by biochars of pine needles with different pyrolytic temperatures. *Environmental Science & Technology*, *42*(14), 5137–5143. doi:10.1021/es8002684 PMID:18754360

Chen, C., Zhou, W., & Lin, D. (2015). Sorption characteristics of N-nitrosodimethylamine onto biochar from aqueous solution. *Bioresource Technology*, *179*, 359–366. doi:10.1016/j.biortech.2014.12.059 PMID:25553566

Chen, X., Chen, G., Chen, L., Chen, Y., Lehmann, J., McBride, M. B., & Hay, A. G. (2011). Adsorption of copper and zinc by biochars produced from pyrolysis of hardwood and corn straw in aqueous solution. *Bioresource Technology*, *102*(19), 8877–8884. doi:10.1016/j.biortech.2011.06.078 PMID:21764299

Chen, Z., Chen, B., & Chiou, C. T. (2012b). Fast and slow rates of naphthalene sorption to biochars produced at different temperatures. *Environmental Science & Technology*, *46*(20), 11104–11111. doi:10.1021/es302345e PMID:22970831

Chen, Z., Chen, B., Zhou, D., & Chen, W. (2012a). Bisolute sorption and thermodynamic behavior of organic pollutants to biomass-derived biochars at two pyrolytic temperatures. *Environmental Science & Technology*, *46*(22), 12476–12483. doi:10.1021/es303351e PMID:23121559

Demirbas, A. (2004). Effects of temperature and particle size on bio-char yield from pyrolysis of agricultural residues. *Journal of Analytical and Applied Pyrolysis*, *72*(2), 243–248. doi:10.1016/j.jaap.2004.07.003

Dong, X., Ma, L. Q., Zhu, Y., Li, Y., & Gu, B. (2013). Mechanistic investigation of mercury sorption by Brazilian pepper biochars of different pyrolytic temperatures based on X-ray photoelectron spectroscopy and flow calorimetry. *Environmental Science & Technology*, *47*(21), 12156–12164. doi:10.1021/es4017816 PMID:24040905

Downie, A., Crosky, A., & Munroe, P. (2009). Physical properties of biochar. In J. Lehmann & S. Joseph (Eds.), Biochar for environmental management: Science and technology (pp. 13–32). Academic Press.

Freddo, A., Cai, C., & Reid, J. (2012). Environmental contextualisation of potential toxic elements and polycyclic aromatic hydrocarbons in biochar. *Environmental Pollution*, *171*, 18–24. doi:10.1016/j.envpol.2012.07.009 PMID:22863991

González-Naranjo, V., Boltes, K., & Biel, M. (2013). Mobility of ibuprofen, a persistent active drug, in soils irrigated with reclaimed water. *Plant, Soil and Environment*, *59*(2), 68–73. doi:10.17221/590/2012-PSE

Hameed, B. H., & El-Khaiary, M. I. (2008). Kinetics and equilibrium studies of malachite green adsorption on rice straw-derived char. *Journal of Hazardous Materials*, *153*(1-2), 701–708. doi:10.1016/j.jhazmat.2007.09.019 PMID:17942219

Hao, F., Zhao, X., Ouyang, W., Lin, C., Chen, S., Shan, Y., & Lai, X. (2013). Molecular structure of corncob-derived biochars and the mechanism of atrazine sorption. *Journal of Agronomy*, *105*(3), 773–782. doi:10.2134/agronj2012.0311

Heberer, T. (2002). Occurrence, fate, and removal of pharmaceutical residues in the aquatic environment: A review of recent research data. *Toxicology Letters*, *131*(1-2), 5–17. doi:10.1016/S0378-4274(02)00041-3 PMID:11988354

Hu, B., Wang, K., Wu, L., Yu, S.-H., Antonietti, M., & Titirici, M.-M. (2010). Engineering carbon materials from the hydrothermal carbonization process of biomass. *Advanced Materials*, *22*(7), 813–828. doi:10.1002/adma.200902812 PMID:20217791

IBI. (2013). *Standardized product definition and product testing guidelines for biochar that is used in soil.* International Biochar Initiative.

Inyang, M., Gao, B., Zimmerman, A., Zhang, M., & Chen, H. (2014). Synthesis, characterization, and dye sorption ability of carbon nanotube–biochar nanocomposites. *Chemical Engineering Journal, 236*, 39–46. doi:10.1016/j.cej.2013.09.074

Kasozi, G. N., Zimmerman, A. R., Nkedi-Kizza, P., & Gao, B. (2010). Catechol and humic acid sorption onto a range of laboratory-produced black carbons (Biochars). *Environmental Science & Technology, 44*(16), 6189–6195. doi:10.1021/es1014423 PMID:20669904

Kimosop, S. J., Getenga, Z. M., Orata, F., Okello, V. A., & Cheruiyot, J. K. (2016). Residue levels and discharge loads of antibiotics in wastewater treatment plants (WWTPs), hospital lagoons, and rivers within Lake Victoria Basin, Kenya. *Environmental Monitoring and Assessment, 188*(9), 532. doi:10.100710661-016-5534-6 PMID:27553946

Kimosop, S. J., Okello, V. A., Orata, F., Getenga, Z. M., & Shikuku, V. O. (2017). Green Remediation of Carbamazepine from Water Using Novel Magnetic Iron Modified Carbonized Baggasse: Kinetics, Equilibrium and Mechanistic Studies. *Chemical Science International Journal, 18*(3), 1–9. doi:10.9734/CSJI/2017/32444

Kumar, S., Loganathan, V. A., Gupta, R. B., & Barnett, M. O. (2011). An assessment of U (VI) removal from groundwater using biochar produced from hydrothermal carbonization. *Journal of Environmental Management, 92*(10), 2504–2512. doi:10.1016/j.jenvman.2011.05.013 PMID:21665352

Lehmann, J. (2007). A handful of carbon. *Nature, 447*(7141), 143–144. doi:10.1038/447143a PMID:17495905

Lehmann, J., Rillig, M. C., Thies, J., Masiello, C. A., Hockaday, W. C., & Crowley, D. (2011). Biochar effects on soil biota—a review. *Soil Biology & Biochemistry, 43*(9), 1812–1836. doi:10.1016/j.soilbio.2011.04.022

Li, Q., Mahendra, S., Lyon, D., Brunet, L., Liga, M., Li, D., & Alvarez, P. (2008). Antimicrobial nanomaterials for water disinfection and microbial control: Potential applications and implications. *Water Resources, 42*(18), 4591–4602. PMID:18804836

Liu, Z., Zhang, F.-S., & Wu, J. (2010). Characterization and application of chars produced from pinewood pyrolysis and hydrothermal treatment. *Fuel, 89*(2), 510–514. doi:10.1016/j.fuel.2009.08.042

Lu, H., Zhang, W., Yang, Y., Huang, X., Wang, S., & Qiu, R. (2012). Relative distribution of $Pb^{2+}$ sorption mechanisms by sludge-derived biochar. *Water Resources, 46*, 854–862. PMID:22189294

Mahmoud, D. K., Salleh, M. A. M., Karim, W. A. W. A., Idris, A., & Abidin, Z. Z. (2012). Batch adsorption of basic dye using acid treated kenaf fibre char: Equilibrium, kinetic and thermodynamic studies. *Chemical Engineering Journal, 181–182*(1), 449–457. doi:10.1016/j.cej.2011.11.116

Michael, I., Rizzo, L., McArdell, C. S., Manaia, C. M., Merlin, C., Schwartz, T., ... Fatta-Kassinos, D. (2013). Urban wastewater treatment plants as hotspots for the release of antibiotics in the environment: A review. *Water Resources, 47*, 957–995. PMID:23266388

Mohan, D., Pittman, C. U., & Steele, P. H. (2006). Pyrolysis of wood/biomass for bio-oil, a critical review. *Energy & Fuels*, *20*(3), 848–889. doi:10.1021/ef0502397

Mui, E. L. K., Cheung, W. H., Valix, M., & McKay, G. (2010). Dye adsorption onto char from bamboo. *Journal of Hazardous Materials*, *177*(1–3), 1001–1005. doi:10.1016/j.jhazmat.2010.01.018 PMID:20097002

Mukherjee, A., Zimmerman, A., & Harris, W. (2011). Surface chemistry variations among a series of laboratory-produced biochars. *Geoderma*, *163*(3–4), 247–255. doi:10.1016/j.geoderma.2011.04.021

Murphy, E., Zachara, J., Smith, S., Phillips, J., & Wietsma, T. (1994). Interaction of hydrophobic organic compounds with mineral-bound humic substances. *Environmental Science & Technology*, *28*(7), 1291–1299. doi:10.1021/es00056a017 PMID:22176321

Ng'eno, E., Orata, F., Lilechi, D. B., Shikuku, V. O., & Kimosop, S. (2016). Adsorption of caffeine and ciprofloxacin onto pyrolytically derived water hyacinth biochar: Isothermal, kinetics and thermodynamics. *Journal of Chemistry and Chemical Engineering*, *10*, 185194.

Nguyen, T., Cho, H., Poster, D., & Ball, W. (2007). Evidence for a pore-filling mechanism in the adsorption of aromatic hydrocarbons to a natural wood char. *Environmental Science & Technology*, *41*(4), 1212–1217. doi:10.1021/es0617845 PMID:17593721

Nikolaou, A., Meric, S., & Fatta, D. (2007). Occurrence patterns of pharmaceuticals in water and wastewater environments. *Analytical and Bioanalytical Chemistry*, *387*(4), 1225–1234. doi:10.100700216-006-1035-8 PMID:17205270

Onay, O. (2007). Influence of pyrolysis temperature and heating rate on the production of bio-oil and char from sunflower seed by pyrolysis, using a well-swept fixed-bed reaction. *Fuel Processing Technology*, *88*(5), 523–553. doi:10.1016/j.fuproc.2007.01.001

Ozcimen, D., & Ersoy-Mericboyu, A. (2010). Characterization of biochar and bio-oil samples obtained from carbonization of various biomass materials. *Renewable Energy*, *35*(6), 1319–1324. doi:10.1016/j.renene.2009.11.042

Park, J. H., Choppola, G. K., Bolan, N. S., Chung, J. W., & Chuasavathi, T. (2011). Biochar reduces the bioavailability and phytotoxicity of heavy metals. *Plant and Soil*, *348*(1-2), 439–451. doi:10.100711104-011-0948-y

Parshetti, G. K., Chowdhury, S., & Balasubramanian, R. (2014). Hydrothermal conversion of urban food waste to chars for removal of textile dyes from contaminated waters. *Bioresource Technology*, *161*, 310–319. doi:10.1016/j.biortech.2014.03.087 PMID:24727353

Pignatello, J., Kwon, S., & Lu, Y. (2006). Effect of natural organic substances on the surface and adsorptive properties of environmental black carbon (char): Attenuation of surface activity by humic and fulvic acids. *Environmental Science & Technology*, *40*(24), 7757–7763. doi:10.1021/es061307m PMID:17256524

Qian, K., Kumar, A., Patil, K., Bellmer, D., Wang, D., Yuan, W., & Huhnke, R. (2013). Effects of biomasss feedstocks and gasification conditions on the physicochemical properties of char. *Energies*, *6*(8), 3972–3986. doi:10.3390/en6083972

Qiu, Y., Zheng, Z., Zhou, Z., & Sheng, G. D. (2009). Effectiveness and mechanisms of dye adsorption on a straw-based biochar. *Bioresource Technology*, *100*(21), 5348–5351. doi:10.1016/j.biortech.2009.05.054 PMID:19540756

Ren, G., Hu, D., Cheng, E., Vargasreus, M., Reip, P., & Allaker, R. (2009). Characterisation of copper oxide nanoparticles for antimicrobial applications. *International Journal of Antimicrobial Agents*, *33*(6), 587–590. doi:10.1016/j.ijantimicag.2008.12.004 PMID:19195845

Schimmelpfennig, S., & Glaser, B. (2012). One step forward toward characterization: Some important material properties to distinguish biochars. *Journal of Environmental Quality*, *41*(4), 1001–1013. doi:10.2134/jeq2011.0146 PMID:22751042

Shikuku, V.O., Zanella, R., Kowenje, C.O., Donato, F.F., Bandeira, N., & Prestes, O.D. (2017). Single and Competitive Removal of Sulfachloropyridazine and Sulfadimethoxine onto Natural Kaolinite Clay: Kinetics, Isotherms and Thermodynamics Studies. *South African Journal of Chemistry*, *70*.

Sohi, S. P. (2012). Carbon storage with benefits. *Science*, *338*(6110), 1034–1035. doi:10.1126cience.1225987 PMID:23180849

Sun, J., Lian, F., Liu, Z., Zhu, L., & Song, Z. (2014). Biochars derived from various crop straws: Characterization and Cd (II) removal potential. *Ecotoxicology and Environmental Safety*, *106*, 226–231. doi:10.1016/j.ecoenv.2014.04.042 PMID:24859708

Sun, K., Keiluweit, M., Kleber, M., Pan, Z., & Xing, B. (2011). Sorption of fluorinated herbicides to plant biomass-derived biochars as a function of molecular structure. *Bioresource Technology*, *102*(21), 9897–9903. doi:10.1016/j.biortech.2011.08.036 PMID:21907572

Sun, L., Wan, S., & Luo, W. (2013). Biochars prepared from anaerobic digestion residue, palm bark, and eucalyptus for adsorption of cationic methylene blue dye: Characterization, equilibrium, and kinetic studies. *Bioresource Technology*, *140*, 406–413. doi:10.1016/j.biortech.2013.04.116 PMID:23714096

Tan, X., Yunguo, L., Guangming, Z., Xin, W., Xinjiang, H., Zhongzhu, Y., & Yanling, G. (2015). Application of biochar for the removal of pollutants from aqueous solutions. *Chemosphere*, *125*, 70–85. doi:10.1016/j.chemosphere.2014.12.058 PMID:25618190

Tsai, W. T., & Chen, H. R. (2013). Adsorption kinetics of herbicide paraquat in aqueous solution onto a low-cost adsorbent, swine-manure-derived biochar. *International Journal of Environmental Science and Technology*, *10*(6), 1349–1356. doi:10.100713762-012-0174-z

Vamvuka, D. (2011). Bio-oil, solid and gaseous biofuels fro biomass pyrolysis processes- an overview. *International Journal of Energy Research*, *35*(10), 835–862. doi:10.1002/er.1804

Xu, R. K., Xiao, S. C., Yuan, J. H., & Zhao, A. Z. (2011). Adsorption of methyl violet from aqueous solutions by the biochars derived from crop residues. *Bioresource Technology*, *102*(22), 10293–10298. doi:10.1016/j.biortech.2011.08.089 PMID:21924897

Yang, Y., Lin, X., Wei, B., Zhao, Y., & Wang, J. (2013). Evaluation of adsorption potential of bamboo biochar for metal-complex dye: Equilibrium, kinetics and artificial neural network modeling. *International Journal of Environmental Science and Technology*. doi:10.100713762-013-0306-0

Yao, Y., Gao, B., Chen, H., Jiang, L., Inyang, M., Zimmerman, A. R., ... Li, H. (2012). Adsorption of sulfamethoxazole on biochar and its impact on reclaimed water irrigation. *Journal of Hazardous Materials*, *209*, 408–413. doi:10.1016/j.jhazmat.2012.01.046 PMID:22321858

Zhang, P., Sun, H., Yu, L., & Sun, T. (2013). Adsorption and catalytic hydrolysis of carbaryl and atrazine on pig manure-derived biochars: Impact of structural properties of biochars. *Journal of Hazardous Materials*, *244*, 217–224. doi:10.1016/j.jhazmat.2012.11.046 PMID:23246958

Zheng, H., Wang, Z., Zhao, J., Herbert, S., & Xing, B. (2013). Sorption of antibiotic sulfamethoxazole varies with biochars produced at different temperatures. *Environmental Pollution*, *181*, 60–67. doi:10.1016/j.envpol.2013.05.056 PMID:23811180

Zhou, Y., Gao, B., Zimmerman, A., & Cao, X. (2014). Biochar-supported zerovalent iron reclaims silver from aqueous solution to form antimicrobial nanocomposite. *Chemosphere*, *117*, 801–805. doi:10.1016/j.chemosphere.2014.10.057 PMID:25461951

Zhu, D., Kwon, S., & Pignatello, J. (2005). Adsorption of single-ring organic compounds to wood charcoals prepared under different thermochemical conditions. *International Journal of Environmental Science and Technology*, *39*(11), 3990–3998. PMID:15984774

# Chapter 3
# Soil Bioremediation Techniques

**P. Senthil Kumar**
*SSN College of Engineering, India*

**Femina Carolin C.**
*SSN College of Engineering, India*

## ABSTRACT

*Soil pollution is rising rapidly due to the existence of pollutants or natural alterations in the soil. It makes the drinking water ineffective and unusable by the human beings. The major cause of the soil contamination is agricultural activities, industrial activities, and inadmissible disposal of waste in the soil. The most common pollutants to accumulate in the soil are petroleum hydrocarbons, solvents, pesticides, lead, and other heavy metals. The important technology to remediate the pollutants or contaminants in the soil is bioremediation. The utilization of bioremediation in the contaminated soil is increasing rapidly due to the presence of toxic pollutants. It is the most advanced technologies which make use of organisms to deteriorate the harmful compounds in order to prevent the soil pollution. The aim of the chapter is to describe the available bioremediation technologies and their application in removing the pollutants exist in the soil.*

## INTRODUCTION

Issues related to the soil contamination is arising due to industrial activities, wastewater discharge and incorrect usage of chemical fertilizers and pesticides (Tang et al., 2015). Soil contamination with hazardous chemicals and toxins results in quick effect to the characteristics of the earth. Technology development for the restore of the polluted environment in an eco-friendly approach is increasing. The major worry in the world is soil contamination. Advancements in the treatment technologies are required because of the presence of contaminants which affect the biological system and human well-being. Soil contamination with toxic compounds rises ecological and health concerns. Site contamination is increasing rapidly that attracts many countries to develop a treatment technology. Nowadays, the hazard of contaminated soil on human health is high, leads to efforts has to be taken to remediate the sites and redeveloped for its use (Zouboulis and Moussas, 2011). The plants grown in the soil absorbs toxins from the soil. Entry of these crops into the human beings causes congenital illnesses and endless medical issues which

DOI: 10.4018/978-1-5225-5754-8.ch003

cannot be cured effectively. Prolonged food poisoning also occurs due to the intake of these plants and crops. Through the inward breath of vaporized contaminants, it specifically influences the individuals. Due to the soil pollutants, ground water also gets affected that represents an extraordinary danger to the individuals by the utilization and moreover contaminated related sick are generated. Neurological problems are provoked to the young children due to the availability of trace quantities of metals in the soil. The arrival of volatile compounds from the air increases air pollution and the substantial metals reach into the ground water, lakes and oceans through run-off which results in water pollution. By consuming non-renewable energy resources like fossil fuels in the soil, the activity and growth of microbes get affected. Soil disintegration is another issue which is made by binding of fungi and bacteria together to decay the soil properties. The land becomes unsatisfactory for agricultural business due to its loss of fertility of the soil. The absence of nutrients in the soil decreases the generation of vegetables and natural products. Consuming lot of chemicals by the creatures changes the natural way of life, expanded death rates and even creature eradication. It demonstrates negative impacts which influence the metabolic exercises of creatures and arthropods. Traditional strategies like the removal of contaminated soil and transfer into the landfill site have some specific disadvantages. Presently, several technologies are usable for the treatment of contaminated soil and the treatments techniques are incineration, soil washing, soil flushing, dechlorination, ultra violet oxidation, thermal desorption etc. These technologies are more expensive and incomplete degradation of contaminants occurs. Hence biological treatment like bioremediation are chosen for the treatment of pollutants and plays a vital role in degrading a wide range of organic contaminants exist in the soil. Soil remediation is a natural depletion method which expels the contaminants by breaking the poisonous natural compounds into non-lethal natural compounds accumulated in the soil. Soil remediation comprises of different methods used to renovate the soil. These species consume contaminants and other pollutants that act as a medium for the growth of organisms in the soil. Microorganisms which are included in the bioremediation procedure are called bioremediators. Bioremediation is a biological conversion of hazardous compounds into non-hazardous compounds. It is a promising green innovation technique with the utilization of microbes like bacteria, fungi, algae, and plants. The few organisms involved in the bioremediation of soil contaminants are described in Table 1. By providing a suitable environment to the microorganisms, microbial degradation gets stimulated and consume contaminants and pollutants as s food for their growth. Bioremediation technology does not need a homogenous environment to degrade the pollutants like other conventional techniques. To carry out bioremediation procedure experience and specific approaches in ecology, microbiology, chemistry, and geology are needed. So far bioremediation usage is increasing worldwide with a great success. The intent of this proposal is to bring out the bioremediation techniques available for the corruption of pollutants and contaminants present in the soil. This information will help to identify the appropriate treatment technology required for the removal contaminants and pollutants from the soil.

## BIOREMEDIATION PROCESS

Interest related to the development of a biological system for the remediation technologies have been increased due to their application in the eradication of pollutants at the contaminated site (Grobelak et al., 2015). Researchers introduced various bioremediation techniques based on the nature and type of the pollutant. Through bioremediation process, the concentration of end product is lower than the concentration limit suggested by government authorities. Bioremediation is a biological mechanism to

*Table 1. Types of microorganism involved in bioremediation of soil*

| S.No | Bacteria | Fungi |
|---|---|---|
| 1. | *Pseudomonas sp.* | *Penicillium sp* |
| 2. | *Alcaligenes sp.* | *Trichoderma sp* |
| 3. | *Enterococcus sp.* | *Aspergillus sp* |
| 4. | *Bacillus sp.* | *Phlebia* |
| 5. | *Thiobacillus sp.* | *Fusarium oxysporum* |
| 6. | *Methosinus sp* | *Claviceps sp* |
| 7. | *Methanogens* | *Rhizopus sp* |
| 8. | *Streptomyces sp* | *Mucor sp* |

decrease the concentration of a pollutant at the pollutant affected site. It employs the living organisms or can be separated from the site and incorporated into the treated material with the help of bioreactors. It is generally accomplished through bioaugmentation and biostimulation based on conditions of soil and microbial population (Lin et al., 2010). Low population of organisms and improper microbial diversities affect the performance of bioremediation process. Thus, expanding the capacity of the microbial degraders by bioaugmentation or advancing the movement of indigenous microbial degraders by biostimulation could move forward bioremediation proficiency towards soil contaminated site. Monitoring of microbial population and its activity helps to find the efficiency of bioremediation. In bioaugmentation process, exogenous microbes are applied to increase the microbial community. In biostimulation nutrient sources like nitrogen and phosphorous are employed in order to develop the bioremediation method. Hence the role of bioaugmentation and biostimulation is high in case of bioremediation efficiency. Bioremediation application is expanding a result of its financial and ecological remediation innovation. It is found most applied method in the removal of toxins from the soil than the physical and chemical techniques. It is reasonable for the treatment of different components like volatile organic compounds, benzene, toluene, ethyl benzene, xylene (BTEX) compounds, phenolic compounds, polycyclic aromatic hydrocarbons (PAHs) (particularly the simpler aromatic compounds), petroleum hydrocarbons, nitroaromatic compounds, metals, complex (high molecular weight) PAHs, chlorinated hydrocarbons. Furthermore, it breaks down the pollutants into carbon-dioxide and methane. The selection principle for the bioremediation process is pollutant region, characteristics of pollutants in the soil, cost, and type of surroundings etc (Frutos et al., 2012; Smith et al., 2015).

## TYPES OF SOIL BIOREMEDIATION STRATEGIES

To satisfy the requirements of bioremediation at each remediation site, a different type of approaches have been developed. Depending on the saturation and contaminated site soil bioremediation is classified into two types for the exclusion of pollutants and contaminants from the soil. The techniques are in-situ and ex-situ bioremediation process. These remediation processes are a good alternative to the traditional cleanup technologies. They have great potential in dealing with the wide range of pollutants present in the soil.

## In-Situ Bioremediation Process

In this procedure, the contaminant which was expected to bioremediate are not required to isolate from the site and hence there is no disturbance in the distribution or concentration of a pollutant in the soil (Azubuike et al., 2016). Consequently, there is a less interference in the surface of the soil. The organisms degrade the pollutants by the inclusion of nutrients towards the soil surface. It is considered as a more affordable strategy for the declination of toxins in the soil. It is a most desirable method because it avoids the circulation of pollutants from one area to another. Also, this process brings out the important activities between the biomass and pollutants and upgrades the microbial degradation at the toxic polluted soil. The major disadvantage in the in-situ process is, the lower layer or the depth of the soil cannot be treated properly and less effective than the ex-situ process. This procedure relies on pH, temperature, the presence of nutrients and moisture content. The utilization of in-situ process expanded because of the porosity of the soil. In-situ remediation includes methods like bioventing, biosparging, bioslurping and phytoremediation. Based on the metabolic activities, the in-situ bioremediation process is categorized into two types. The high-level categories of metabolism are an aerobic and anaerobic process. The contaminants like fuel hydrocarbons are degraded through the aerobic pathway, some contaminants like carbon tetrachloride undergo anaerobic process and trichloroethene are degraded either by the aerobic or anaerobic process. Soil porosity enhances the usage of in-situ bioremediation process at different polluted sites. In-situ remediation process provides good technical, economic and environmental advantages and the cost of operation is cheap than the ex-situ process (Velimirovic et al., 2014).

### Bioventing

Bioventing is a general method in in-situ treatment and employed in large scale because of its compatibility. In order to rise degradation rate of pollutants, this method joins with the physical venting process. This procedure builds the bioremediation procedure through the supply of oxygen in the unsaturated soil zone to stimulate the bioremediation by raising the activities of microorganisms. Bioventing is typically connected in the high-temperature regions. The achievement of bioventing process relies on air flow rates. It is an integration of volatilization with bioremediation process. This method can be resolved by two techniques like air injection or by vapor extraction in which vapor treatment is required and costlier. These two techniques supply the oxygen to the indigenous organisms and stimulate bioremediation of toxins (Hohener and Ponsin, 2014). The compounds which undergo aerobic degradation are more suitable for bioventing process. The presence of oxygen in the soil determines the metabolism that takes place in the soil. Aerobic metabolism gives more benefits than another metabolic process because it produces more energy (Meneghetti et al., 2012). Due to the pressure gradient on the surface of the soil, atmospheric oxygen flows on the subsurface and leads to aerobic decomposition. Nitrogen salts and phosphorous are additionally included as an added substance in the contaminated zone to transfer the polluted substance into a harmless form. Permeability and size of the particle allow air flow into the soil. The conveyance of these supplement sources into the soil depends on the arrangement of the soil. The nutrient dispersion is constrained in the earth soil as a result of its low penetrability. The existence of moisture content in the clay is also high because the high surface area allows the water to retain in it. In the passive technology of bioventing process exchange of gas takes place through the atmospheric pressure and in the active technology of bioventing process the air pumped into the site through blower or pump. The bioventing process is reasonable to the coarse finished soil. The utilization of this kind

*Figure 1. Bioventing of contaminated soil*

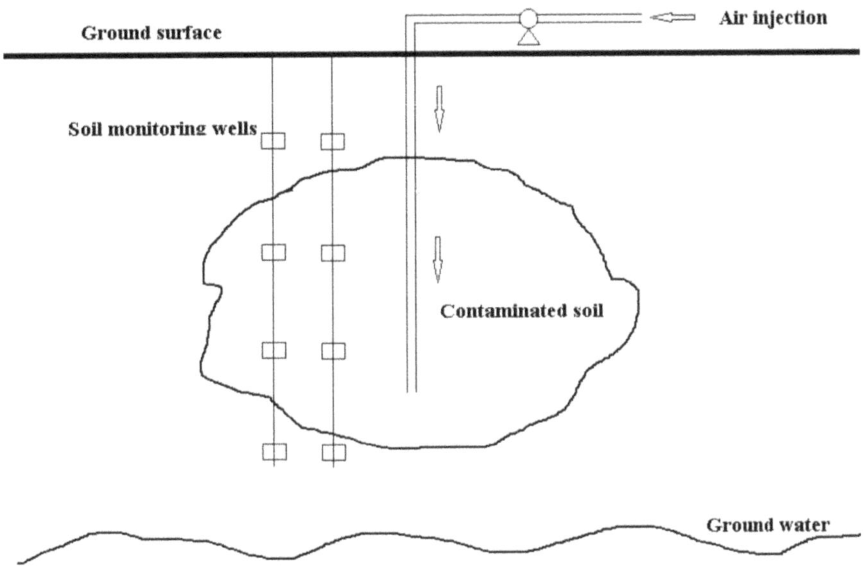


of in-situ process has been expanded to the restoring sites polluted with petroleum products, aromatic hydrocarbons, and non-volatile hydraulic oils. The schematic representation of bioventing process was depicted in Figure 1.

## Biosparging

This technique was established in the year 1980 in order to give enhanced maintenance of moisture content. Like bioventing strategy, biosparging system supplies the air into the immersed and unsaturated soil zone which causes the decrease of energy consumption (Kao et al., 2008). Because of the infusion of air at the immersed zone, the unpredictable contaminants in the soil are treated at the surface of the soil to empower the biodegradation procedure. Additionally, oxygen concentration gets increased leads to the stimulation of degradation rate by the existence of microbial populations. It is more identical to the process like in-situ air sparging which depends on air flow rates and biosparging increases bioremediation of pollutants in the soil. The elements which impact the biosparging procedure are soil penetrability and contamination biodegradability. Blowers/compressors, piping, and air injection wells are the components required for the sir sparging system (Johnson et al., 2010). In biosparging method, there is a controlled release of nutrients, oxygen, and microbes to regulate the moisture content. By injecting pressure into the soil mass, the diffusion of material takes place in the macropores and micropores of the soil. Wells are created around the soil and soil vapor extracted. This soil vapor is treated for the biodegradation of volatile compounds and nutrients are added to the treated vapor stream. Then the vapor stream is administered into the contaminated soil below the ground water level. It is widely used in the soil contaminated site because it can easily adapt with other remediation technologies. This treatment is commonly applied in the treatment of petroleum products like diesel and kerosene. Forecasting the airflow is difficult in this process. It is less expensive in the installation of air injector, monitoring, sampling and easy to design and construct due to its flexibility.

## Bioslurping

Bioslurping is an exclusive strategy for an in-situ process which implement vacuum to isolate the soil vapor and free product, for example, light non-fluid stage fluids (LNAPLs) from the polluted zone. Sometimes LNAPLs hold in the soil within the unsaturated zone. They can also present in the form of soil gas in the zone. Bioslurping systems implement atmospheric pressure to extract soil gas from the soil pores. The segregated vapor and product are treated and this method demonstrates very successfully because the only small amount of soil vapor is pumped at once henceforth the requirement of the treatment plant is low to treat the extracted soil. This method act as a conventional bioventing process in order to remove the free products completely so that complete remediation takes place (Kim et al., 2014). It is applicable to the soil polluted with the petroleum hydrocarbons. The bioslurping process additionally controlled in the soil harmed with volatile and semi-volatile organic compounds. The drawback of this procedure is, the microbial movement is low because of the low oxygen exchange rate. Also, enormous humidity in the soil decreases microbial activity and restricts the permeability into the soil. But this approach needs to be selected after analyzing the characteristics of the contaminated site.

## Phytoremediation

Phytoremediation is a moderately later field of research and application because of its low cost effective and promising methodology to clean up the pollutants at different polluted sites. Phytoremediation is less expensive which utilized solar energy to clean up the sites polluted with fewer contaminants. It is an in-situ method which utilizes the plants to remediate the toxins from the polluted soil. The plant based remediation process has been shown as a powerful innovation for the treatment of soil. The plants growing in the contaminates soil are called good phytoremediators. The plants can be grown, harvested and disposed of easily in a natural manner. Usually, the plants have connections between soil, water, and microorganisms over millions of years. They produce enzymes or uptake the contaminants from the soil for the degradation process. Plants act as natural surroundings to aerobic and anaerobic microorganisms to rapidly increase the degrading capacity. The phytoremediation process strongly depends on the factors like nutrients, soil pH, contaminant concentration and so on. Based on capacity, plants are chosen to tolerate and accumulate pollutants (Mesa et al., 2015). It is an accessible technique and additionally, recovery and reuse of products are possible. Deep rooted trees, grasses, legumes, and aquatic plants are convoluted in the phytoremediation strategy. Mostly in phytoremediation process, hyper accumulators are used for absorbing the pollutants with their roots. These hyperaccumulators absorb metals than the normal plants because it has growth rate, rapid biomass production and fights against pests and diseases. About 400 species were identified as hyper accumulators. Sometimes pre-treatment of the polluted site is required that stimulates the activity of microbes and decreases toxic effect to plants. Plants diminish toxicity and uptake soil pollutants through various mechanisms. The mechanisms involving in the phytoremediation process are phytoextraction, rhizofiltration, phytodegradation, immobilization, phytovolatilization etc and these mechanisms depend on the kind of toxin present in the soil. Description of phytoremediation mechanisms is explained in Table 2. The inorganic pollutants are degraded by the mechanisms like phytoextraction, transformation and sequestration methods and organic pollutants are remediated by the mechanisms like phytovolatilization, phytostabilization, and mineralization process. Both rhizospheric microorganisms and endophytic bacteria play a major role in the degradation of harmful impurities in the soil. Rhizospheric bacteria stimulates biomass growth and patience towards the heavy metals available in the soil. Some metals get accumulated in the plants and regained after remediation process which is known as phytomining (Wu et al., 2015). Table 3 represents some of the inorganic pollutants removed by plant based bioremediation process. Plants moderates the development of soil contaminants stimulates evapotranspiration and uptake the contaminants through roots. The upsides of phytoremediation process are less expensive, eco-friendly, low maintenance cost, protection of soil, low installation, avoidance of erosion, transfer of biomass for energy, assisting biodiversity and so on (Ali et al., 2013, Thijs et al., 2016). Though it has some advantages, this technique has certain limitations like metal accumulators in plants reduces biomass growth rate, tightly bound part of metals in the soil are difficult to isolate from the soil, plant development is not good in the site contaminated with highly concentrated metal pollutants, absence of appropriate care leads to risk of food chain. Plants are autotrophic hence the enzymes need to degrade the organic pollutants into carbon-di-oxide and water which was found to be a major drawback in the phytoremediation process (Lee, 2013). Another approach involved in the detoxification of pollutants in the phytoremediation process are transgenic plants. Using genetic engineering method, transgenic plants are produced which is used to decontaminate the pollutants in the soil. It is necessary to increase the activity of microorganisms present in soil and plants, contaminant convenience and their availability and need to understand response between the plants and microbes to stimulate the efficiency of phytoremediation process (Becerra-Castro et al., 2013). Modifications in the growth of microorganisms influence the organisms to respond to the plants for the detoxification process. It is widely applied to clean up the pollutants like metals, pesticides, solvents, explosives, unrefined petroleum, polyaromatic hydrocarbons and landfill leachates.

## Ex-Situ Bioremediation Process

In this process, the soil is removed from the polluted site, then treated and returned back to the treated area. Ex-situ bioremediation process is costly and more complicated than in-situ method. Without damaging top soil plants remediate the pollutants and moderates its usage and fertility of the soil. This is generally applied process for soil remediation because of its appropriate usage in the environment, easiness, and consent with the governmental protocols. The disadvantages in ex-situ method are health risk which is

*Table 2. Different type of phytoremediation mechanisms and their description*

| S.NO | Mechanisms | Description |
|---|---|---|
| 1. | Phytovolatilization | This process volatilizes the contaminants to the surroundings |
| 2. | Phyto-stabilization | It destabilizes the contaminants by the compounds produced from the plants |
| 3. | Rhizosphere biodegradation | Natural substances released from the plant roots act as a nutrient to the soil microbes and enhances biodegradation |
| 4. | Phytoaccumulation | Using roots, the organic contaminants in the soil are transferred from the soil to the plants |
| 5. | Phytodesalination | This method involved to exclude salts from the saline soils by using halophytes |

Source: Archaya et al., 2014; Ali et al., 2013

*Table 3. Different types of plants involved in the phytoremediation of inorganic pollutants*

| S.NO | Name of inorganic pollutant | Plant name | References |
|---|---|---|---|
| 1. | Cadmium | *Allium cepa* | (Chen et al., 2014) |
| 2. | Cobalt | *Trifolium nigriscens* | (Bani et al., 2014) |
| 3. | Chromium | *Mylopharyngodon* | (Li et al., 2015) |
| 4. | Copper | *Karelinia caspica* | (Liu et al., 2016) |
| 5. | Nickel | *Daucus carotta* | (Szolnoki and Farsang, 2013) |
| 6. | Zinc | *Spinacia oleraceae* | (Ferri et al., 2015) |
| 7. | Lead | *Alcea aucheri* | (Ravanbakhsh et al., 2016) |

related because of the exchange of contaminated site. Simple monitoring this system and time requirement is low to clear the contaminants. It is incorporated with composting, landfarming, biopiling and bioreactors. The most part of the ex-situ bioremediation process occurs by solid phase and slurry phase bioremediation process. The slurry stage bioremediation happens in the reactor where the nutrients and water get blended to keep the contact with the nutrients and organisms. The solid phase bioremediation treats the contaminants over the ground surface. This procedure takes a long time to remediate the poisons and requires substantial space for the treatment. The major upside of the bioremediation process is pre-treatment of polluted soil before the remediation are not required which makes less expensive than the in-situ bioremediation. In the ex-situ process, the uneven distribution of pollutant in the soil can be easily solved by optimizing environmental conditions.

## Composting

Composting is characterized as a conservative and environment-friendly biological procedure of microbial degradation of pollutant by the group of indigenous microorganisms (Zhao et al., 2017). The types of microorganisms' rich in the composting process are mesophilic and thermophilic microorganisms. It is a natural process in which organic wastes are changed over by microorganisms through aerobic and anaerobic conditions to harmless. Utilization of indigenous microorganisms in the composting process heat results in degradation natural material in the waste. During composting process, appropriate thermophilic conditions must be maintained properly. Composting happens at a temperature range from 55° to 65°C. These temperature ranges are mostly maintained by the indigenous microorganisms. Since it is a temperature dependent process, the metabolic movement and contaminants solubility get expanded in the manures. Organic wastes like wood, hay, manure, and vegetative waste get involved with the contaminated soil to improve the porosity of the blend to be disintegrated. In order to stabilize the end product like an organic waste, compost is generally used which increases characteristics of the soil, provides nutrients to the plants and reduces soil erosion. This method uses rotating drums, enclosed tunnels, and trenches to carry out the bioremediation process. The compost plays a dual role like bioremediation and soil amendment. After getting the nutrients from the compost by microbes, degradation of soil contaminants takes place (Taiwo et al., 2016). These natural sources give porosity, carbon and nitrogen sources which improve the degradation or breakdown of contaminants. By the process of composting, organic compounds get transformed into organic constituents like humic substances in the soil (Lashermes et al., 2012). It has successfully used to remediate the soil contaminated with oils, PAHs, organic compounds etc.There are three types of process designs are used in composting process. They are aerated static pile composting, mechanically agitated vessel composting and windrow composting. Windrow composting is considered to be a most effective alternative technique for treating the contaminated soil. After the excavation of contaminated soil, rocks and other debris are eliminated. Materials like manure and agricultural waste are used and layered into piles which are known as windrows. Subsequently, the windrows are dismantled and the manure is taken to the last transfer region. Most extreme degradation productivity is accomplished through looking after oxygenation, irrigation and checking moisture content and temperature. This method has certain limitations and concerns like the requirement of space and costs, uncontrolled release of volatile organic compounds, increase in volume due to the addition of inclusion of bulking agents and incomplete degradation leads to careful disposal of end products.

## Landfarming

It is the simplest strategy due to its less equipment required for the operation. The soil is expelled and isolated through the mechanical operation called sieving and it is applied to the prepared area for the natural degradation to occur. The contaminated soil layer is put over the cleaned soil of around 0.4 meters thick and aerobic degradation takes place by autochthonous microorganisms (Silva-Castro et al., 2015). It does not provide large surface area for the treatment, instead, the contaminated soil part is removed and kept in the treatment beds with the supply of aeration. The sheets like synthetic, concrete, or clay membrane are utilized to cover the polluted soil layer. Oxygen, nutrients and moisture content is provided to stimulate the aerobic microorganisms to stimulate remediation process. Metabolic activity of the aerobic microorganisms varies based on the factor like pH, temperature, humidity, soil characteristics and so on (Silva-Castro et al., 2013). Crushed limestone or agricultural lime are used to control pH during the landfarming procedure. The guideline behind this procedure is to advance and quicken the natural degradation of contaminants in the soil. As per the government regulations, the landfarming can be used at any climatic conditions and locations. Mostly landfarming occurs naturally but sometimes require biostimulation and bioaugmentation to enhance this method. The heterotrophic organisms showed best degradation capacity in remediating the soil pollutants. Based on biodegradation, the parameters like soil nutrients, pH, buffer capacity and humidity are controlled and optimized. Although it is the simplest technique, it is restricted in certain applications because of its large operating space, low microbial activity, low efficiency in inorganic pollutant etc. It is also not adequate in the hot regions of soil. It is highly useful for treating the soil contaminated with petroleum hydrocarbons, volatile and semi-volatile organic compounds, wood wastes, sludges etc.

## Biopiling

This procedure happens around few weeks to several months to degrade the pollutants in the soil. It relies on the elements like oxygen, nutrients, pH, temperature, moisture content and so on. This technique is often viewed because of its low cost, and more sustainable than the other techniques (Lors et al., 2012). Soil amendments combined with scraped soil and form into piles enclosed for the bioremediation to occur. Depending upon the sieving process and aeration, biopiling efficiency increases. The biopiling procedure comprises of layers like air circulation system with contaminated soil, nutrient system, and leachate collection system. This leachate system used to regulate the petroleum constituents in the soil through the biodegradation process. Through the vacuum or pressure, the aeration system transfers the air into the soil. Covering the piles with the material prevents the exhaust of vapors into the atmosphere. Plastics which was used to cover the soil in this procedure used to control evaporation, volatilization, and entry of contaminants into the uncontaminated site. Since the temperature, air circulation and nutrients are regulated in the system which provides efficiency towards this particular ex-situ technique (Whelan et al 2015). If necessary, additional vacuum aeration system can be incorporated in order to collect the volatile compounds. The reduction of moisture content in the soil helps to prevent the soil constituents from leaching. At high-temperature soil, biopiling is highly efficient but the application of biopiling in the cold zones are remained untested. The remediation time required in the biopiling is less due to its adaptability and high rate of degradation occurs in this system in which heating system incorporated that raises the microbial activity. Materials like straw, sawdust, wooden barks stimulate the remediation method in biopiling system. It also utilized in pilot scale with the same performance achieved in lab scale studies (Chemlal et al., 2013). It is commonly applied to decrease petroleum constituents in polluted soil. It is applicable for the treatment of volatile organic compounds, pesticides and hydrocarbons and so on.

## Bioreactors

For quick and protected bioremediation process, mostly bioreactors are chosen for the treatment of contaminated soil (Pino- Herrera et al., 2017). The bioreactor is a system in which the contaminants are treated in the media with the help of microorganisms. By series of biological mechanisms, pollutants are converted into a specific product which takes place in a closed container or vessel. In this process, microorganisms play a major role in the degradation of pollutants by mixing in the closed compartment. It overcomes the disadvantages of other ex-situ methods. Continuous bioremediation of pollutants can be obtained in the bioreactor by regulating the conditions of pH, moisture, temperature, nutrients, and oxygen. Extended mixing in the bioreactor leads to the increase in diffusion, contact, and desorption of contaminants from the soil. The in-situ methods provide poor mixing that causes low permeability and creates a struggle in the inclusion of nutrients and oxygen for the remediation takes place. But this problem can be overcome by the utilization of aeration systems which enhances the mixing process during the remediation time. Conditions in the reactor enhance the optimum growth of microbes. The mixing takes place in the reactor between the nutrients, contaminated soil, water, and microbes. It is generally utilized for the treatment of mud soils and it was observed to be a quick process. Generally, the reaction takes place in the vessel which is joined with the mixing system, aeration pumps for influent and effluent. The different performing methods of bioreactors are batch, fed-batch, sequencing batch, continuous and multistage. Capital investment and market economy decide the selection of bioreactor for bioremediation. This method is contradictory to the in-situ method because controlled environment is maintained in the bioreactor which results in the success of ex-situ method. The types of bioreactors are plug flow, fluidized bed, slurry bioreactors, vapor phase bioreactors and so on. Figure 2 represents pictorial representation of soil remediation using bioreactors. The productivity of the bioreactor relies on the elements like the presence of nutrients, pollutant bioavailability, and mass transfer between the organisms and the pollutant exist in the soil. It can also be used to treat the soil contaminated with volatile organic compounds. Moreover, it provides a high rate of degradation than the in-situ technologies due to the controlled environmental conditions. Since it is an ex-situ process, high manpower and safety measures are needed to transfer the contaminated soil to the treatment site which depicts that this technique is a cost ineffective process. Control of processing conditions and regular monitoring influences the microbial growth which leads to the decrease of remediation time (Nikakhtari et al., 2009). Improper maintenance of parameter in the bioreactor brings about the diminishment of microbial action and less effective. The bioreactor is inappropriate for the usage in pilot scale study because the volume of pollutant treated may be too large, high manpower requirement and highly expenses. Bioreactors are applied to the treatment of halogenated and non-halogenated hydrocarbon, pesticides, petrochemicals, solvents, wood preservatives and volatile organic compounds. It is appropriate to treat the soil with low permeability and soil with heterogeneous characteristics. Pollutants removed by different types of bioreactors are depicted in Table 4.

*Table 4. Selected reports on the use of bioreactors for the remediation of contaminated soil*

| S.NO. | Bioreactor used | Name of the pollutant | References |
|---|---|---|---|
| 1. | Slurry bioreactor | 4-nitrophenol (4NP) | (Tomei et al., 2013) |
| 2. | Solid–liquid two-phase partitioning bioreactor | Polycyclic aromatic hydrocarbons (PAHs) | (Rehmann et al., 2008) |
| 3. | Two-liquid-phase bioreactor (TLPB) | Polycyclic aromatic hydrocarbons (PAHs) | (Wang et al., 2010) |
| 4. | Sequence batch reactor | Carbofuran | (Plangklang and Reungsang, 2010) |
| 5. | Aerobic soil slurry bioreactors | Lindane (herbicide) | (Varo-Arguello et al., 2012) |

*Figure 2. Remediation of soil using bioreactors*

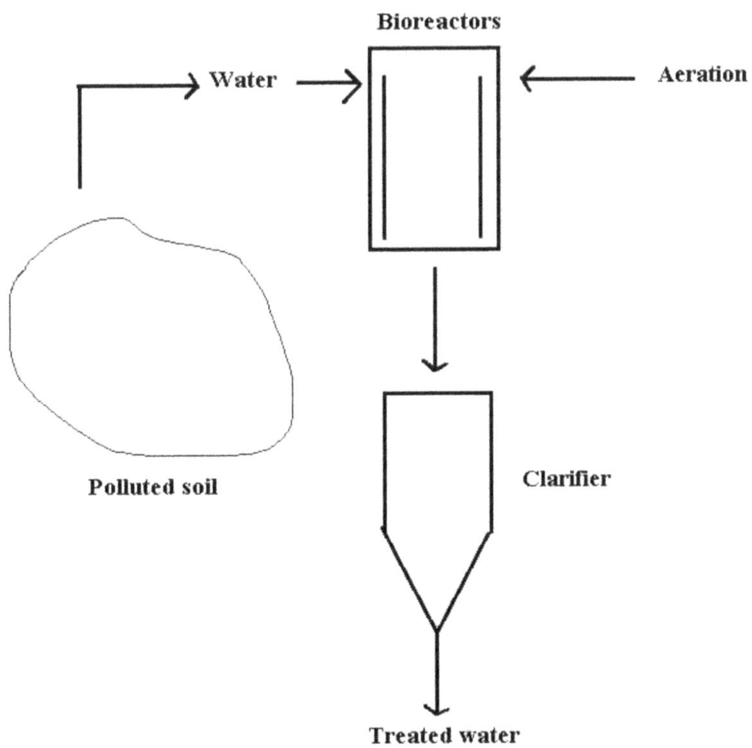

## FUTURE PROSPECTS

It is clear from the above-mentioned remediation technologies are found to be more effective in removing the pollutants from a contaminated site. Additionally, the role of organisms in bioremediation process is more in restoring toxins from the site. Hence low nutrients, their prosperity, degrading capacity, pollutant availability impede the success of bioremediation process. So, it is necessary to understand the metabolic process and activities of soil microorganisms to increase their response towards the bioremediation process. Like other treatment techniques, bioremediation has certain disadvantages such as it can be applied to the compounds which are biodegradable. Problems related to the bioremediation products are increasing because the end products after bioremediation are sometimes more hazardous than the parent compounds. Hence it needs to be analyzed frequently and monitoring of bioremediation is required in order to obtain non-hazardous end products from bioremediation process. In the remediation procedure, environmental effect needs to be evaluated that should be included in the installation of remediation objectives. Consequently, more consideration ought to be paid to the strategies for evaluating remediation efficiency while growing new remediation innovations in future.

## CONCLUSION

This chapter represents the current technologies available in in-situ and ex-situ method and the knowledge regarding the remediation procedure for the degradation of pollutant in the soil. Bioremediation methods have been explored for solving the issues related to the soil contaminants. A vast number of remediation procedures are accessible to clean up the pollutants and are intensively applied in practice. But they are capable of degrading the all the pollutants present in the soil. The increment of issues because of the soil contamination raised the advancement of new remediation advances like bioremediation. The bioremediation practices are chosen based on the type of organic and inorganic pollutant present in the soil. The determination of bioremediation procedure is additionally in view of the critical factor like the cost of remediation technique. After site characterization, most appropriate and practical bioremediation techniques like in-situ or ex-situ methods are selected. Ex-situ bioremediation technique is costlier because of the transportation of contaminated site. Yet at the same time, they treat the toxins in the controlled form. The absence of excavation in the in-situ method found to be more affordable but inefficient in treating the dangerous compounds than the ex-situ method. Characteristics of the soil, soil depth, types of soil, contaminated soil location and characteristics and performance of bioremediation technique influences to decide the most applicable technique for treating polluted soil. The techniques mentioned in this chapter has been successfully applied to degrade the soil polluted with pesticides, organic solvents, inorganic compounds and other chemicals.

## REFERENCES

Ali, H., Khan, E., & Sajad, M. A. (2013). Phytoremediation of heavy metals—concepts and applications. *Chemosphere*, *91*(7), 869–881. doi:10.1016/j.chemosphere.2013.01.075 PMID:23466085

Archaya, S., Gopinath, L. R., & Bhuvaneswari, R. (2014). A Review on Advances in Phytoremediation of Pesticide Polluted Soils. *International Journal of Advances in Interdisciplinary Research, 1*, 1–6.

Azubuike, C. C., Chikere, C. B., & Okpokwasili, G. C. (2016). Bioremediation techniques–classification based on site of application: Principles, advantages, limitations and prospects. *World Journal of Microbiology & Biotechnology, 32*(11), 180. doi:10.100711274-016-2137-x PMID:27638318

Bani, A., Echevarria, G., Sulce, C., & Morel, J. L. (2015). Improving the agronomy of *Alyssum murale* for extensive phytomining: A five-year field study. *International Journal of Phytoremediation, 17*(2), 117–127. doi:10.1080/15226514.2013.862204 PMID:25237722

Becerra-Castro, C., Kidd, P. S., Rodriguez-Garrido, B., Monterroso, C., Santos-Ucha, P., & Prieto-Fernandez, A. (2013). Phytoremediation of hexachlorocyclohexane (HCH)-contaminated soils using Cytisus striatus and bacterial inoculants in soils with distinct organic matter content. *Environmental Pollution, 178*, 202–210. doi:10.1016/j.envpol.2013.03.027 PMID:23583940

Chemlal, R., Abdi, N., Lounici, H., Drouiche, N., Pauss, A., & Mameri, N. (2013). Modeling and qualitative study of diesel biodegradation using biopile process in sandy soil. *International Biodeterioration & Biodegradation, 78*, 43–48. doi:10.1016/j.ibiod.2012.12.014

Chen, H., Arocena, J. M., Li, J., Thring, R. W., & Zhou, J. (2014). Assessments of chromium (and other metals) in vegetables and potential bio-accumulations in humans living in areas affected by tannery wastes. *Chemosphere, 112*, 412–419. doi:10.1016/j.chemosphere.2014.04.091 PMID:25048934

Ferri, R., Hashim, D., Smith, D. R., Guazzetti, S., Donna, F., Ferretti, E., ... Lucchini, R. G. (2015). Metal contamination of home garden soils and cultivated vegetables in the province of Brescia, Italy. *Implications for human exposure. The Science of the Total Environment, 518-519*, 507–517. doi:10.1016/j.scitotenv.2015.02.072 PMID:25777956

Frutos, F. J. G., Perez, R., Escolano, O., Rubio, A., Gimeno, A., Fernandez, M. D., ... Laguna, J. (2012). Remediation trials for hydrocarbon-contaminated sludge from a soil washing process: Evaluation of bioremediation technologies. *Journal of Hazardous Materials, 199–200*, 262–271. doi:10.1016/j.jhazmat.2011.11.017 PMID:22118850

Grobelak, A., Napora, A., & Kacprzak, M. (2015). Using plant growth-promoting rhizobacteria (PGPR) to improve plant growth. *Ecological Engineering, 84*, 22–28. doi:10.1016/j.ecoleng.2015.07.019

Hohener, P., & Ponsin, V. (2014). In situ vadose zone bioremediation. *Current Opinion in Biotechnology, 27*, 1–7. doi:10.1016/j.copbio.2013.08.018 PMID:24863890

Johnson, P. C., Johnson, R. L., Bruce, C. L., & Leeson, A. (2010). Advances in In Situ Air Sparging/Biosparging. *Bioremediation Journal, 5*(4), 251–266. doi:10.1080/20018891079311

Kao, C. M., Chen, C. Y., Chen, S. C., Chien, H. Y., & Chen, Y. L. (2008). Application of in situ biosparging to remediate a petroleum-hydrocarbon spill site: Field and microbial evaluation. *Chemosphere, 70*(8), 1492–1499. doi:10.1016/j.chemosphere.2007.08.029 PMID:17950413

Kim, S., Krajmalnik-Brown, R., Kim, J. O., & Chung, J. (2014). Remediation of petroleum hydrocarbon-contaminated sites by DNA diagnosis-based bioslurping technology. *The Science of the Total Environment, 497-498*, 250–259. doi:10.1016/j.scitotenv.2014.08.002 PMID:25129160

Lee, J. H. (2013). An overview of phytoremediation as a potentially promising technology for environmental pollution control. *Biotechnology and Bioprocess Engineering; BBE, 18*(3), 431–439. doi:10.100712257-013-0193-8

Li, N., Kang, Y., Pan, W., Zeng, L., Zhang, Q., & Luo, J. (2015). Concentration and transportation of heavy metals in vegetables and risk assessment of human exposure to bio accessible heavy metals in soil near a waste-incinerator site. *South China. The Science of the Total Environment, 521-522*, 144–151. doi:10.1016/j.scitotenv.2015.03.081 PMID:25829292

Lin, T. C., Pan, P. T., & Cheng, S. S. (2010). Ex situ bioremediation of oil-contaminated soil. *Journal of Hazardous Materials, 176*(1-3), 27–34. doi:10.1016/j.jhazmat.2009.10.080 PMID:20053499

Liu, Z., Hamuti, A., Abdulla, H., Zhang, F., & Mao, X. (2016). Accumulation of metallic elements by native species thriving in two mine tailings in Aletai, China. *Environmental Earth Sciences, 75*(9), 781. doi:10.100712665-016-5594-5

Lors, C., Damidot, D., Ponge, J. F., & Perie, F. (2012). Comparison of a bioremediation process of PAHs in a PAH-contaminated soil at field and laboratory scales. *Environmental Pollution, 165*, 11–17. doi:10.1016/j.envpol.2012.02.004 PMID:22390976

Meneghetti, L. R. R., Thome, A., Schnaid, F., Prietto, P. D. M., & Cavelhao, G. (2012). Natural attenuation and biostimulation of biodiesel contaminated soils from Southern Brazil with different particle sizes. *Journal of Environmental Science & Engineering, 1*(2B), 155–162.

Mesa, J., Rodriguez-Llorente, I. D., Pajuelo, E., Piedras, J. M., Caviedes, M. A., Redondo-Gomez, S., & Mateos-Naranjo, E. (2015). Moving closer towards restoration of contaminated estuaries: Bioaugmentation with autochthonous rhizobacteria improves metal rhizoaccumulation in native Spartina maritima. *Journal of Hazardous Materials, 300*, 263–271. doi:10.1016/j.jhazmat.2015.07.006 PMID:26188869

Nikakhtari, H., Kumar, P., Nemati, M., & Hill, G. A. (2009). Biodegradation of diesel oil in a baffled roller bioreactor. *Journal of Chemical Technology and Biotechnology (Oxford, Oxfordshire), 84*(4), 525–532. doi:10.1002/jctb.2074

Plangklang, P., & Reungsang, A. (2010). Bioaugmentation of carbofuran by *Burkholderia cepacia* PCL3 in a bioslurry phase sequencing batch reactor. *Process Biochemistry, 45*(2), 230–238. doi:10.1016/j.procbio.2009.09.013

Ravanbakhsh, M., Ronaghi, A.-M., Taghavi, S. M., & Jousset, A. (2016). Screening for the next generation heavy metal hyperaccumulators for dryland decontamination. *Journal of Environmental Chemical Engineering, 4*(2), 2350–2355. doi:10.1016/j.jece.2016.04.013

Rehmann, L., Prpich, G. P., & Daugulis, A. J. (2008). Remediation of PAH contaminated soils: Application of a solid–liquid two-phase partitioning bioreactor. *Chemosphere, 73*(5), 798–804. doi:10.1016/j.chemosphere.2008.06.006 PMID:18640698

Silva-Castro, G. A., Rodelas, B., Perucha, C., Laguna, J., Gonzalez-Lopez, J., & Calvo, C. (2013). Bioremediation of diesel-polluted soil using biostimulation as post treatment after oxidation with Fenton-like reagents: Assays in a pilot plant. *The Science of the Total Environment*, *445–446*, 347–355. doi:10.1016/j.scitotenv.2012.12.081 PMID:23354375

Silva-Castro, G. A., Uad, I., Rodrıguez-Calvo, A., Gonzalez-Lopez, J., & Calvo, C. (2015). Response of autochthonous microbiota of diesel polluted soils to land- farming treatments. *Environmental Research*, *137*, 49–58. doi:10.1016/j.envres.2014.11.009 PMID:25486545

Szolnoki, Z., & Farsang, A. (2013). Evaluation of metal mobility and bioaccessibility in soils of urban vegetable gardens using sequential extraction. *Water, Air, and Soil Pollution*, *224*(10), 1737. doi:10.100711270-013-1737-4

Taiwo, A. M., Gbadebo, A. M., Oyedepo, J. A., Ojekunle, Z. O., Alo, O. M., Oyeniran, A. A., ... Taiwo, O. T. (2016). Bioremediation of industrially contaminated soil using compost and plant technology. *Journal of Hazardous Materials*, *304*, 166–172. doi:10.1016/j.jhazmat.2015.10.061 PMID:26551220

Tang, Z., Zhang, L., Huang, Q., Yang, Y., Nie, Z., Cheng, J., ... Chai, M. (2015). Contamination and risk of heavy metals in soils and sediments from a typical plastic waste recycling area in North China. *Ecotoxicology and Environmental Safety*, *122*, 343–351. doi:10.1016/j.ecoenv.2015.08.006 PMID:26318969

Thijs, S., Sillen, W., Rineau, F., Weyens, N., & Vangronsveld, J. (2016). Towards an Enhanced Understanding of Plant–Microbiome Interactions to Improve Phytoremediation: Engineering the Metaorganism. *Frontiers in Microbiology*, *7*, 341. doi:10.3389/fmicb.2016.00341 PMID:27014254

Tomei, M. C., Angelucci, D. M., Annesini, M. C., & Daugulis, A. J. (2013). Ex situ remediation of polluted soils by absorptive polymers, and a comparison of slurry and two-phase partitioning bioreactors for ultimate contaminant degradation. *Journal of Hazardous Materials*, *262*, 31–37. doi:10.1016/j.jhazmat.2013.08.026 PMID:24007996

Varo-Arguello, W. E., Camacho-Perez, B., Rios-Leal, E., Vazquez-Landaverde, P. A., Ponce-Noyola, M. T., Barrera-Cortes, J., ... Poggi-Varaldo, H. M. (2012). Triphasic slurry bioreactors for the bioremediation of lindane-impacted soil under aerobic and anaerobic conditions. *Environmental Engineering and Management Journal*, *11*(10), 1811–1823.

Velimirovic, M., Tosco, T., Uyttebroek, M., Luna, M., Gastone, F., De Boer, C., ... Bastiaens, L. (2014). Field assessment of guar gum stabilized microscale zerovalent iron particles for in-situ remediation of 1,1,1-trichloroethane. *Journal of Contaminant Hydrology*, *164*, 88–99. doi:10.1016/j.jconhyd.2014.05.009 PMID:24963597

Wang, C., Wang, F., Wang, T., Bian, Y., Yang, X., & Jiang, X. (2010). PAHs biodegradation potential of indigenous consortia from agricultural soil and contaminated soil in two-liquid-phase bioreactor (TLPB). *Journal of Hazardous Materials*, *176*(1-3), 41–47. doi:10.1016/j.jhazmat.2009.10.123 PMID:19954884

Whelan, M. J., Coulon, F., Hince, G., Rayner, J., McWatters, R., Spedding, T., & Snape, I. (2015). Fate and transport of petroleum hydrocarbons in engineered biopiles in polar regions. *Chemosphere*, *131*, 232–240. doi:10.1016/j.chemosphere.2014.10.088 PMID:25563162

Wu, Z., Banuelos, G. S., Lin, Z.-Q., Liu, Y., Yuan, L., Yin, X., & Li, M. (2015). Biofortification and phytoremediation of Selenum in China. *Frontiers in Plant Science*, *6*, 1–8. doi:10.3389/fpls.2015.00136 PMID:25653664

Zhao, Y., Wei, Y., Zhang, Y., Wen, X., Xi, B., Zhao, X., ... Wei, Z. (2017). Roles of composts in soil based on the assessment of humification degree of fulvic acids. *Ecological Indicators*, *72*, 473–480. doi:10.1016/j.ecolind.2016.08.051

Zouboulis, A. I., & Moussas, P. A. (2011). Groundwater and Soil Pollution: Bioremediation. Encyclopedia of Environmental Health, 1037–1044.

## KEY TERMS AND DEFINITIONS

**Bioremediation:** It is the process of utilization of microbes to remove the pollutants from the contaminated site.

**Bioventing:** This method delivers the nutrients to the soil organisms through wells for the stimulation of degradation.

**Composting:** Organic constituents in the soil gets converted into humus like product.

**Ex-Situ Bioremediation Process:** Treating the excavated contaminants or pollutants of the soil away from the site.

**In-Situ Bioremediation Process:** Treating the contaminants or pollutants of the soil at the site.

**Phytoaccumulation:** Plants involved in the accumulation of pollutants present in the soil are called phytoaccumulation.

**Phytoremediation:** Contaminants break down by plants is called phytoremediation.

**Pollutants:** It is a compound which provides undesirable effects to the resources like human, surroundings, organisms, etc.

# Chapter 4
# Potential Applications of Nanomaterials in Wastewater Treatment:
## Nanoadsorbents Performance

**Hamidreza Sadegh**
*West Pomeranian University of Technology, Szczecin, Poland*

**Gomaa A. M. Ali**
*Al-Azhar University, Egypt*

## ABSTRACT

*High-quality water is one of the most important challenges around the world. Conventional techniques of wastewater treatment need to be developed. Therefore, finding sustainable, environmentally friendly, and efficient treatment techniques is required. In this regard, due to the extraordinary potential of nanotechnology resulted from nanoscale size characteristics, recently nanomaterials have been the subject of novel research and development worldwide. In this chapter, the authors review recent development of the direct applications of nanomaterial as an adsorbent in adsorption systems for integrating nanoparticles into conventional treatment technologies for wastewater treatment, especially a wide range of candidate nanomaterials and its properties. In addition, advantages and limitations as compared to existing processes are discussed.*

## INTRODUCTION

Nowadays, industrial technology is most important matters in quality of life. So, rapid population growth in developing region will continue to intensify the clean water demand from industry and energy perspective (Gupta et al., 2015a). The larger the industry is, the more waste and pollution it usually generates, and the bigger the impact on the environment. Recently, many physical and chemical methods have been developed to reduce the levels of pollution in the wastewater (Gupta et al., 2017b; Habeeb et al., 2017a; Habeeb et al., 2017c; Sadegh et al., 2016). Each method has certain advantages and disadvantages

DOI: 10.4018/978-1-5225-5754-8.ch004

(Table 1). Example, coagulation method reported as a low-cost method but, utilization of chemicals was high and membrane filtration reported as high performance for removal of heavy metal ions with lower space requirement but is very expensive (Ahmed & Ahmaruzzaman, 2016). Among of all, adsorption is considered as an efficient and economical technique and it is easy to operate for pollution removal from wastewater effluents (Gupta et al., 2015b; Habeeb et al., 2017b; Rajabi et al., 2016; Zare et al., 2016). The general advantages and disadvantages of adsorption methods are given in Figure 1. In addition, since the adsorption is reversible in some cases, adsorbents can be regenerated. There are three types of adsorption, the interaction between adsorbent-adsorbate (physisorption), adsorbent-solvent (chemisorption) and adsorbate-solvent (electrostatic interactions) (Qiu et al., 2009; Zare et al., 2015a; Zare et al., 2015b). In addition, underlying mechanisms involved in the adsorption material surface have been studied. As reported, many factors affect the efficiency and performance of adsorbents e.g., initial ion concentration, adsorbent dosage, temperature, pH value, contacting time, ionic strength, coexisting ions, and stirring speed (Ambashta & Sillanpää, 2010; Yu et al., 2017). It is important to understand how adsorbent interact with different adsorbate (i.e. heavy metal ions and dyes) in the laboratory scale to determine their potential for application in wastewater treatment and their contribution to large scale (Wan Ngah et al., 2011). However, adsorption has certain limitations also, such as it could not achieve a good status at commercial levels due to the lack of suitable adsorbents of high adsorption capacity and unavailability of commercial scale columns. Moreover, the design of suitable adsorbent, which has good performance, should be considered as a crucial stage. To fulfill the objective of the development of an effective adsorbent that is active and rapid for pollution removal, the advancement in nanotechnology has proved an incredible potential for the developing novel adsorbents were developed by nanomaterials (Sadegh et al., 2017b).

Nanomaterials have a small size structures of a few nanometers (less than 100 nm) (Zarbin, 2007). In the field of wastewater treatment, a variety of materials have been developed having unique functionalities for potential decontamination of industrial effluents, surface water, groundwater and drinking water (Liao et al., 2003).

In this chapter, the application of nanomaterials as adsorbents for removal of pollution from wastewater has been reviewed. It also includes a summary of the nanoadsorbents for the removal of heavy metal ions and dyes. The recent output of adsorption systems and isotherm modeling of adsorption is also reviewed.

*Figure 1. Advantages and disadvantages of the adsorption method*

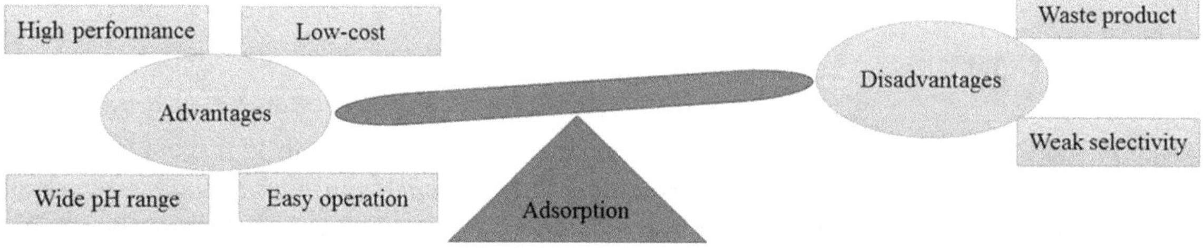

*Table 1. Comparison of treatment methods*

| Method | Advantage | Disadvantage | Reference |
|---|---|---|---|
| Coagulation | Low-cost | High chemicals utilization | (Ahmed & Ahmaruzzaman, 2016) |
| Membrane filtration | High removal of heavy metals | High-cost | (Ahmed & Ahmaruzzaman, 2016) |
| Adsorption | Easy operation, less sludge production<br>Utilization of low-cost adsorbents | Desorption | (Sadegh, 2017) |
| Electrochemical | Efficient for the removal of low concentrations metal ions<br>Low chemical usage | The initial investment is high<br>Need high electrical supply | (Ahmed & Ahmaruzzaman, 2016) |
| Electrodialysis | High segregation of metals | Clogging and energy loss | (Nguyen et al., 2013) |
| Ion exchange | High transformation of components | Removes only limited metal ions<br>High-cost operation | (Farooq et al., 2010) |
| Photocatalysis | Removal of both the metal ions and organic pollutants concurrently | It takes a prolonged time to remove the metals | (Ihsanullah et al., 2016) |
| Biological treatment | Removal of heavy metal ions | Needs developments | (Ahmaruzzaman, 2009) |
| Oxidation | No need for electricity | Rusting occurs in the system due to the usage of oxidation | (Patil et al., 2016) |

## NANOADSORBENTS

There is a key criteria that an adsorbent material must satisfy, for the sequestration of contaminants to be both economical and operational, including; high adsorption capacity to reduce both adsorbent quantity and equipment size, fast adsorption kinetics, stability in environments and regenerability. Several types of nanoadsorbents have been developed over the recent years which are capable of removing pollutants such as carbon nanotubes (CNTs), graphene and nanoparticles. These three groups are applicable for adsorption of heavy metal ions, dyes and many other compounds (Figure 2).

### Carbon Nanotubes

In 1991, CNTs were discovered by Injima (Iijima, 1991). CNTs can be seen as a curled graphene sheet. Synthetic techniques of CNTs mainly include arc discharge vaporization, laser ablation, chemical vapor deposition, electrolysis and flame synthesis, etc. CNTs are categorized as single-walled (SWCNTs), double-walled (DWCNTs) and multi-walled (MWCNTs) as shown in Figure 3 (Moradi et al., 2015b). The DWCNTs and MWCNTs have one or more outer tubes successively enveloping SWCNTs. CNTs are utilized in wastewater treatment; one of the good reviews in this matter was written by Gupta et al. (Gupta et al., 2016; Sadegh, 2017; Zare et al., 2015a; Zare et al., 2015b) in which CNTs was admired as an adsorbent for heavy metal ions because of unique properties like unique bonding configurations, ultra-high surface area, relative ease of attachment for numerous chemical functional groups, and the

*Figure 2. Types of nanoadsorbents*

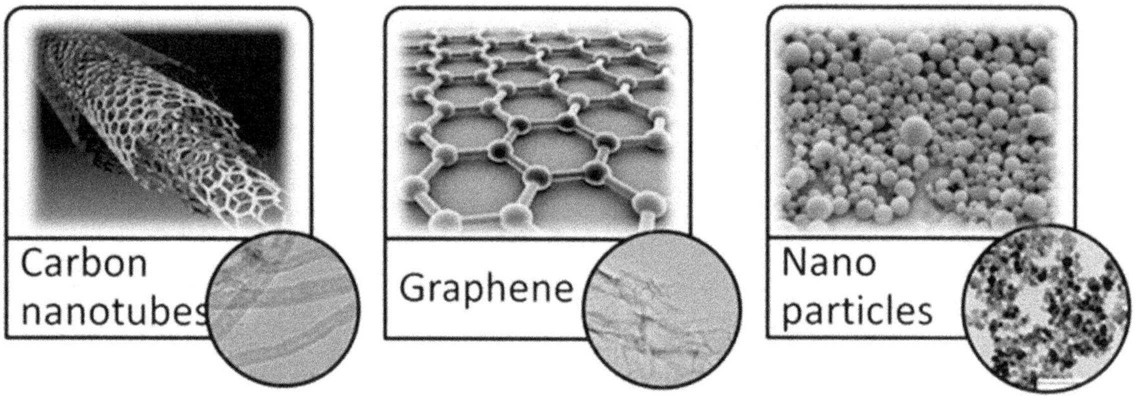


ability to decorate CNTs with nanoparticles. CNTs have a highly hydrophobic and nature, which it can be a problem for the application. So, to solve this problem, covalent and non-covalent strategies developed both as different methods for surface modification of CNTs.

Focusing on the application in wastewater, some articles have reported the functionalization effects of CNTs. In 2015, Sadegh et al. have focused on researching the utilization of CNTs for adsorption of dye and suggested microwave method in well-functionalized nanotubes (Sadegh et al., 2015). While some others covered the adsorption studies, Rajabi et al. have achieved the studies regarding analyzing the efficiency of the modified CNTs with oxidation process as an easy method on the adsorption of malachite green, introduced hydroxyl as a functional group to the sidewalls of CNTs (Rajabi et al., 2016). Thus, CNTs present the selectivity ability by being functionalized with different kinds of groups making them sufficient for adsorption system systems (Ali et al., 2017). However, only a few studies on the application of CNTs for pollution removal from aqueous solution have been reported and the CNTs were typically directly used without further treatment. In addition, some studies have been shown that oxidation method can be effective in the removal of dyes and heavy metal ions from aqueous solutions (Gupta et al., 2016; Gupta et al., 2017a; Sadegh et al., 2016).

## Graphene

Graphene is a single atomic sheet of $sp^2$-bonded carbon atoms densely packed in a honeycomb crystal lattice, has become a rapidly rising star among carbon materials, owing to its excellent chemical, mechanical, electrical and optical properties (Figure 3) (Novoselov, 2004). Graphene and its derivative e.g. graphene oxide (GO) which can be obtained by chemical oxidation following either the Hummers, Brodie, Staudenmaier, etc. have a highly hydrophilic nature, despite the excellent properties (Ali et al., 2015; Marina et al., 2016). So, it can be difficult to be used in separation in case of wastewater application. Hence, functionalization methods by introduction of functional groups are an effective way for making excellent properties. Therefore, increases adsorption performance of pollution removal from wastewater. For example, Gupta et al. modified GO using β-cyclodextrin for the adsorption of toxic Bisphenol A from aqueous solution (Gupta et al., 2017b). In a similar study, modified GO using β-Cyclodextrin capped graphene-magnetite nanocomposite for the removal of Bisphenol-A (Ragavan & Rastogi, 2017).

*Figure 3. Structure of graphene and CNTs types*

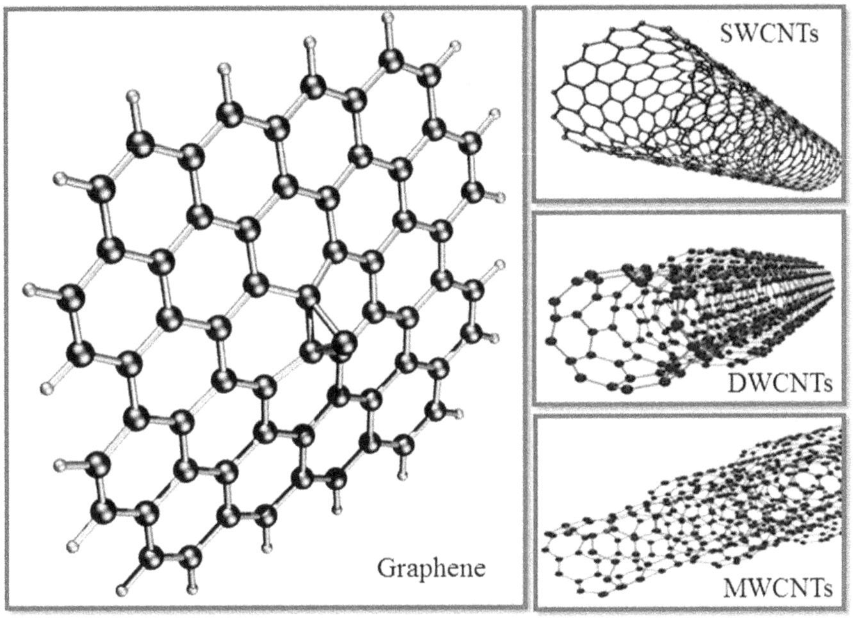


The modified GO exhibited a higher performance compared to the pure graphene and GO (Moradi et al., 2015a).

## Nanoparticles

Nanoparticles (NPs) including single component materials such as: $TiO_2$, $SiO_2$, $MnO_2$, ZnO, and $Fe_3O_4$, or nanocomposite materials such as $Co_3O_4/SiO_2$, $B_2O_3/TiO_2$, are currently the most widely used adsorbents for the removal of pollution from wastewater (Abdel Ghafar et al., 2015; Lee & Yang, 2012; Wang et al., 2013; Zhang et al., 2013). In addition, nanocomposites of NPs with carbon materials were shown a great ability toward pollutants removal as they possess both system component characteristics (Habeeb et al., 2017b). NPs are effective and economical for the rapid removal and recovery of metal ions from wastewater effluents due to their large surface area and optimal magnetic properties. NPs are more effective adsorbents than the others for the removal of various pollutants. Moreover, their surface modification by the attachment of inorganic shells and organic molecules stabilizes and prevents the oxidation of nanoparticles. In addition, these surface functionalities provide sites for the uptake of specific/selective ions and, thus, enhance the efficiency of their removal. Example, Zare et al., have observed that the adsorption efficiency of ions by $Fe_3O_4$ nanoparticles is strongly dependent on pH, temperature, the amount of the adsorbent and the incubation time (Zare et al., 2016). Singh et al. reported the removal of toxic metal ions from wastewater by using carboxyl-, amine- and thiol-functionalized $Fe_3O_4$ nanoparticles (Singh et al., 2011). Comparison study in term of adsorption capacity of some metal ions and dyes is given in Table 2 for the above discussed nanoadsorbents.

*Table 2. Comparison of the adsorption capacity of some nanoadsorbents*

| Type | Adsorbents | Adsorption Capacity (mg g$^{-1}$) | | Reference |
|---|---|---|---|---|
| | | Metal ions | Dyes | |
| CNTs | MWCNTs | Ammonium ions (129) | ---- | (Sadegh et al., 2017a) |
| | MWCNTs | ---- | Acid red 18(166.67) | (Shirmardi et al., 2012) |
| | MWCNTs/$Fe_3O_4$ | $Ni^{+2}$ (9.18) | | (Chen et al., 2009) |
| | MWCNT-COOH-Cysteamine | ---- | Amido black 10B (90) | (Sadegh et al., 2016) |
| Graphene | Graphene oxide | $Co^{+2}$ (21.28) | ---- | (Lingamdinne et al., 2016) |
| | $CoFe_2O_4$-functionalized graphene sheets | ---- | Methyl orange (71.54) | (Li et al., 2011) |
| | Graphene oxide–β-cyclodextrin | ---- | Bisphenol A (373.4) | (Gupta et al., 2017b) |
| | $Fe_3O_4$/graphene | $As^{+4}$ (5.83) $As^{+3}$ (13.10) | ---- | (Chandra et al., 2010) |
| Nanoparticles | $Co_3O_4/SiO_2$ | ---- | Methelene blue (53.87) | (Abdel Ghafar et al., 2015) |
| | $Fe_3O_4$ | Ammonium ions (133.21) | ---- | (Zare et al., 2016) |
| | ZnO | $Cd^{+2}$ (28.1) | ---- | (Wang et al., 2013) |
| | $B_2O_3/TiO_2$ | $Cd^{+2}$ (49) | ---- | (Kalfa et al., 2009) |

## ECONOMIC FEASIBILITY OF NANOMATERIALS AND FUTURE PERSPECTIVES

Adsorption of toxic substances can increase the toxicity of nanomaterials and finally affects the behavior of these substances in the environment. A good adsorbent should be stable and able to undergo an appropriate number of reuse with maintaining some level of its capacity. Therefore, adsorbent must be regenerated for further using. Common regeneration procedures include thermal treatment, chemical extraction, bio-regeneration, supercritical regeneration, ultrasonic regeneration, and Fenton and electro-Fenton regeneration. It is the most important criteria for the selection of an adsorbent for the removal of pollution. The cost of adsorption process is mainly dependent on the cost of adsorbent used for the removal of pollutions from wastewater. The most common nanoadsorbent in lab scale used for the adsorption is synthesis which is quite expensive. Hence, low-cost materials are sorely needed which are comparable to some nanomaterials in terms of adsorption capacity, economic feasibility and should be locally available.

## CONCLUSION

There is a significant need for advanced water technologies to ensure a high quality of water, eliminate chemical and biological pollutants, and intensify industrial production processes of wastewater. The presence of pollution in wastewater is a major concern for environmental conservation and human health. So far, the process of removal of these toxic pollutions has not reached the optimum conditions. To increase stringent environmental regulations more and more, a wide range of treatment technologies such as membrane filtration, adsorption, electrochemical treatment, electrodialysis, ion exchange, photocatalysis, biological treatment and oxidation have been used for wastewater treatment. Many reports confirmed that adsorption method was most effective than other methods. The adsorbent is an important part of adsorption. In this regard, nanotechnology is one of the ideal options for developing adsorbent. Nanomaterials used as adsorbents include CNTs, graphene and NPs. There is no doubt of efficiency of nanomaterials adsorption as an adsorbent; however, this technology needs developing.

## REFERENCES

Abdel Ghafar, H. H., Ali, G. A. M., Fouad, O. A., & Makhlouf, S. A. (2015). Enhancement of adsorption efficiency of methylene blue on $Co_3O_{4/}SiO_2$ nanocomposite. *Desalination and Water Treatment*, *53*(11), 2980–2989. doi:10.1080/19443994.2013.871343

Ahmaruzzaman, M. (2009). Role of fly ash in the removal of organic pollutants from wastewater. *Energy & Fuels*, *23*(3), 1494–1511. doi:10.1021/ef8002697

Ahmed, M. J. K., & Ahmaruzzaman, M. (2016). A review on potential usage of industrial waste materials for binding heavy metal ions from aqueous solutions. *Journal of Water Process Engineering*, *10*, 39–47. doi:10.1016/j.jwpe.2016.01.014

Ali, G. A. M., Lih Teo, E. Y., Aboelazm, E. A. A., Sadegh, H., Memar, A. O. H., Shahryari-Ghoshekandi, R., & Chong, K. F. (2017). Capacitive performance of cysteamine functionalized carbon nanotubes. *Materials Chemistry and Physics*, *197*, 100–104. doi:10.1016/j.matchemphys.2017.05.019

Ali, G. A. M., Makhlouf, S. A., Yusoff, M. M., & Chong, K. F. (2015). Structural and electrochemical characteristics of graphene nanosheets as supercapacitor electrodes. *Reviews on Advanced Materials Science*, *40*(1), 35–43.

Ambashta, R. D., & Sillanpää, M. (2010). Water purification using magnetic assistance: A review. *Journal of Hazardous Materials*, *180*(1-3), 38–49. doi:10.1016/j.jhazmat.2010.04.105 PMID:20488616

Chandra, V., Park, J., Chun, Y., Lee, J. W., Hwang, I.-C., & Kim, K. S. (2010). Water-dispersible magnetite-reduced graphene oxide composites for arsenic removal. *ACS Nano*, *4*(7), 3979–3986. doi:10.1021/nn1008897 PMID:20552997

Chen, C., Hu, J., Shao, D., Li, J., & Wang, X. (2009). Adsorption behavior of multiwall carbon nanotube/iron oxide magnetic composites for Ni (II) and Sr (II). *Journal of Hazardous Materials*, *164*(2), 923–928. doi:10.1016/j.jhazmat.2008.08.089 PMID:18842337

Farooq, U., Kozinski, J. A., Khan, M. A., & Athar, M. (2010). Biosorption of heavy metal ions using wheat based biosorbents – A review of the recent literature. *Bioresource Technology*, *101*(14), 5043–5053. doi:10.1016/j.biortech.2010.02.030 PMID:20223652

Gupta, V., Moradi, O., Tyagi, I., Agarwal, S., Sadegh, H., Shahryari-Ghoshekandi, R., ... Garshasbi, A. (2016). Study on the removal of heavy metal ions from industry waste by carbon nanotubes: effect of the surface modification: a review. *Critical Reviews in Environmental Science and Technology*, *46*(2), 93–118. doi:10.1080/10643389.2015.1061874

Gupta, V., Sadegh, H., Yari, M., Ghoshekandi, R. S., Maazinejad, B., & Chahardori, M. (2015a). Removal of ammonium ions from wastewater: A short review in development of efficient methods. *Global Journal of Environmental Science and Management*, *1*(2), 149.

Gupta, V. K., Agarwal, S., Bharti, A. K., & Sadegh, H. (2017a). Adsorption mechanism of functionalized multi-walled carbon nanotubes for advanced Cu (II) removal. *Journal of Molecular Liquids*, *230*, 667–673. doi:10.1016/j.molliq.2017.01.083

Gupta, V. K., Agarwal, S., Sadegh, H., Ali, G. A. M., Bharti, A. K., & Hamdy, A. S. (2017b). Facile route synthesis of novel graphene oxide-β-cyclodextrin nanocomposite and its application as adsorbent for removal of toxic bisphenol A from the aqueous phase. *Journal of Molecular Liquids*, *237*, 466–472. doi:10.1016/j.molliq.2017.04.113

Gupta, V. K., Tyagi, I., Agarwal, S., Sadegh, H., Shahryari-ghoshekandi, R., Yari, M., & Yousefi-nejat, O. (2015b). Experimental study of surfaces of hydrogel polymers HEMA, HEMA–EEMA–MA, and PVA as adsorbent for removal of azo dyes from liquid phase. *Journal of Molecular Liquids*, *206*, 129–136. doi:10.1016/j.molliq.2015.02.015

Habeeb, O. A., Kanthasamy, R., Ali, G. A. M., Sethupathi, S., & Yunus, R. B. M. (2017a). Hydrogen sulfide emission sources, regulations, and removal techniques: A review. *Reviews in Chemical Engineering*, *0*(0). doi:10.1515/revce-2017-0004

Habeeb, O. A., Ramesh, K., Ali, G. A. M., & Yunus, R. M. (2017b). Experimental design technique on removal of hydrogen sulfide using CaO-eggshells dispersed onto palm kernel shell activated carbon: Experiment, optimization, equilibrium and kinetic studies. Journal of Wuhan University of Technology-Mater. *Science Editor*, *32*(2), 305–320. doi:10.100711595-017-1597-7

Habeeb, O. A., Ramesh, K., Ali, G. A. M., & Yunus, R. M. (2017c). Low-cost and eco-friendly activated carbon from modified palm kernel shell for hydrogen sulfide removal from wastewater: Adsorption and kinetic studies. *Desalination and Water Treatment*, *84*, 205–214. doi:10.5004/dwt.2017.21196

Ihsanullah, A., Abbas, A., Al-Amer, A. M., Laoui, T., Al-Marri, M. J., Nasser, M. S., ... Atieh, M. A. (2016). Heavy metal removal from aqueous solution by advanced carbon nanotubes: Critical review of adsorption applications. *Separation and Purification Technology*, *157*, 141–161. doi:10.1016/j.seppur.2015.11.039

Iijima, S. (1991). Helical microtubules of graphitic carbon. *Nature*, *354*(6348), 56–58. doi:10.1038/354056a0

Kalfa, O. M., Yalçınkaya, Ö., & Türker, A. R. (2009). Synthesis of nano $B_2O_3/TiO_2$ composite material as a new solid phase extractor and its application to preconcentration and separation of cadmium. *Journal of Hazardous Materials*, *166*(1), 455–461. doi:10.1016/j.jhazmat.2008.11.112 PMID:19135793

Lee, Y.-C., & Yang, J.-W. (2012). Self-assembled flower-like $TiO_2$ on exfoliated graphite oxide for heavy metal removal. *Journal of Industrial and Engineering Chemistry*, *18*(3), 1178–1185. doi:10.1016/j.jiec.2012.01.005

Li, N., Zheng, M., Chang, X., Ji, G., Lu, H., Xue, L., ... Cao, J. (2011). Preparation of magnetic $CoFe_2O_4$-functionalized graphene sheets via a facile hydrothermal method and their adsorption properties. *Journal of Solid State Chemistry*, *184*(4), 953–958. doi:10.1016/j.jssc.2011.01.014

Liao, M.-H., Wu, K.-Y., & Chen, D.-H. (2003). Fast Removal of Basic Dyes by a Novel Magnetic Nano-adsorbent. *Chemistry Letters*, *32*(6), 488–489. doi:10.1246/cl.2003.488

Lingamdinne, L. P., Koduru, J. R., Roh, H., Choi, Y.-L., Chang, Y.-Y., & Yang, J.-K. (2016). Adsorption removal of Co (II) from waste-water using graphene oxide. *Hydrometallurgy*, *165*, 90–96. doi:10.1016/j.hydromet.2015.10.021

Marina, P. E., Ali, G. A. M., See, L. M., Teo, E. Y. L., Ng, E.-P., & Chong, K. F. (2016). In situ growth of redox-active iron-centered nanoparticles on graphene sheets for specific capacitance enhancement. *Arabian Journal of Chemistry*. doi:10.1016/j.arabjc.2016.02.006

Moradi, O., Gupta, V. K., Agarwal, S., Tyagi, I., Asif, M., Makhlouf, A. S. H., ... Shahryari-ghoshekandi, R. (2015a). Characteristics and electrical conductivity of graphene and graphene oxide for adsorption of cationic dyes from liquids: Kinetic and thermodynamic study. *Journal of Industrial and Engineering Chemistry*, *28*, 294–301. doi:10.1016/j.jiec.2015.03.005

Moradi, O., Sadegh, H., Shahryari-Ghoshekandi, R., & Norouzi, M. (2015b). Application of carbon nanotubes in nanomedicine: New medical approach for tomorrow. *Handbook of Research on Diverse Applications of Nanotechnology in Biomedicine. Chemical Engineering (Albany, N.Y.)*, 90–128.

Nguyen, T. A. H., Ngo, H. H., Guo, W. S., Zhang, J., Liang, S., Yue, Q. Y., ... Nguyen, T. V. (2013). Applicability of agricultural waste and by-products for adsorptive removal of heavy metals from wastewater. *Bioresource Technology*, *148*, 574–585. doi:10.1016/j.biortech.2013.08.124 PMID:24045220

Novoselov, K. S. (2004). Electric Field Effect in Atomically Thin Carbon Films. *Science*, *306*(5696), 666–669. doi:10.1126cience.1102896 PMID:15499015

Patil, D. S., Chavan, S. M., & Oubagaranadin, J. U. K. (2016). A review of technologies for manganese removal from wastewaters. *Journal of Environmental Chemical Engineering*, *4*(1), 468–487. doi:10.1016/j.jece.2015.11.028

Qiu, H., Lv, L., Pan, B., Zhang, Q., Zhang, W., & Zhang, Q. (2009). Critical review in adsorption kinetic models. *Journal of Zhejiang University. Science A*, *10*(5), 716–724. doi:10.1631/jzus.A0820524

Ragavan, K. V., & Rastogi, N. K. (2017). β-Cyclodextrin capped graphene-magnetite nanocomposite for selective adsorption of Bisphenol-A. *Carbohydrate Polymers*, *168*, 129–137. doi:10.1016/j.carbpol.2017.03.045 PMID:28457432

Rajabi, M., Mirza, B., Mahanpoor, K., Mirjalili, M., Najafi, F., Moradi, O., ... Tyagi, I. (2016). Adsorption of malachite green from aqueous solution by carboxylate group functionalized multi-walled carbon nanotubes: Determination of equilibrium and kinetics parameters. *Journal of Industrial and Engineering Chemistry*, *34*, 130–138. doi:10.1016/j.jiec.2015.11.001

Sadegh, H. (2017). Development of graphene oxide from graphite: A review on synthesis, characterization and its application in wastewater treatment. *Reviews on Advanced Materials Science*, *49*, 38–43.

Sadegh, H., Ali, G. A. M., Abbasi, Z., & Nadagoud, M. N. (2017a). Adsorption of Ammonium Ions onto Multi-Walled Carbon Nanotubes. *Studia Universitatis Babes-Bolyai. Chemia*, *62*(2), 233–245. doi:10.24193ubbchem.2017.2.18

Sadegh, H., Ali, G. A. M., Gupta, V. K., Makhlouf, A. S. H., Shahryari-Ghoshekandi, R., Nadagouda, M. N., ... Megiel, E. (2017b). The role of nanomaterials as effective adsorbents and their applications in wastewater treatment. *Journal ofNanostructure in Chemistry*, *7*(1), 1–14. doi:10.100740097-017-0219-4

Sadegh, H., Shahryari-ghoshekandi, R., Agarwal, S., Tyagi, I., Asif, M., & Gupta, V. K. (2015). Microwave-assisted removal of malachite green by carboxylate functionalized multi-walled carbon nanotubes: Kinetics and equilibrium study. *Journal of Molecular Liquids*, *206*, 151–158. doi:10.1016/j.molliq.2015.02.007

Sadegh, H., Zare, K., Maazinejad, B., Shahryari-Ghoshekandi, R., Tyagi, I., Agarwal, S., & Gupta, V. K. (2016). Synthesis of MWCNT-COOH-Cysteamine composite and its application for dye removal. *Journal of Molecular Liquids*, *215*, 221–228. doi:10.1016/j.molliq.2015.12.042

Shirmardi, M., Mesdaghinia, A., Mahvi, A. H., Nasseri, S., & Nabizadeh, R. (2012). Kinetics and equilibrium studies on adsorption of acid red 18 (Azo-Dye) using multiwall carbon nanotubes (MWCNTs) from aqueous solution. *Journal of Chemistry*, *9*(4), 2371–2383.

Singh, S., Barick, K., & Bahadur, D. (2011). Novel and efficient three dimensional mesoporous ZnO nanoassemblies for environmental remediation. *International Journal ofNanoscience, 10*(4-5), 1001-1005.

Wan Ngah, W. S., Teong, L. C., & Hanafiah, M. A. K. M. (2011). Adsorption of dyes and heavy metal ions by chitosan composites: A review. *Carbohydrate Polymers*, *83*(4), 1446–1456. doi:10.1016/j.carbpol.2010.11.004

Wang, X., Cai, W., Liu, S., Wang, G., Wu, Z., & Zhao, H. (2013). ZnO hollow microspheres with exposed porous nanosheets surface: Structurally enhanced adsorption towards heavy metal ions. *Colloids and Surfaces. A, Physicochemical and Engineering Aspects*, *422*, 199–205. doi:10.1016/j.colsurfa.2013.01.031

Yu, L., Han, M., & He, F. (2017). A review of treating oily wastewater. *Arabian Journal of Chemistry*, *10*, S1913–S1922. doi:10.1016/j.arabjc.2013.07.020

Zarbin, A. J. G. (2007). Química de (nano)materiais. *Quimica Nova*, *30*(6), 1469–1479. doi:10.1590/S0100-40422007000600016

Zare, K., Gupta, V. K., Moradi, O., Makhlouf, A. S. H., Sillanpää, M., Nadagouda, M. N., ... Kazemi, M. (2015a). A comparative study on the basis of adsorption capacity between CNTs and activated carbon as adsorbents for removal of noxious synthetic dyes: A review. *Journal of Nanostructure in Chemistry*, *5*(2), 227–236. doi:10.100740097-015-0158-x

Zare, K., Sadegh, H., Shahryari-ghoshekandi, R., Asif, M., Tyagi, I., Agarwal, S., & Gupta, V. K. (2016). Equilibrium and kinetic study of ammonium ion adsorption by $Fe_3O_4$ nanoparticles from aqueous solutions. *Journal of Molecular Liquids*, *213*, 345–350. doi:10.1016/j.molliq.2015.08.045

Zare, K., Sadegh, H., Shahryari-ghoshekandi, R., Maazinejad, B., Ali, V., Tyagi, I., ... Gupta, V. K. (2015b). Enhanced removal of toxic Congo red dye using multi walled carbon nanotubes: Kinetic, equilibrium studies and its comparison with other adsorbents. *Journal of Molecular Liquids*, *212*, 266–271. doi:10.1016/j.molliq.2015.09.027

Zhang, J., Zhai, S., Li, S., Xiao, Z., Song, Y., An, Q., & Tian, G. (2013). Pb (II) removal of Fe3O4@ SiO2–NH2 core–shell nanomaterials prepared via a controllable sol–gel process. *Chemical Engineering Journal*, *215*, 461–471. doi:10.1016/j.cej.2012.11.043

# Chapter 5
# Plastic Waste Pollution and Its Management in India: A Review

**Athar Hussain**
*Chaudhary Brahm Prakash Government Engineering College, India*

**Ayushman Bhattacharya**
*Chaudhary Brahm Prakash Government Engineering College, India*

**Arfat Ahmed**
*Chandigarh College of Engineering and Technology, India*

## ABSTRACT

*Plastic, one of the most preferred materials in today's industrial world, is posing a serious threat to the environment and consumer health in many direct and indirect ways. The global plastic production increased over years due to the vast applications of plastics in many sectors. More than 50% of the plastic waste generated in the country is recycled and used in the manufacture of various plastic products. The remaining half is disposed of at landfill sites or simply burned in incinerators. The burning of plastics, especially PVC, releases this dioxin and also furan into the atmosphere. In this chapter, the authors examine the environmental and health effects and harm caused by the burning of plastics in detail. It focuses on the current status of plastic waste management in India and industries working under the extended producer responsibility. Therefore, an attempt has been made to review the current practices prevalent in India to deal with this plastic waste and problems associated with it.*

## 1. INTRODUCTION

Plastic, with its exclusive qualities is now a serious worldwide environmental and health concern, essentially due to its non-biodegradable nature. More than 50% of the plastic waste generated in the country is recycled and used in the manufacture of various plastic products. Dioxin is a highly carcinogenic and toxic by-product of the manufacturing process of plastics. Burning of plastics, especially PVC,

DOI: 10.4018/978-1-5225-5754-8.ch005

releases this dioxin and also furan into the atmosphere. Plastics are so versatile in use that their impact on the environment is extremely wide ranging. Careless disposal of plastic bags chokes drains, blocks the porosity of the soil and causes problems for groundwater recharge. Dioxin is the common name for a class of 75 chemicals. It is a toxic waste product formed when waste containing chlorine is burned or when products containing chlorine are manufactured. Dioxins are among the most potent synthetic chemicals ever tested, causing cancer and harming our immune and reproductive systems even at very low concentrations (Pavani and Rajeswari, 2014). Plastic is a relatively cheap, durable, and versatile material. These properties have led to its use in the creation of thousands of products, which have brought benefits to society in terms of economic activity, jobs, and quality of life. However, plastic waste can also impose negative externalities such as greenhouse gas emissions or ecological damage. It is usually non-biodegradable and, therefore, can remain as waste in the environment for a very long time; it may pose risks to human health and the environment; in some cases, it can be difficult to reuse and/or recycle (Toxic link, 2016). Plastic disturbs the soil microbial activity. Plastic bags can also contaminate food-stuffs due to leaching of toxic dyes and transfer of pathogens. Furthermore, flame-retardant materials such as polyvinyl chloride (PVC) are known to cause corrosion in incinerators during combustion due to their constituent halogen substances; these materials also produce halogen compounds such as dioxins.

Moreover, $CO_2$ discharged from the combustion of polymers causes environmental problems such as global warming and acid rain. In addition, suspected endocrine disrupting chemicals, typically Bisphenol A, can dissolve out of polycarbonate (Fatima, 2014). As per the government report more than 15,000 tonnes of plastic waste are generated in India everyday, of which 6,000 tonnes remain uncollected and littered. As per a report of a Task Force constituted by erstwhile Planning Commission in 2014 indicates that 62 million tonnes of municipal solid waste is generated in India annually in urban areas. However, as per the CPCB report of 2014-15, 51.4 million tonnes of solid waste were generated in the country, of which 91 per cent was collected, and 27 per cent was treated and remaining 73 per cent disposed of at dump sites. Central Pollution Control Board has estimated that the generation of 15,342 tonnes of plastic waste in the country, out of which, 9,205 tonnes were reported to be recycled and leaving 6,137 tonnes uncollected and littered (Press Trust of India, 2016). The growth of the Indian plastic industry has been phenomenal equal to17% is higher than for the plastic industry elsewhere in the world (Panda et tal, 2010). Petrochemical products permeate the entire spectrum of daily use items and cover almost every sphere of life like clothing, housing, construction, furniture, automobiles, household items, agriculture, horticulture, irrigation, packaging, medical appliances, electronics and electrical etc. These industries hence drive the demand growth of petrochemicals. Current low per capita consumption level of plastic products as compared to developed countries per capita consumption as shown in Figure 1 suggests that India offers a huge opportunity over long term (FICCI-2014).

To manufacture finished products, polymers are processed through various types of techniques namely extrusion, injection moulding, blow moulding and roto moulding. Various products manufactured through these processes are highlighted in the Table 1. In India, extrusion-based methods account for 75% of the total amount of plastics processed; this is very similar to the situation in Western Europe.

Over 27% of all extruders in the plastics processing industry are operated for the manufacture of blown films. Injection moulding is the second largest processing technique (19%) and is mainly used for the manufacture of household products, packaging, and in the electrical sector. Blow moulding and rotational moulding constitute a small share; these processes serve to manufacture products like bottles, drums, tanks etc. (Muth et al, 2006).

*Figure 1. Per capita plastic products consumption (Kg/person) (Source: A report on plastics industry, FICCI-2014)*

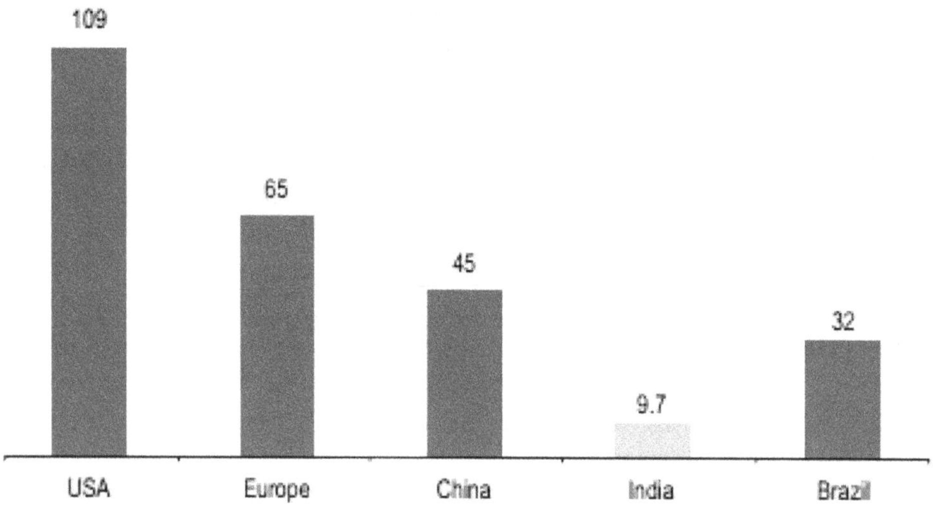


*Table 1. Manufacturing processes of plastics*

| | |
|---|---|
| Extrusion | Films and Sheets, Fibre and Filaments Pipe, Conduits and profiles, miscellaneous applications |
| Injection Moulding | Industrial Injection Moulding, Household Injection and Thermo-ware/ Moulded luggage |
| Blow Moulding | Bottles, Containers, Toys and Housewares |
| Roto Moulding | Large circular tanks such as water tanks |

Source: A report on plastics industry - ficci

However plastics can be categorized as Recyclable Plastics (Thermoplastics) like PET, HDPE, LDPE, PP, PVC, PS etc. and Non-Recyclable Plastic (Thermoset & others) like Multilayer & Laminated plastics, PUF, Bakelite, Polycarbonate, Melamine, Nylon etc.

Technologies for plastics separation are Triboelectric separation, Froth flotation methods, Speed accelerators, Solvent-based, Hyperspectral imaging, Ultrasound technology, Laser-induced breakdown spectroscopy, X-ray fluorescence and infrared spectroscopy(Rahimi and Garcí, 2017). A triboelectric separator sorts materials on the basis of a surface charge transfer phenomenon. When materials are rubbed against each other, one material becomes positively charged, and the other becomes negatively charged or remains neutral.

Particles are mixed and contact one another in a rotating drum to allow charging. Materials with a particle size of approximately 2–4 mm is having the highest in both purity and recovery in the tribo-electric process. Plastic waste can also be sorted by a speed accelerator technique, developed by Result Technology (Switzerland). This technique uses a high-speed accelerator to delaminate shredded waste, and the delaminated material is separated by air classification, sieves, and electrostatics. Using X-ray fluorescent (XRF) spectroscopy, different types of flame-retardants (FRs) can be identified. No matter how efficient the recycling scheme is, sorting is the most important step in the recycling loop. (Salem et al, 2009).

## 2. ENVIRONMENTAL IMPACTS

The carbon footprint of plastic (Low density polythylene or Polyethylene terepthalate, polyethylene) is about 6 kg $CO_2$ per kg of plastic. The production of 1 kg of polyethylene (PET or LDPE), requires the equivalent of 2 kg of oil for energy and raw material .Polyethylene (PE) is the most commonly used plastic for plastic bags. Burning 1 kg of oil creates about 3 kg of carbon dioxide. In other words: Per kg of plastic, about 6 kg carbon dioxide is created during production and incineration (Times for Change, 2018). The combustion of plastics as presented in Table 2 infers that several chemical compounds cause a serious risk to health and the environmental elements, among which, especially, the incineration of PVC poses the greatest threats. The use of poly (vinyl chloride) is highly restricted or banned in many countries of the European Union. During the incineration of PVC, up to 2 milligrams per gram of phosgene is generated, which is one of the most dangerous gases, and a serious risk to health. Phosgene ($COCl_2$) is an organic compound, colourless, its odour is reminiscent of musty hay, an asphyxiant gas it is a strong poison. It was used as a chemical weapon during World War I. It hydrolyses easily and is freely soluble in non-polar solvents. This also reacts with the condensing vapour during combustion, whereby hydrogen chloride is formed, which is also a toxic compound. The equation 1 presents the chemical reaction (NAGY and KUTI, 2016):

$$COCl_2 + H_2O \rightarrow CO_2 + 2HCl \quad (1)$$

In terms of quantities, it can be said that every five grams of burned PVC derivatives pollute a cubic meter of air to such an extent that it will surely damage our health. The compounds generated by the incineration of PVC are acetone, toluene, xylene, acetaldehyde, benzaldehyde, phosgene etc.

Need of Biodegradable Plastics- Although traditional plastics can be reused or recycled, it take many decades and even more than two hundred years to decompose, leading to serious environmental problems. The solution for this problem is to produce oxo-biodegradable plastics (oxo-bio) by using the degradable to water (d2w) additive during the standard production process of plastics. The impact

*Table 2. Combustion characteristics of plastics used in households*

| Name of substance | Ignition temperature (°C) | Heat of combustion (MJ/kg) |
|---|---|---|
| Polyethylene(PE) | 350 | 46.3 |
| Polyethylene terephthalate (PET) | 500 | 22.7 |
| Polystyrene(P5) | 470 | 41.6 |
| Polypropylene(PP) | 410 | 46.6 |
| Polyamide(PA) | 500 | 31.4 |
| Poly(vinyl chloride)(PVC) | 760 | 19.26 |

Source: (NAGY & KUTI, 2016)

of the additive is therefore negligible and there is no special training required for workers. One of the common constituents of biodegradable plastics is polyhydroxyalkanonate (Pavani and Rajeswari, 2014).

## 3. PLASTIC WASTE MANAGEMENT RULES, 2016

As per the Plastic Waste Management Rules 2016 in India: Increasing the thickness of plastic carry bags from 40 to 50 micron and stipulation of 50 micron thickness for plastic sheets is likely to increase the cost by about 20%. Hence, the tendency to provide free carry bags will come down and collection by the waste-pickers also increase to some extent. The producers, importers and brand owners who introduce the plastic carry bags, multi-layered plastic sachet, or pouches, or packaging in the market within a period of six months from the date of publication of these rules, need to establish a system for collecting back the plastic waste generated due to their products. All institutional generators of plastic waste, shall segregate and store the waste generated by them in accordance with the Solid Waste Management Rules, and handover segregated wastes to authorized waste processing or disposal facilities or deposition centers, either on its own or through the authorized waste collection agency. All waste generators shall pay such user fee, or charge, as may be specified in the bye-laws of the local bodies for plastic waste management, such as waste collection, or operation of the facility thereof, etc. The shopkeepers and street vendors willing to provide plastic carry bags for dispensing any commodity shall register with local body. The introduction of provision to collect fee from the producers, importers of plastic carry bags / multilayered packaging and vendors selling the same, will strengthen the financial status of local authorities and improve Plastic Waste Management System. The options on reuse of plastic in various applications namely, road construction, waste to oil, waste to energy will enhance the recycling of plastic. The responsibility to provide land for establishing waste management facility has been made to the Department with business allocation of land allotment in the State Government. This would eliminate the issue of getting land for the waste management facility (MOEF&CC notification, 2016).

## 4. PLASTIC WASTE MANAGEMENT: INDIAN SCENARIO

An attempt has been made to summarize the CPCB study on assessment and quantification of plastics waste generation in MSW in 60 major cities of India (Central Pollution Control Board, 2015) suggests that out of total plastics Waste, thermoplastics content is about 94% (Recyclable) and rest 6% belong to family of others including thermoset plastics (Non-Recyclable). The data mentioned in Table 3 indicates that the majority of the plastics waste (PW) obtained about 66% generated by HDPE/LPDE materials which is of mixed plastic wastes like Polybags, Multilayer pouches used for packing food items, Ghutkas etc. The households are the biggest source of plastics waste. The mechanical recycling requires extensive sorting is necessary to separate the packaging waste and the isolation of pure plastics was found to be too difficult.

The containers, films and other oversized items which consist of Polyethylene, Polypropylene and Polystyrene can undergo mechanical recycling. But the mixed plastics, any blended material or other multilayer films should be prepared for chemical or feedstock recycling. In the chemical recycling process, contamination and heterogeneity are not a problem. The multilayer components which consists of 2-3 layers of different plastics, will find difficult on the mechanical recycling. Hence the monomer recovery

*Table 3. The consolidated details of classification of different constituents of plastics*

| S.No | Description | Total Percentage Obtained |
|---|---|---|
| 1 | PET | 8.66 |
| 2 | HDPE/LDPE | 66.91 |
| 3 | PVC | 4.14 |
| 4 | PP | 9.9 |
| 5 | PS | 4.77 |
| 6 | Others | 6.43 |

(Source: CPCB, 2015)

makes logistics and economic sense for the multilayer pouches which are littered in more quantity and keep piling up on garbage leads unhygienic condition. It is also observed that these multilayer/ Metalized pouched are not lifted by the rag pickers, because collecting them is not profitable and non-recyclable hence thrown/dumped in the dumpsite.

## 5. MANAGEMENT OF PLASTIC WASTE IN INDIA

House-to-house collection is system is adapted in few cities. In those cities that use house-to-house collection the waste that are collected from houses, offices, small shops and small markets. Here people are required to deposit their wastes in communal containers/ community bins (stationary or haul types), from which it is collected by municipal crews, handcarts and tricycles are used for waste collection from individual houses at a specific time in the morning, when residents deposit the stored waste into the handcarts. The waste in the handcarts is either transferred to community bins or directly transferred to vehicles going to the disposal site. Transportation of waste from collection point to disposal sites is carried out by using different types of vehicles depending on the distances to be covered by them. Larger vehicles carry the waste from the collection points to the disposal sites. Comparatively small vehicles discharge waste at transfer stations where the wastes are loaded into larger vehicles for transportation to the disposal sites. In metro cities transfer stations located at different places to support intermediate transfer of waste from the surrounding areas up to the dumping grounds.

## 6. TREATMENT AND DISPOSAL

There is no processing of waste being done in most of the Indian cities. The entire waste, which is collected, is taken for dumping to the disposal site. At present there is no sanitary landfill site in most of the cities. The disposal is carried out following the method of crude dumping where the waste is neither spread nor covered. In some areas the garbage waste is recklessly burnt in open dump yards placed on the main highway road. Land filling of mixed waste like non-biodegradable, inert waste and other waste that are not suitable either for recycling or for biological processing is dumped together with recyclable material. Waste processing facilities are not available with proper capacity except few cities (CPCB, 2015).

## 6.1 Incineration

The characteristics of plastic waste incinerated is shown in Table 4. In India, incineration is not a common practice, since the garbage tends to be low in calorific value and volumes are generally low for a central facility. The technology for incineration is not available domestically and import options are highly capital intensive. An incinerator plant was established in Delhi during the 1980s and was expected to generate power for the local grid. However, the operational experience was not satisfactory (Tapan Narayana, 2009). The combustible percentage is found to be near in both the plastic waste and the liquid fuel but percentage of moisture content and ash content is found to be higher in PVC plastic waste and HDPE plastics waste (Chun-Teh Li, 2001).

# 7. OPTIONS FOR PLASTIC WASTE MANAGEMENT

Recycling of plastics through environmentally sound manner is carried in order to minimize the pollution during the process and as a result to enhance the efficiency of the process and conserve the energy. Plastics recycling technologies have been historically divided into four general types -primary, secondary, tertiary and quaternary. Primary recycling involves processing of a waste/scrap into a product with characteristics similar to those of original product. Secondary recycling involves processing of waste/scrap plastics into materials that have characteristics different from those of original plastics product. Tertiary recycling involves the production of basic chemicals and fuels from plastics waste/scrap as part of the municipal waste stream or as a segregated waste. Quaternary recycling retrieves the energy content of waste/scrap plastics by burning / incineration. This process is not in use in India.

# 8. UTILIZATION OF PLASTICS WASTE IN DIFFERENT FIELD

## 8.1 Road Construction

The plastic waste (Bags, Cups, Thermocloe) made out of PE, PP, & PS are separated, cleaned and shredded into small pieces by passing through 4.35mm sieve. The aggregate (granite) is heated to 170 C in the mini hot mix plant and shredded plastic waste is added, it's get softened and coated over the aggregate. Immediately the hot bitumen of 160 C is added and mixed well. As the polymer and the bitumen are in the molten state they get mixed and the blend is formed at the surface of the aggregate. The mixture is transferred to the road for laying.

*Table 4. Components of three plastic waste and liquid fuel*

| Component | PVC plastics | PP plastics | HDPE plastics | Liquid diesel |
|---|---|---|---|---|
| Combustible (%) | 97.3 | 99.4 | 98.4 | 99.9 |
| Moisture (%) | 1.67 | 0.27 | 0.05 | 0.05 |
| Ash (%) | 1.08 | 0.37 | 1.51 | 0.04 |

(Source: Chun-Teh Li, 2001)

## 8.2 Cement Kilns

The recycling of plastics waste for recovery of energy is the use of plastics waste as an alternative to fusible fuel in Cement Kilns. Any material having calorific value of at least 2,500 kcals are accepted as an alternative fuel in cement kilns, provided it is available at a cost less than the normal fossil fuel: plastics waste, which have quite high calorific values, offer a viable alternative fuel. In fact, Cement Kilns can be utilized for burning of some hazardous waste and for recovering precious energy out of it for production of cement. (CPCB, 2015).

## 8.3 Conversion of Plastic Waste Into Liquid Refuse Derived Fuel, RDF (Oil)

The pyrolysis by direct heating was adopted to produce the paraffin and crude oil from the plastic wastes in the 1990s. The small-scaled process is featured by facilitation, convenience and low equipment investment However, the temperature caused by pyrolysis is higher and all the reactive time is longer than the other methods. The octane number of gasoline gained is relatively low and the pour point of diesel oil is high. More paraffin is produced in the process of pyrolysis. Although this process is simple and convenient, the converting rate and yield is still lower. Since the total yield of fuel oil with pyrolysis is still lower and the quality of oil is not satisfied as gasoline and diesel oil, the upgrade by catalytic cracking for the crude products gained with pyrolysis can be used. Having improved the quality of finished oil, this process has been widely used in many factories. The system consists of the knapper, extrusion machine, pyrolysis reactor, catalytic cracking reactor, fractionating tower, heating and temperature controller, separator of oil and water, and oilcan. The catalyst for the process is prepared by using Faujasite zeolite(05 - 35 wt%), Pseudoboehmite alumina(10 - 40 wt%), Polyammonium silicate (01 - 10 wt%),Kaolin clay(15 - 60 wt%).(Source: Raja and Murali, 2011).

# 9. INDIAN CITIES SCENARIO OF PLASTICS WASTE GENERATION

In almost all the cities the MSW generated from different sources like door to door, community bin at the street corners and from markets are collected by the municipal trucks or by the firms under contract to the government and dumped in the open dump yard. The plastic waste generation in Indian cities as per the survey done by CPCB are summarized below in Table 5.Perusal of data as summarized in Table 5 indicates that the average waste generation of PET plastics is higher in Faridabad city. Similarly the average waste generation of HDPE/LDPE plastics is found to be higher in Faridabad. Also the average waste generation of polyvinyl chloride plastics is higher in Varanasi. The average waste generation of polypropylene and polystyrene plastics is found to be higher in Amritsar and Delhi.

## 9.1 Study of Delhi City: PW (Kg/Mt)

In Delhi, the quantity of plastic waste has been assessed as 10.14%, which comprises of 76% of HDPE/LDPE, 6% of Polypropylene and 10% of Polystyrene material. The total MSW generated in this city was with an average of 6800 MT/ Day. During the survey on different weekdays, the minimum plastic waste generation was observed to be 87.23 Kg/MT and maximum was 118.74Kg/MT. It has been observed no post-treatment operation for MSW is carried out in the city and 100% of MSW is dumped as land-filling. Approximately, 200-250 rag pickers were voluntarily collecting the valuable plastics waste that was generated throughout the day. Further the compost fertilizer plant having the capacity of about 500MT was established in the city, but currently the plant produces with an output of about only 200 MT. Delhi, being the capital city requires urgent attention for efforts to recycle the recyclable plastic waste such as PET, PE & PP & PVC etc. since the quantum of waste would increase further in future. With over 10% of Plastics Waste in MSW, waste management authorities/civic bodies should set-up Plastics Waste Management Cell (PWMC) exclusively to deal with plastics waste by adopting prevailing technologies (CPCB, 2015).

Also the data obtained for Patna city reveals that the total MSW generated was about 220 MT/Day. The plastic waste was observed as 57.25 Kg/MT, out of which the majority of plastics waste found as carry bags and packaging pouches which is of 65.15% made up of HDPE/LDPE material. The field study reveals that the minimum plastic waste generation 35.98 Kg/MT and maximum of 76.57Kg/MT (CPCB, 2015).

## 10. WORKING OF INDUSTRIES UNDER EXTENDED PRODUCERS RESPONSIBILITY

### 10.1 Dabur India Limited

On 3rd January 2018 India's largest Science-based Ayurveda major Dabur India Limited has decided to form a consortium with other responsible organizations to collect and help Multilayer Plastic (MLP) Waste Collection and sending for co-processing as RDF (Refuse-Derived Fuel) in cement kilns for a Pilot Project in Gujarat state covering 11 major cities. This is also part of Gujrat first mega Environment Protection initiative called 'EPR Connect: Towards a Sustainable Future". Under Gujarat 1st Extended Producers Responsibility (EPR) consortium, Dabur is part of processing the first batch of collected waste of MLPs for recycling from 11 cities (Ahmedabad, Sanand, Chattral, Changodar, Mehsana, Kadi, Kalol, Hathijan, Gandhinagar, Rakanpur and Santej) in Gujarat State. This joint initiative is being implemented by Nepra Resource Management Private Limited, a waste management company, based in Ahmedabad duly guided and supported by Gujrat Pollution Control Board (GPCB), and Gujrat URBAN Development Department (Dabur Media, n.d.).

### 10.2 Colgate-Palmolive Company

The Colgate-Palmolive Company is an American multinational consumer products company focused on household, health care, and personal hygiene products. Personal care products giant Colgate-Palmolive has committed to making 100% of its packaging for three of four product categories completely recyclable by 2020. In addition, Colgate has committed to work towards developing a fully recyclable toothpaste tube or package, which would bring its fourth product category close to the same sustainability standard. The company also agreed to increase the average recycled content of its packaging to 50%. Increasing the recycled content of key materials such as polyethylene (PE), polyethylene terephthalate (PET), high-density polyethylene (HDPE), polypropylene (PP), and paper pulp (fiber) is crucial to improving the sustainability of consumer packaging. Colgate's commitments includes that by 2020 achieve 100 percent recyclable portfolio of Home, Pet, and Personal Care packaging - three out of four of its product categories. Develop a fully recyclable toothpaste tube or other packaging in the fourth category, Oral Care. Increase recycled content of packaging from 40 percent to 50 percent. Reduce or eliminate use of PVC in packaging.

## 10.3 Coca-Cola Company

Coca-Cola, or Coke, is a carbonated soft drink produced by The Coca-Cola Company.

As a responsible corporate citizen, [Coca-Cola] welcome[s] the introduction of 'Extended Producer Responsibility' as part of the new Plastic Waste Management Rules, 2016.As a system, coca-cola have always been a proponent of circular thinking and are committed to improving the recycling rates of PET waste in India. In the waste guzzling metropolis of Gurugram, coca-cola has launched a source segregation programme – 'Alag Karo – Har Din Teen Bin' in partnership with Saahas, Tetra Pak and GIZ. Through this initiative, [they] are taking the practice of segregating waste at source to households, schools and commercial establishments across Gurugram. Coca-Cola's bottling partners continue to work together with consumers, collection agencies and authorised recycling partners to ensure that PET waste is collected and recycled to the extent possible. *(Coca-Cola, n.d.)*

## 11. CONCLUSION

Improved waste management systems and regulatory discipline are required in India because volumes of plastic waste will clearly rise in the future as the per capita consumption of plastic products increases. The ill effects of Climate Change have already begun to be felt. Toxic substances are released via burning from plastics, open combustion, incineration, posing a threat to the surrounding areas including vegetation and health of individuals. Proper development of policy with respect to chemical exposure caused by plastic must be set in place with encouraging research in this direction. A sustainable step towards tomorrow's cleaner and healthier environment is the need of the hour. For this reason Extended Producer Responsibility (EPR) concept is introduced for all producers. The concept of Extended Producer Responsibility (EPR) is that the responsibility of the producer of a product is extended beyond conven-

*Figure 2. Collection & Transportation of MSW (Source: CPCB, 2015)*

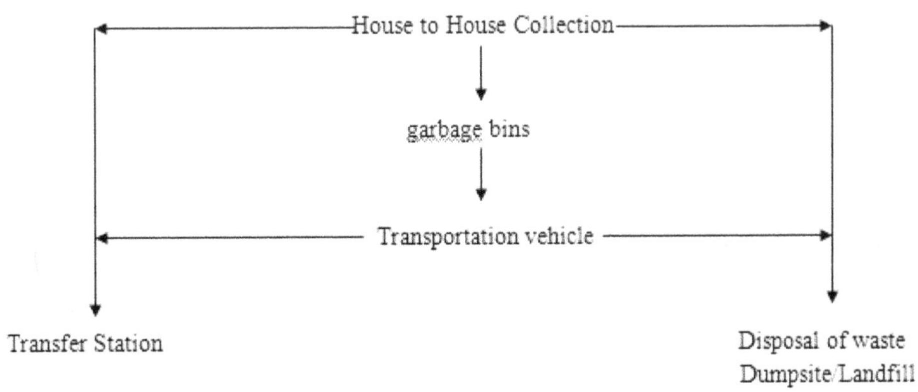

*Table 5. Plastic waste generation of Indian cities*

| City | PET | HDPE/LDPE | PVC | PP | PS | OTHERS |
|---|---|---|---|---|---|---|
| Lucknow | 2.4-9.1 | 19.42-48.16 | 1.42-4.77 | 4.14-13.93 | 0.66-6.31 | 55.59-62.93 |
| Allahabad | 3.5-6.0 | 8.6-40.40 | 1.06-6.78 | 7.78-17.43 | 0.0-1.72 | 4.06-23.90 |
| Chandigarh | 0.0-0.8 | 8.0-21.60 | 0.2-5.6 | 0.32-9.6 | 0.8-3.2 | 4.0-13.6 |
| Delhi | 1.41-5.98 | 85.24-67.34 | 0.85-2.63 | 2.92-9.33 | 6.42-17.22 | 1.0-9.29 |
| Faridabad | 8.73-15.36 | 74.4-104.54 | 0.32-1.65 | 1.14-7.79 | 4.3-6.99 | 0.0-1.0 |
| Jammu | 0.32-2.56 | 16.32-36.88 | 0.64-14.16 | 12.16-28 | 4.48-8.16 | 7.2-34.08 |
| Srinagar | 2.32-10.4 | 16.68-25.06 | 4.0-5.68 | 9.11-10.56 | 2.32-5.76 | 1.52-9.04 |
| Shimla | 1.56-3.68 | 13.64-26.64 | 0.56-1.32 | 8.04-20.24 | 1.0-3.36 | 4.2-13.44 |
| Amritsar | 0.0-03.04 | 15.04-28.32 | 0.0-2.26 | 9.92-23.2 | 1.44-5.6 | 0.0-15.52 |
| Dehradun | 5.10-10.35 | 36.21-43.15 | 2.57-7.41 | 6.14-9.61 | 1.56-4.34 | 2.17-5.33 |
| Agra | 8.74-10.33 | 42.68-45.0 | 5.97-8.22 | 6.74-10.16 | 0.18-1.24 | 7.67-11.01 |
| Meerut | 1.55-6.49 | 41.10-42.26 | 3.022-6.92 | 4.77-7.11 | 0.62-2.48 | 3.40-7.24 |
| Varanasi | 2.84-12.37 | 13.03-42.89 | 2.58-15.27 | 3.98-10.05 | 0.54-2.21 | 3.53-10.2 |
| Kanpur | 3.36-8.36 | 40.19-53.15 | 4.26-8.17 | 2.08-7.02 | 0.34-0.95 | 1.03-3.44 |
| Patna | 1.0-1.67 | 33.25-46.56 | 0.42-6.93 | 1.70-9.63 | 1.68-6.8 | 2.12-10.76 |
| Ranchi | 0.65-1.02 | 35.64-42.54 | 0.36-1.19 | 5.71-7.0 | 3.36-5.89 | 6.02-9.31 |

(Source: CPCB, 2015)

tional sales to its post-consumer or end-of-life (EOL) stage. This means that the producer is responsible for collection of the used products or packaging material and ensure its safe recycling or disposal. In line with the principle of 'extended producer's responsibility' the new rules for the first time in India has underlined the role of municipal bodies in not only ensuring safe collection and disposal of plastic wastes but also in engaging agencies or groups working in waste management including waste pickers.

## REFERENCES


Achyut, K. P., Singh, R. K., & Mishra, D. K. (2010). Thermolysis of waste plastics to liquid fuel A suitable method for plastic waste management and manufacture of value added products—A world prospective. Renewable and Sustainable Energy Reviews, 233–248.

Central Pollution Control Board. (2015). *Assessment & Quantification of Plastics Waste Generation in Major cities –CPCB Report, 2015.* Retrieved from http://www.indiaenvironmentportal.org.in/files/file/Assessment%20and%20Quantification%20of%20Plastics%20Waste.pdf

Coca-Cola. (n.d.). *Sustainability Update 2016/17. Message from our leadership.* Retrieved from https://www.coca-colaindia.com/sustainability-report/sustainability-message

Fatima, S. (2014). Chemical Recycle of Plastics. *American Journal of Engineering Research,* 3(11), 93-108.

Li, C.-T., Zhuang, H.-K., Hsieh, L.-T., Lee, W.-J., & Tsao, M.-C. (2001). PAH emission from the incineration of three plastic wastes. *Environment International*, *27*(1), 61–67. doi:10.1016/S0160-4120(01)00056-3 PMID:11488391

Media, D. (n.d.). *Dabur Joins EPR Connect: Towards a Sustainable Future Project*. Retrieved from https://www.dabur.com/in/en-us/media/dabur-joins-epr-connect-towards-a-sustainable-future-project

Muth, P., & Premnath. (2006). Plastics materials flow analysis for India. *Resources, Conservation and Recycling, 47*, 222–244.

Narayana, T. (2009). Municipal solid waste management in India: From waste disposal to recovery of resources. Waste Management, 29, 1163–1166.

Panda, A., Singh, R. K., & Mishra, D. K. (2010). *Thermolysus of Waste Plastics to Liquid Fuel: A Suitable Method for Plastic Waste Management and Manufacture of Value Added Products—A World Prospective*. Retrieved from https://www.researchgate.net/publication/227421226_Thermolysis_of_waste_plastics_to_liquid_fuel_A_suitable_method_for_plastic_waste_management_and_manufacture_of_value_added_products--A_world_prospective

Pavani, P., & Rajeswari, T. R. (2014). Impact of Plastics on Environmental Pollution. *Journal of Chemical and Pharmaceutical Sciences*, (3).

Press Trust of India. (2016). *15,342 tn plastic waste generated in India everyday: Dave*. Retrieved from http://www.business-standard.com/article/pti-stories/15-342-tn-plastic-waste-generated-in-india-everyday-dave-116080200866_1.html

Rahimi, A., & García, J.M. (2017). Chemical recycling of waste plastics for new materials production. *Nature Reviews Chemistry, 1*.

Salem, S. M., Lettieri, P., & Baeyens, J. (2009). Recycling and recovery routes of plastic solid waste (PSW): A review. Waste Management, 29, 2625–2643.

The Federation of Indian Chambers of Commerce and Industry (FICCI). (2014). *Potential Of Plastics Industry In Northern India With Special Focus On Plasticulture And Food Processing: A report on plastics industry*. Retrieved from http://ficci.in/spdocument/20396/Knowledge-Paper-ps.pdf

Toxic Link. (2016). *WEEE plastic and Brominated Flame Retardants: A Report on WEEE Plastic Recycling*. Retrieved from http://toxicslink.org/docs/Report-WEEE-plastics.pdf

# Chapter 6
# Industrial Wastewater Pollution and Advanced Treatment Techniques

**Smita Chaudhry**
*Kurukshetra University, India*

**Shivani Garg**
*Kurukshetra University, India*

## ABSTRACT

*Industry creates more pressure on water resources by wastewater discharge than the quantity used in production. The wastewater produced by industries may be either excessively acidic or alkaline or may contain high or low concentrations of colored matter, organic or toxic materials, and possibly pathogenic bacteria. It is necessary to pre-treat the wastes prior to release to the sewer or a full treatment is necessary when this is discharged directly to surface or ground waters and it must be within the effluent standard limits provided by the environmental protection organizations. The management and control of liquid wastes in the industry as well as the selection of the different possible treatments for the wastewater prior to its discharge to the sewer system was studied. These would protect the environment and also benefits from the waste materials can be gained. Opportunities for introducing pollution prevention measures for different types of pollutants produced by different industries are discussed in this chapter.*

## BACKGROUND

The main wastewater collector, the Cloaca Maxima, in Rome presumably follows the course of an old ditch which was used at about 500 BC as a collector for wastewater ((Lamprecht, 1988). By 1880, scientists began to understand pathogenic bacteria and their association with specific disease. For example, calcium chloride was used to treat faeces from typhoid patients before disposal to sewers (White, 1972). Frankland (1869) proposed ten parameters to analyses the river water quality. The first activated sludge wastewater treatment plant was taken into operation in 1932, Kyläsaari, Finland. The first modern treatment plant functioned properly built after the Second World War in 1957 in Tali. An Industrial ecology

DOI: 10.4018/978-1-5225-5754-8.ch006

was recognized as one step in "waste utilization" rather than "waste treatment" in the late 1990s. In 1996, the construction of a new blower building steel air lines, and provision of new sanitaire diffusers in the aeration chamber was established. The replacement of the original gas chlorination equipment was done with a liquid chlorination facility to meet *E. coli* effluent limitations in 2012. According to historical studies it has been observed that wastewater treatment techniques have improved with the development of economic conditions, civilization and awareness towards public health. The pollution due to industrial wastewater and the modern treatment techniques are discussed in this chapter.

## INTRODUCTION

Wastewater can be defined as 'used water discharged from homes, industry, cities and agriculture'. Industry creates more pressure on water resources from the impacts of wastewater discharges and their pollution potential than by the quantity used in production. Industrial effluents are the most important sources of toxic contaminants in any environment (Mohana et al., 2008). Industries such as agricultural based, forest based and marine based on organic raw materials are the largest contributors to organic pollution, while oil, steel, and mining industries represent the major risk for heavy metal release. 70% of industrial wastes in developing countries are disposed off untreated into water supplies (UN- Water 2009). Wastewater discharges from industrial and commercial sources may contain pollutants at levels that could affect the quality of receiving waters. The AQUASTAT database of the Food and Agriculture Organization of the United Nations (FAO) estimates that global freshwater withdrawals rate is 3,928 $km^3$ per year (Raza et al., 2012). An estimated 44% (1,716 $km^3$ per year) of this water is consumed, mainly by agriculture through evaporation in irrigated cropland. The remaining 56% (2,212 $km^3$ per year) is released into the environment as wastewater. It has been reported that irrigation with sewage or sewage mixed with industrial effluents results in saving of 25 to 50% of N and P fertilizer and increase 15-27% higher crop productivity, over the normal waters (Anonymous, 2004). An estimated 38354 million litres per day (MLD), sewage is generated in major cities of India, but the sewage treatment capacity is only of 11786 MLD (~30%) Similarly, only 60% of industrial waste water, mostly of large scale industries, is treated (CPCB, 2009). Much industrial waste water is discharged without treatment to open water resources, reducing the quality of larger volumes of water and sometimes infiltrating aquifers and contaminating groundwater resources. The wastewater discharge from industries varies in flow rate and pollution strength. So, it is impossible to assign fixed values to their constituents. The treatment capacity is strongly correlated with the countries' income: in lower-middle-income countries on average 28% of the generated wastewater is reported to be treated, and in low-income countries, only 8% is treated, while in high income countries the ratio is closer to 70% (Sato *et al.*, 2013). India is a World leader in dairy sector with 15% of global share. The milk production has increased significantly from 137.7 to 164 million tonnes from 2013-14 to 2016-17. Dairy industry contributes market share of Pasteurized Milk (73%), Milk Power (8%), and Ghee (8%), Ice cream (4%), Butter (3%), and Rest (4%-Flavoured milk, Cheese etc.). It is very water intensive industry in terms of consumption & waste water generation (i.e. 4.0 Litter water per Litter of milk). Surat is one of the wealthiest city in India where dairy industries generates 800 $m^3$ wastewater effluent having high effluent COD, oil and grease content. The waste water treatment capacity in India has increased by about 2.5 times since 1978-79 yet hardly 10% of the sewage generated is treated effectively, while the rest discharge into the environment and is responsible for large-scale pollution of rivers and ground waters (Trivedy and Nakate, 2001). Industries are the manufacturing or technically productive enterprises that produce many goods but are big user of water. This chapter describes different types of industries that produce wastewater and their treatment methods.

## CATEGORIES OF INDUSTRY

In most industries, wastewater effluents result from water use in cooling (from disposing of excess heat to the environment), sanitary wastewater (from washing, drinking, etc.), processed wastewater (including both water used for production and washing products and for removal and transport of waste and by-products) and cleaning (including wastewater from cleaning and maintenance of industrial areas). The wastewater produced from urban areas is about evenly divided between industrial and municipal sources, excluding the large volumes of cooling water discharged by the electric power industry. The quality and quantity of wastewater vary from industry to industry and these variations are brought about by bath discharges, operation start-ups and shutdowns, working-hour distribution and so on.

In Figure 1, the classification system is based on the raw material different industries used for production. These industries are the big contributor of water pollution. The values of typical concentration parameters ($BOD_5$, COD, suspended solids) for different industrial effluents are given in Table 1. Industrial wastewater characteristics vary industries to industries and are also much more qualitatively and quantitatively different than domestic wastewater in its composition.

The Ministry of Environment, Forest and Climate Change (MoEFCC) has given the criteria of categorization of industrial sectors based on the Pollution Index (PI) which is a function of the emissions (air pollutants), effluents (water pollutants), hazardous wastes generated and consumption of resources (Table 2). For this purpose, the references are taken from the Water (Prevention and Control of Pollution) Cess (Amendment) Act, 2003, Standards so far prescribed for various pollutants under Environment (Protection) Act, 1986 and Doon Valley Notification, 1989 issued by MoEFCC. The Pollution Index

*Figure 1. Classification of industries on the basis of products produced*

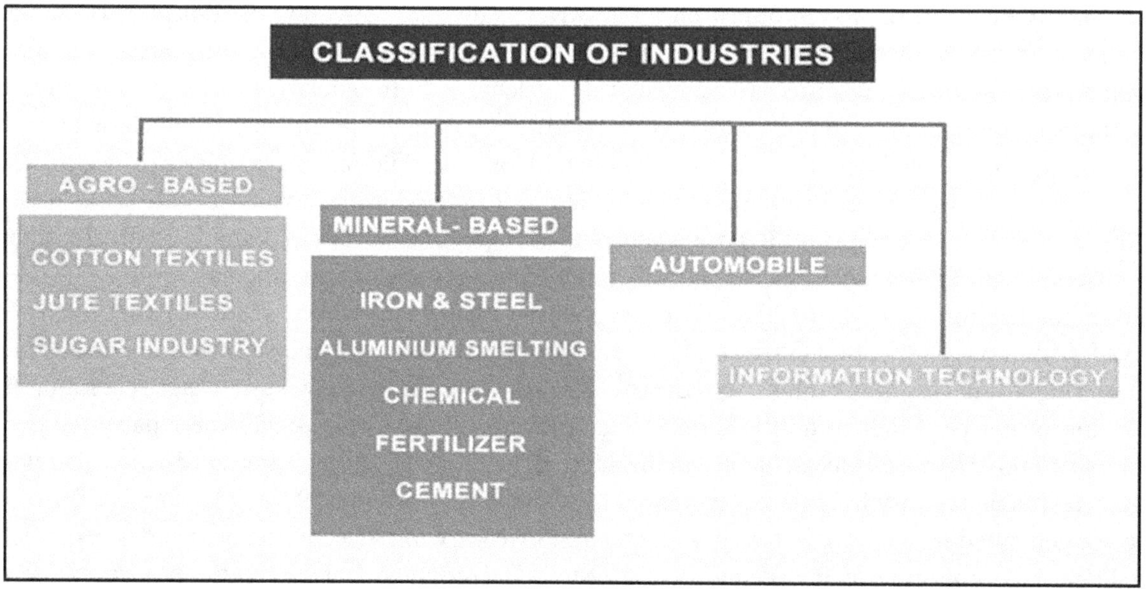

*Table 1. List of comparative strengths of Industrial wastewater and their principal pollutants.*

| Industries | Principal Pollutants | $BOD_5$ (mg/l) | COD (mg/l) | SS (mg/l) |
|---|---|---|---|---|
| Canning | Carbohydrates, proteins | 2000 | - | 7000 |
| Tannery | Carbohydrates, proteins | 1000-2000 | 2000-4000 | 2000-3000 |
| Household chemicals | soap, oven cleaners etc. | 1600 | 2700 | 250-500 |
| Dairy industry | Carbohydrates, Fat, Proteins | 600-1000 | - | 200-400 |
| Brewery | Colours, liquors | 850 | 1700 | 90 |
| Oil Refinery | Oil, petrochemicals | 100-500 | 150-800 | 130-600 |

(Source: Abdallh et al., 2016)

*Table 2. Categories of the industrial sector with pollution index score given by MoEFCC*

| Industrial Sector With Pollution Index (PI) Score | Category |
|---|---|
| 60 and above | Red Category |
| 41 to 59 | Orange Category |
| 21 to 40 | Green Category |
| up to 20 | White Category |

(Source: CPCB, 2016)

(PI) is a number from 0 to 100 given to any industrial sector and the increasing value of PI indicates the increasing degree of pollution load from the industrial sector.

According to Ministry of Environment, Forests and Climate Change, Govt. of India, as mentioned in Table 2, Industries identified as Red category with PI score 60 and above are heavily polluting such as sugar, fertilizers, pulp and paper, oil refinery, cement etc., industries identified as orange category with PI score from 41 to 59 are manufacture of mirror from sheet glass and photo framing, cotton spinning and weaving, automobile servicing and repair stations, hotels and restaurants etc., and all those industries or processes which are not covered under the "Red" and/or "Orange" category; identified as Green category are mineralised water, bakery products, chilling plants and cold storage, tailoring and garment making, etc., and industries under white category with PI score up to 20 are Bicycles, baby carriages, bailing of waste papers, bio fertilizer /bio-pesticides, biscuits trays, blending / packing of tea, block making of printing, chalk making, compressed oxygen gas etc.

## Approaches for Choosing Suitable Technology for Treatment of Wastewater From Different Industries

The main goal of design an industrial wastewater technology process is to minimize the volume and toxicity and the final treatment residue, since final disposal can sustain significant cost and liability. The evaluation and selection of waste treatment technologies typically follows a logical series of steps that help to meet the goal of minimizing waste toxicity and volume. Multiple process alternatives which suit all related law and regulations should be considered and the technical and economic aspects of these alternatives should be evaluated by detailed comparative cost analysis, which should include investment, operation, in order to determine optimum treatment for the industry (Buyukkamaci and Koken, 2010).

*Figure 2. Approach for developing a well-operating, cost-effective treatment system for Industrial Waste Water*

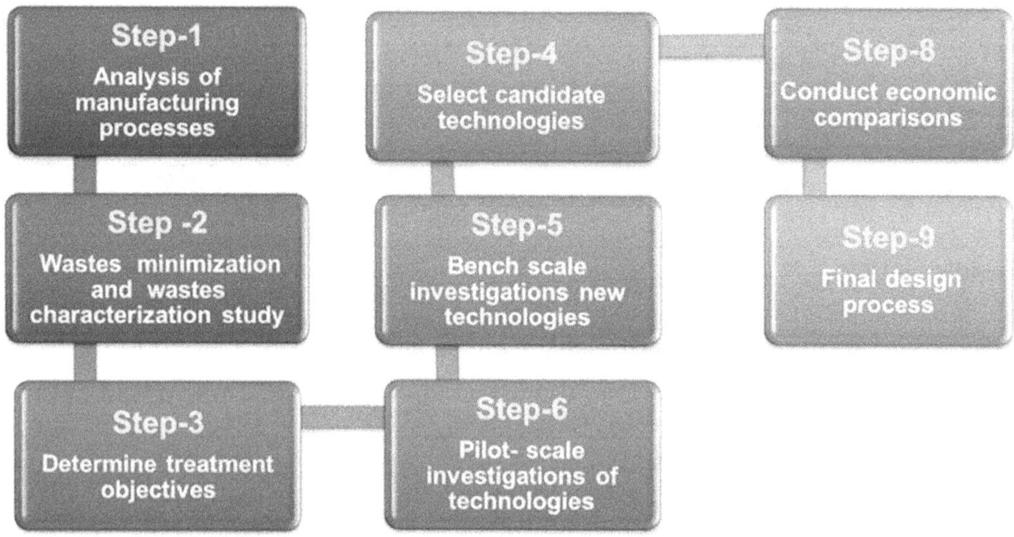


Industrial waste water (IWW) characteristics vary not only with the industry that generates them, but also within the industry itself. Treatment of IWW is a complex due to the wide variety of compounds and concentrations it may contain. So, treatment of different kinds of industrial wastewater specific technologies is necessary.

*Table 3. Treatment of Different Kinds of Industrial Wastewater with Different Technologies*

| Name of Industries | Wastewater Quality | Methods of Treatment | References |
|---|---|---|---|
| **Power industry and cooling water** | Ammonia, cyanide, thiocyanate, sulfides, and a wide variety of complex hydrocarbons | A combination of Electro Dialysis Reversal and Reverse Osmosis (RO) | Judd and Jefferson, 2003 |
| **Pulp and paper industry** | Coloration, non-biodegradable organic materials | Chemical, Aerobic and Anaerobic Biological Processes; Adsorption, Advanced Oxidation and Membrane Filtration | Ko and Fan, 2010. |
| **Textile industry** | Dyes, oil and grease, organic compounds | Fenton process, PAC Coagulation and Ion exchange | Ustun, Solmaz and Birgul, 2007 |
| **Food and beverage industry** | Oil and grease (O&G), organic hydrocarbons | MBRs of Microfiltration, Coagulation/ Flocculation | Amuda and Amoo, 2007 |
| **Pharmaceutical industry** | Pharmaceuticals and personal care products of 20 compounds including antibiotics, hormones, | Physicochemical and Modified Activated Sludge with Co-existence of anoxic- oxic condition, combination of Coagulation –Fenton Process | Perdigon-Melon et al., 2010 |

## STANDARDS AND GUIDELINES FOR EFFLUENT

Discharge standards vary from country to country and in many cases, from state to state, reflecting their specificities, development stage, economical level, commitment of environmental protection and various other factors. It is easier for the environmental agencies to control, monitor and enforce point discharges. Discharge standards play an important role in most countries in the world to prevent water body pollution and protect environment. It sets limits for quality of wastewater from the different industries. Standards must be compiled with, through enforcement of the legislation, by the entities involved the discharges and use of the water.

The World Health Organization (1989) has recognized both the potential and risk of untreated wastewater use and so has developed guidelines for policy makers attempting to legislate permission for the safe use of wastewater. WHO (2003) is currently revising their guidelines on wastewater reuse.

## TREATMENT TECHNIQUES OF INDUSTRIAL WASTEWATER

Many processes for treating industrial wastewater involve varieties technologies for reduction of toxicity in order to meet environmental-based treatment standards. Subsequent to the selection of appropriate treatment technology, more pilot-plant scale experiments with real industrial wastewater must be performed on a larger scale. However, there are a very wide variety of technologies for choosing a suitable individual to treat specified wastes; intergradations of more than one technology could be more economic for removal of high strength and toxic industrial. New directives and regulations demand the necessity for finding the optimal solution of treatment in order to decrease hazardous compounds bellow the maximal concentration prescribed by the regulations. The constituents of the Industrial Waste Water (IWW) are highly variable and specific. The approach used to treat and dispose of IWW is distinctly different from that used for municipal wastes (Woodard, 2005).

Technologies for treating IWW according its characteristics are generally divided into three broad categories:

*Table 4. Standards given by Central Pollution Control Board (2006) for effluent discharge*

| Name of Parameters | Permissible Limit |
|---|---|
| **pH** | 5.5 -9.0 |
| **$BOD_5$ (mg/l)** | 100 |
| **COD (mg/l)** | 250 |
| **TDS(mg/l)** | 2100 |
| **TSS (mg/l)** | 100 |
| **Temperature (°C)** | ≤40 |
| **Appearance** | Normal |
| **Odour** | - |

- **Biological Methods:** The principal applications of the biological processes are: carbonaceous organic material representing in biochemical oxygen demand (BOD) removal, nitrification, denitrification, stabilization, and phosphorous removal. Aerobic biological processes are commonly achieving high degree of treatment efficiency, while anaerobic uses the concept of resource recovery and utilization with achieving objective of pollution control. In recent years, high rate anaerobic/aerobic bioreactors have been increasingly employed for degradation high strength IWW with advantage of minimal space requirements, low capital cost and good COD removal efficiencies (in excess of 83%) (Chan et al., 2009).
- **Chemical Methods:** This treatment can be used for removing substances through producing insoluble solids and gas, coagulation of a colloidal suspension, biological degradable substances from non-biodegradable, destroying or deactivating chelating agents, and Non objectionable substances that can be removed more easily. Coagulants allow the colloidal particles to join together by slow stirring. Combination of advanced oxidation processes or biological technologies for treatment of certain difficult (or impossible) industrial wastewater were used and have greatly increased the removal of difficult pollutants. The ability to increase biodegradability and detoxify effluent streams containing polar and hydrophilic chemicals by alternative treatment with advance oxidation processes, such as photo-catalysis, ozonation, and ultrasound-oxidation has been extensively studied in the recent years
- **Physical Methods:** This treatment accomplishes removal of substances by use of naturally occurring forces, such as gravity, electrical attraction, and van der Waal forces, as well as physical barriers. Physical methods of WWT include sedimentation, flotation, and adsorption, as well as barriers such as bar racks, screens, deep bed filters, membranes, electro dialysis and ion exchange. Segregation of total suspended solids from any industry's waste plays an important part in its overall waste treatment program for elimination a major portion of the contaminant and for separation of one type of contaminant for more easily and economically further treatment. Membrane processes can enhance conventional processes by concentration of components in the reactor, such as in a MBR. Membrane processes have a good efficiency of water recycle and reuse to meet water reusable request in many industries (foodstuff, leather, textile, electronic industry).

There are essentially two types of wastewater collection and treatment systems:

1. **Off-Site Systems:** Where waste is transported through a sewerage network to a treatment plant or disposal point.
2. **On-Site Systems:** Where waste is accumulated in a pit or septic tank. This tank can be periodically emptied or a new pit/septic tank can be opened in another location. Certain on-site systems have leaching beds that infiltrate the partiality treated water from septic tanks into the ground (old and overstressed systems are a significant cause of pollution in some areas). In the case of emptying, waste is transported for treatment and/or disposal. On-site systems can also include small-scale sewerage systems that convey wastewater to treatment plants located nearby.

The toxicity, mobility, and loading of industrial pollutants have potentially more significant impacts on water resources, human health and the environment than actual volumes of wastewater. The first step is to keep the volumes and toxicity of pollution to a minimum at the point of origin, from concept to design and in operations and maintenance. This includes

- Substitution with more environmentally friendly raw materials and biodegradable process chemicals, as well as staff education and training to address pollution-related issues.
- The second step is to recycle as much water as possible within a plant, thus minimizing discharge. Small- and medium-sized enterprises (SMEs) and informal industries often discharge their wastewater into municipal systems or directly into the environment. Industries discharging into municipal systems or surface water have to meet discharge regulations to avoid fines, so in many cases, end-of-pipe treatment is required at the plant before release. In some situations, however, industries may find it more economical to pay fines than to invest in treatment to meet regulations.
- One notable opportunity for industrial wastewater use and recycling is the cooperation between plants through industrial symbiosis. This is best seen in eco-industrial parks that locate industries adjacent to one another in such a way as to take advantage of various wastewater flows and water and by-product recycling. For SMEs, this can be a significant way to save on wastewater treatment costs.

The use of water by industry can significantly affect the water quality of receiving waters. The level of wastewater loading from industrial sources varies markedly with the water quality objectives enforced by the regulatory agencies. There are many possible in-plant changes, process modifications, and water-saving measures through which industrial wastewater loads can be significantly reduced. Planed use of reclaimed water is today a common pattern in countries of the Middle East and North Africa, Australia, the Mediterranean, and the United States of America (Mateo-Sagasta et al., 2015). Up to 90% of recent wastewater reductions have been achieved by industries employing such methods as recirculation, operation modifications, effluent reuse or more efficient operation. As a rule, treatment of an industrial effluent is much more expensive without water-saving measures than the total cost of in-plant modifications and residual effluent treatment. There are different levels of wastewater treatment depending upon treatment process adopted.

1. Pre-treatment techniques.
2. Primary Treatment.
3. Secondary treatment or biological treatment.
4. Tertiary treatment.

The whole process of treatment of wastewater in industries is divided into different methods according to type and quality of wastewater (Figure 2). MoEFCC started the common effluent treatment plants scheme in 1991.

## A. Pre-Treatment Techniques

This is the initial stage of wastewater treatment; wastewater constituents such as rags, sticks, floatable grit, and grease that may cause maintenance or operational problem with the treatment operations, processes have been removed. It constitutes following steps:

1. **Screening:** The first unit operation encountered in wastewater treatment plants is screening. According to the method of cleaning, screens are designated as hand cleaned or mechanically cleaned. A screen is a device with openings, generally of uniform size usually coarse screens have

*Figure 3. Classification of wastewater treatment process*

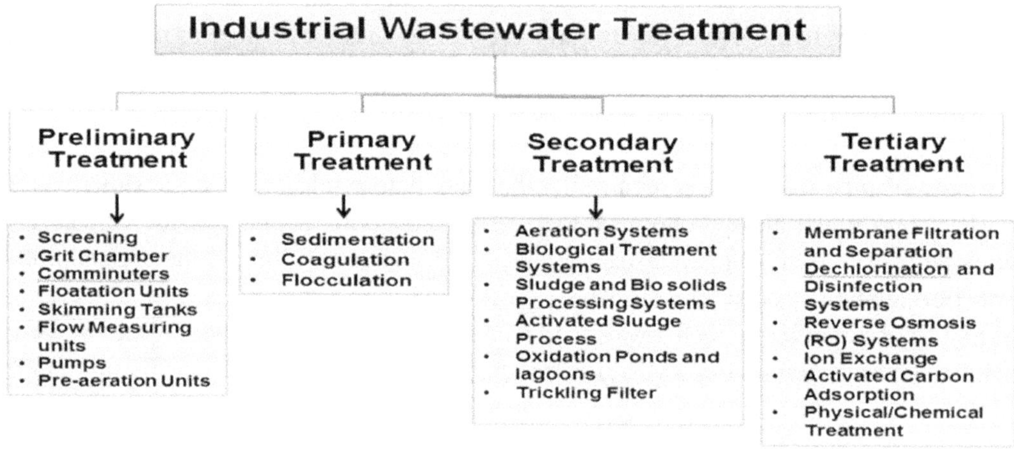


openings of ¼ inch or more, and fine screens have openings of less than ¼ inch that is used to retain the coarse solids found in wastewater. These screens are set at 45° to 60° with vertical. The screens are cleaned manually or by mean of automatically operated rakes.

2. **Grit Chamber:** Grit chamber is the second unit operation used in the treatment of wastewater and it is intended to remove suspended inorganic particles such as sandy and gritty matter from the wastewater. The grit chamber is used to remove grit, consisting of sand, gravel, cinder, or other heavy solids materials that have specific gravity much higher than those of the organic solids in wastewater. Separate removal of suspended inorganic solids in grit chamber and suspended organic solids in primary sedimentation tank is necessary due to different nature and mode of disposal of these solids. Grit can be disposed off after washing, to remove higher size organic matter settled along with grit particles; whereas, the suspended solids settled in the primary sedimentation tank, being organic matter, requires further treatment before disposal.
    a. **Grit Chambers Are of Two Types:** (a) mechanically cleaned and (b) manually cleaned. In mechanically cleaned grit chamber, scraper blades collect the grit settled on the floor of the grit chamber. The grit so collected is elevated to the ground level by several mechanisms such as bucket elevators, jet pump, and air lift. The grit washing mechanisms are also of several designs most of which are agitation devices using either water or air to produce washing action. Manually cleaned grit chambers should be cleaned at least once a week. The simplest method of cleaning is by means of the shovel.
3. **Comminutors:** This is used to grind large size suspended particles. Its role is to protect the pumps from damage. Once the solids are reduced to smaller and more uniformed sizes, it can be returned and move towards subsequent processes without requiring the needs for manual jobs and manpower intervention to clear up the trapped waste materials.
4. **Floatation Units:** Floatation units are used in place of sedimentation, primarily for treating industrial wastewaters containing finely divided suspended solids and oily matter. Floatation technique is used in paper industry to recover fine fibres from the screened effluent and in the oil industry for the clarification of oil-bearing waste. It is also used for treating effluents from tanneries, metal finishing, and cold-rolling and pharmaceutical industries. Particles of density very close to that of

water are very difficult to settle in normal sedimentation tanks and take a long time for separation. In such cases, the separation can be speeded up by aerating the effluent whereby air bubbles are attached to the suspended matter. This has the effect of increasing the buoyancy of the particles; as a result, the particles float to surface from where they can be readily removed. To aid in the floatation process, chemical coagulants such as aluminium and ferric salts or polymer coagulant-aids are often used. These chemicals increase the flocculent structure of the floated particles so that they can easily entrap the air bubbles. Two methods of floatation are currently available such as dispersed-air floatation and dissolved air floatation. In dispersed-air floatation, air is introduced directly into the liquid through a revolving impeller or through diffusers. In dissolved-air floatation, air is intimately brought into contact with the wastewater at a pressure of several atmospheres when the air is dissolved. The pressure on the liquid is reduced to atmospheric level through a back-pressure valve, thereby releasing micron-sized bubbles.

5. **Skimming Tank:** A skimming tank is a chamber so arranged that the floating matter like oil, fat, grease etc., rise and remain on the surface of the wastewater until removed, while the liquid flows out continuously under partitions or baffles walls. A theoretical detention period of 3 minutes is enough. The floating matter can be either hand or mechanically removed. It is mostly present in the industrial wastewater treatment plants.
6. **Flow Measuring Units and Pumps:** Metering and instrumentation devices in numerous sections of a wastewater treatment facility are necessary for adequate plant control and operating flexibility. It checks the flow rate of wastewater in the pipelines.
7. **Pre-Aeration Units:** The addition of air at the initial stages of treatment to freshen the wastewater, remove gases, add oxygen, promote floatation of grease, and aid coagulation.

## B. Primary Treatment

The objective of primary treatment is the removal of settling able organic and inorganic solids by sedimentation, and the removal of material that float (scum) by skimming. Approximately 25 to 50% off incoming biochemical oxygen demand (BOD), 50 to 70% of the total suspended solids (SS), and 65% of the oil and grease are removed during primary treatment. Some organic nitrogen, organic phosphorus, and heavy metals associated with solids are also removed during primary sedimentation.

1. **Sedimentation:** It is a physical water treatment process used to settle out suspended solids in water under the influence of gravity (Cheremisinoff, 2002). Sedimentation is recommended as simple pre-treatment of water prior to application of other purification treatments such as filtration and disinfection methods. It removes undesirable small particulate suspended matters (sand, silt, and clay) and some biological contaminants from water under the influence of gravity. The longer the water is sediment, the more the suspended solids and pathogens will settle to the bottom of the container. Adding special chemicals or some natural coagulants can accelerate sedimentation. Three common chemicals used are aluminium sulphate, poly aluminium chloride (also known as PAC or liquid alum) and ferric sulphate. 'PUR' is a powder product containing both coagulants and disinfectant.
2. **Coagulation:** Non-settlable solids are allowed to settle by the addition of certain chemicals, this process is called as coagulation (Latifossglu et al., 1997) Coagulation destabilizes the particles' charges. Coagulants with charges opposite to those of the suspended solids are added to the water to

neutralize the negative charges on dispersed non-settable solids such as clay and organic substances. Once the charge is neutralized, the small-suspended particles are capable of sticking together. The slightly larger particles formed through this process are called micro flocs and are still too small to be visible to the naked eye. A high-energy, rapid-mix to properly disperse the coagulant and promote particle collisions is needed to achieve good coagulation and formation of the microflocs. Over-mixing does not affect coagulation, but insufficient mixing will leave this step incomplete. Proper contact time in the rapid-mix chamber is typically 1 to 3 minutes.

3. **Flocculation:** Following coagulation, flocculation, a gentle mixing stage, increases the particle size from sub-microscopic micro floc to visible suspended particles. The micro flocs are brought into contact with each other through the process of slow mixing. Collisions of the microfloc particles cause them to bond to produce larger, visible flocs. The floc size are continues to build through additional collisions and interaction with inorganic polymers are formed by the coagulant or with organic polymers added and macroflocs are formed. High molecular weight polymers, called coagulant aids, may be added during this step to help bridge, bind, and strengthen the floc, add weight, and increase settling rate. Once the floc has reached its optimum size and strength, the water is ready for the separation process (sedimentation, floatation or filtration). Design contact times for flocculation range from 15 or 20 minutes to an hour or more.

In water treatment, coagulation and flocculation are practically always applied subsequently before a physical separation. The Coagulation-Flocculation process consists of the following steps:

- **Coagulation-Flocculation:** The use of chemical reagents to destabilize and increase the size of the particles, mixing and increasing of flogging size.
- A physical separation of the solids from the liquid phase. This separation is usually achieved by sedimentation (decantation), floatation or filtration.

The common reagents are mineral and/or organic coagulants (typically iron and aluminium salt, organic polymers), flocculation additives (activated silica, talcum, activated carbon), anionic or cationic flocculants and pH control reagents such as acids or bases. Certain heavy metal chelating agents can also be added during the coagulation step.

## C. Secondary Treatment

The objective of secondary treatment is the further treatment of the effluent from primary treatment to remove the residual organics and suspended solids. Aerobic biological treatment is performed in the presence of oxygen by aerobic microorganisms (principally bacteria) that metabolize the organic matter in the wastewater, thereby producing more microorganisms and inorganic end-products (principally $CO_2$, $NH_3$, and $H_2O$). Several aerobic biological processes are used for secondary treatment differing primarily in the manner in which oxygen is supplied to the microorganisms and in the rate at which organisms metabolize the organic matter.

- **Aeration System:** Secondary treatment is called the activated sludge process. This is because air and "seed" sludge from the plant treatment process are added to the wastewater to break it down further. Air pumped into large aeration tanks mixes the wastewater and sludge that stimulates the growth of oxygen-using bacteria and other tiny organisms that are naturally present in the plant. These beneficial microorganisms consume most of the remaining organic materials that are polluting the water and this produces heavier particles that will settle later in the treatment process. The aeration tanks, which are used to hold the wastewater while oxygen is mixed into it, are made of reinforced concrete and are left open to the atmosphere at the top.

- **Biological Treatment Systems:** The objective of the biological treatment of wastewater is to remove organic matter from the wastewater which is present in the soluble and colloidal form or to remove nutrients such as nitrogen and phosphorous from the wastewater. The microorganisms (principally bacteria) are used to convert the colloidal and dissolved carbonaceous organic matter into various gases and into cell tissue. Cell tissue having high specific gravity than water can be removed in settling tank. Hence, complete treatment of the wastewater will not be achieved unless the cell tissues are removed. Biological removal of degradable organics involves a sequence of steps including mass transfer, adsorption, absorption and biochemical enzymatic reactions. Stabilization of organic substances by microorganisms in a natural aquatic environment or in a controlled environment of biological treatment systems is accomplished by two distinct metabolic processes: respiration and synthesis, also called as catabolism and anabolism, respectively.

The biological processes occur in two ways:

- Aerobic Biological Treatment
    - Activated-Sludge Process
        - Conventional Activated Sludge
        - Extended Aeration
    - Bioreactors
        - Sequential Batch Reactor (SBR)
        - Trickling Filter (Biological Air Filters)
        - Membrane bioreactors
    - Anaerobic biological treatment.
        - Up flow anaerobic sludge biological reactor (UASBR)
        - Septic Tank (anaerobic filters)
        - Anaerobic reactor- post treatment

## Aerobic Biological Treatment

Biological wastewater treatment is primarily used to remove dissolved and colloidal organic matter in a wastewater. It is a natural process occurs in the presence of microorganisms. Bacteria are the primary agents of treatment in the biological treatment process. Aerobic processes are usually operated at low dissolved oxygen (DO) concentrations and the microorganisms are subjected to varying periods of time when no DO is present.

### *Activated-Sludge Process*

In activated sludge process wastewater containing organic matter is aerated in an aeration basin in which micro-organisms metabolize the suspended and soluble organic matter. Aerobic activated sludge reactors have been used on a limited scale as bio-scrubbers for the treatment of odorous air (Bowker, 2000). This process is of two types:

- Conventional Activated Sludge

The conventional activated sludge system contains a tank for wastewater aeration followed by a settler and solids recycle line (Figure 3). The wastewater flows through under constant aeration in the presence of activated sludge and exits at the end of the tank after 4-8 hours of residence time. The oxygen concentration in the reactor should be 0.5-2 mg/l throughout, where values over 2 mg/l are considered lost energy. Conventional chemical additives include aluminum sulfate, ferric chloride, polyelectrolyte, enzymes and surfactants etc. are used for chemical treatment of sludge. The most common and economical is alum ($Al_2 (SO_4).nH_2O$) and widely used as a coagulant in the water treatment (Sahu et al, 2013). Wastewater particles are normally charged negatively, low molecular weight, cationic polyelectrolyte's can act as a coagulant that neutralizes or reduces the negative charge of the particles, similar to the effect of alum or ferric chloride (Sahu et. al., 2013). The coke WWTP at a steel making company in Korea is a conventional activated sludge process for the removal of toxic organic pollutants (Woo et al., 2009). Reuse of composted sludge as a soil conditioner returns C, N, P and elements essential for plant growth back to the soil. The wet digested sludge may be sprayed on to a cropland where it functions as fertilizer. The potential of leaching of the nutrients ground or surface water by rainfall run-off is much reduced. Dried sludge may be used a land fill or soil conditioner. Reuse of composted sludge as a soil conditioner returns C, N, P and elements essential for plant growth back to the soil. The wet digested sludge may be sprayed on to a cropland where it functions as fertilizer. The potential of leaching of the nutrients ground or surface water by rainfall run-off is much reduced. Dried sludge may be used a land fill or soil conditioner.

- Extended *Aeration*

This is the modified form of a conventional activated sludge process in which the production of excess sludge is minimized by oxidation and an increase in residence time, i.e. through the larger size

*Figure 4. The flow diagram of conventional activated sludge system*

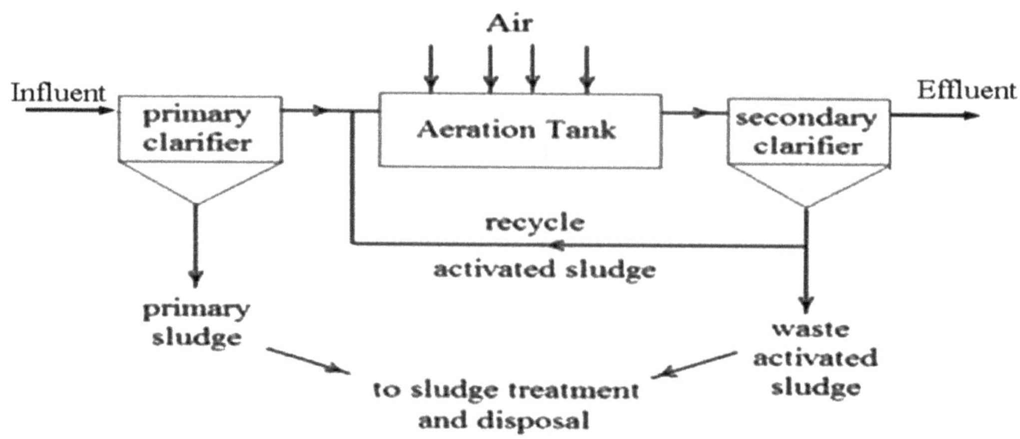

of the aeration tank. The retention time is extended to 1-2 days, which results in a very low net yield of sludge due to its consumption of endogenous respiration. The main advantage of the extended aeration system is in having the minimum of sludge handling facilities as compared with other conventional activated sludge processes. The sludge in extended aeration effluents is very light, of non-degradable nature, and settles with difficulty. Therefore, settling tanks are provided with a longer retention time of approximately 4 hours versus 2 hours for the conventional treatment process.

### *Bioreactor*

Bioreactors are used for carrying out biochemical processes which employ microbes, fungus, plant cells or mammalian cell systems for production of biological products. The bioreactors provide a controlled environment for the production of metabolites which can help to achieve the optimal growth of microbes. The term fermentor is used as synonym to bioreactors.

- Sequential Batch Reactor (SBR)

The unit processes involved in the SBR and conventional activated sludge systems are identical. Aeration and sedimentation/clarification are carried out in both systems. However, there is one important difference. In conventional plants, the processes are carried out simultaneously in separate tanks, whereas in SBR operation the processes are carried out sequentially in the same tank. A unique feature of the SBR system is that there is no need for a return activated-sludge (RAS) system. Because both aeration and settling occur in the same chamber, no sludge is lost in this step, and none has to be returned from the clarifier to maintain the sludge content in the aeration chamber.

- Oxidation Ponds and Lagoons

Wastewater flows continuously through a pond especially constructed for wastewater treatment. The wastewater remains in the ponds for many days. The energy introduced per unit volume of the pond is high, that makes the solids remain dispersed in the liquid medium with complete mixing. The resulting higher biomass concentration in the liquid medium increases the BOD removal efficiency. However, the effluent contains high levels of solids that need to be removed before being discharged. The sedimentation pond downstream provides conditions for the removal. The sludge of the sedimentation pond must be removed after time interval.

- Trickling Filter (Biological Air Filters)

The trickling filter consists of a bed of a highly permeable medium to which microorganisms are attached and through which wastewater is percolated or trickled. The filter media usually consists of either rock (slag is also used) or a variety of plastic packing materials. Rock filter beds are usually circular and the liquid wastewater is distributed over the top of the bed by a rotary distributor, and the wastewater is allowed to contact the media for a short time. The collected liquid is passed to a settling tank where the solids are separated from the treated wastewater. In practice, portion of the liquid collected in the under-drain system or the settled effluent is recycled, usually to dilute the strength of the incoming wastewater and to maintain the biological slime layer in a moist condition. The limitations of the trickling filter include a relatively high incidence of clogging, the long rest period required, and the relatively low loading that can be used.

- Membrane Bioreactors

Membrane separation techniques are used to remove very fine particles from water, to desalinate water, and recently, membranes have been developed to remove organics from water, such as oil and other organics that have clogged and degraded membranes in the past. Membranes are made of various materials but all have a consistent pore size that will permit particles or molecules of a given size to pass through the membrane and will prevent molecules or particles of a size larger than the pore size from passing through. Membrane bioreactor is a better and efficient way to treat wastewater especially that from food industry (Garg & Chaudhry, 2017). Membrane Separation includes ultrafiltration (UF) and reverse osmosis (RO).

## Anaerobic Biological Treatment

### *Up Flow Anaerobic Sludge Blanket Reactor (UASBR)*

The performance of upflow anaerobic fixed-bed reactors filled with low density supports of varying size and specific surface area for the treatment of wastewater and efficiency of reactors increased with decrease in size and increase in specific surface area of the supports (Ganesh et al., 2010). BOD is converted anaerobically by bacteria dispersed in the reactor. The liquid flow is upwards. The upper part of the reactor is divided into settling and gas collection zones. The settling zone allows the exit of clarified effluent in the upper part and the return of solids by gravity to the system, increasing its concentration in the reactor. Amongst the gases formed is methane. The system has no primary sedimentation tank. The sludge production is low, and the excess sludge wasted is already thickened and stabilised.

### *Anaerobic Filters*

BOD is converted anaerobically by bacteria that grow attached to a support medium in the reactor. The tank works submerged, and the flow is upwards. The system requires a primary sedimentation tank (commonly called septic tanks). The sludge production is low and the excess sludge is already stabilised.

### *Anaerobic Reactor- Post Treatment*

UASB reactors produce an effluent that has difficulty in complying with most existing discharge standards. Therefore, some form of post treatment is frequently necessary. The post treatment may be biological or physical- chemical. Generally the post treatment used for wastewater treatment is anaerobic reactors. Sludge production is also low in this treatment.

## D. Tertiary Treatment

Tertiary or advanced wastewater treatment is employed when specific wastewater constituents which cannot be removed by secondary treatment. Tertiary treatment is rare in developing countries. Removals of specific organic and inorganic constituents from industrial effluents make it suitable for reuse. This treatment also removes pathogens from secondary treated effluents.

## Membrane Process

The membrane processes can be classified into sub-processes such as electro dialysis (ED) or electrodialysis reversal (EDR), microfiltration (MF), ultrafiltration (UF), nanofiltration (NF), reverse osmosis (RO). MF, UF, NF and RO use pressure to transport water across the membrane.

## Reverse Osmosis

This is the most widely used membrane filtration system that removes large molecules and ions from effluents by applying pressure to the effluents when it is passing through a selective membrane. RO is used to remove specific dissolved organic constituents from wastewater remaining after advanced treatment with depth filtration or MF. RO system can operate at 90% efficiency or better with respect to total dissolved solids. In addition, it also removes residual organic molecules, turbidity, bacteria and viruses. The solvent of the solution is transferred through a dense membrane tailored to retain salts and low molecules weight solutes. The demineralization of water can be accomplished using thin, micro porous membranes. There are two basic modes of operation one is pressure to drive water through the membrane against the force of osmotic pressure and another system applied pressure in several orders of magnitude in excess of natural osmotic pressure. The membrane commonly used is made up of cellulose acetate and is about 100 μm thick. The film contains microscopic openings that allow water molecules to pass through but retain dissolved solids by either molecular sieving (or by some other mechanism which is not yet completely understood). The process results in a concentrated solution of the ions on the pressure side of the membrane and product water relatively free of ions on the other side of the membrane.

- Reverse Osmosis Desalination Projects

Unlike conventional technology, the market for membrane based desalination plants is growing. Since the early 1990s, hundreds of membrane based brackish water desalination plants with capacities of 10 - 100 cubic metres per day were set up to provide safe drinking water to villages in the states of Andhra Pradesh, Gujarat, Rajasthan and Tamil Nadu (Marwas, 2010).

Reverse osmosis desalination plants were installed also in the industry sector:

- Gujarat Electricity Board, Sikka, state of Gujarat (4.5 million litres per day sea water reverse osmosis plant developed by Ion Exchange, India).
- NIRMA, Bhavnagar, state of Gujarat (10 million litres per day sea water reverse osmosis plant developed by Thermax, India).
- Rashtriya Chemical Factory, Mumbai, state of Maharashtra (12 million litres per day waste water reverse osmosis plant developed by Aquatec, India).
- IFFCO, Phulpur, state of Uttar Pradesh (3 million litres per day wastewater reverse osmosis plant developed by Ion Exchange, India).

## Nanofiltration (NF)

Nanofiltration (NF) is an advance membrane separation technique for wastewater treatment. This process is also called as low-pressure RO or membrane softening. The osmotic back pressure is much lower than that in RO. The operating pressure used in NF is typically 0.5 to 1.5 MPa. The pore size of the membrane is in the range of 0.5-1 nm. Due to the lower operating pressure and higher flow rates, nanofiltration is inexpensive when compared to reverse osmosis. NF allows partial infiltration of monovalent salts such as sodium chloride excluding bivalent salts. NF can reduce total dissolved solids and hardness, reduce colour and odour, and remove heavy metal ions from groundwater. NF process is used for raw water containing TDS in the range sufficient for permeation of essential salts while keeping the total dissolved solids (TDS) in accordance with BIS drinking water standards.

### Ultrafiltration (UF)

The term "ultrafiltration" is forcing solutions at pressures up to several atmospheres through membranes. All types of microorganisms such as viruses and bacteria and all types of particles can be removed by this process. It is a clarification and disinfection membrane operation. More than 50 UF plants for producing drinking water from surface water are in operation in the world (Delgrange-Vincent, et. al., 2000). In most cases, the ultrafiltration membranes represented a reliable barrier achieving the complete removal of *E. coli* and total coliforms (Falsanisi, Liberti and Notarnicola, 2010). The application of UF membrane processes has proved to be competitively estimated and the installing of the UF equipment into the existing wastewater treatment facilities is a further major concern (Bachmann, 2005).

## E. Advanced Wastewater Treatment

For advanced wastewater treatment, treatment methods are dependent upon the characteristics of effluent to be obtained by secondary treatment in order to satisfy further use or disposal of treated wastewater.

### Ultraviolet (UV) Disinfection

The land-based tertiary treatment process is carried out in the filtration and UV disinfection plants. The UV disinfection facility at Mangere, New Zealand is one of the largest in the Southern Hemisphere opened in April, 2003. The plant, which contains the latest UV disinfection technology, is designed to reduce pathogens in the effluent stream significantly. This ensures a very high quality treated effluent discharge. Prior to the disinfection process; secondary treated effluent from the clarifiers will pass through large anthracite coal filters. Filtration down to a level of 15 microns removes any remaining larger particles to clean the treated wastewater, further enabling maximum UV light penetration. At the heart of the disinfection system is a high performance mercury vapour lamp (rated 300 watts) covered with a hard quartz tube. There are 12 channels each with three banks of multiple UV lamps – altogether 7,776 UV lamps. The filtered treated wastewater will pass through these channels to kill off remaining pathogens in the water with UV light.

### Granular Media Filters

Granular media filters are widely used in wastewater treatment for the removal of both organic and inorganic suspended solids. Granular media filters can operate either by gravity flow (gravity filters) or by pressure (pressure filters).The most common types of filters are two and three media filters. A common design for a two media filter would have a bed of 0.5 mm sand layer below a 0.9 mm anthracite layer. A common design for a three media filter would have a 30 to 40 mesh garnet layer below the sand layer. Specialty filters could use different media with different effective sizes.

## Ion Exchange

Ion exchange is a unit process in which ions are removed out of an aqueous solution and is replaced by another ionic species. Toxic ions present in wastewater are exchanged with the non-toxic ions from a solid material called an ion exchanger (Xu, 2005). The ion exchange worked on basic principle is that a weakly bound ion can preferably be displaced by a stronger binding ion. This effect is called the principle of selectivity. The effect of selectivity can be used to remove distinct ions from water and to replace them with others. It can be operated in a batch or continuous mode and has been used for removal of nitrogen, heavy metals and TDS in wastewater applications. It has also been used selectively to remove specific impurities and to recover valuable trace metals like chromium, nickel, copper, lead and cadmium from industrial waste discharges. Some minerals that have ion exchange properties among them are aluminium silicate minerals, which are called zeolites and synthetic zeolites have been prepared using solutions of sodium silicate and sodium aluminate. Alternatively synthetic ion-exchange resins composed of organic polymer with attached functional groups such as $–SO_3\ H^+$ (strongly acidic cation exchange resins), or - $COO^-$ - $H^+$ (weakly acidic cation exchange resins or $-N^+(CH_3)_3OH^-$ (strongly basic anion exchange resins) can be used. Synthetic and industrially produced ion exchange resins consist of small, porous beads that are insoluble in water and organic solvents. The most widely used base-materials are polystyrene and polyacrylate.

## Carbon Adsorption

Carbon adsorption is used to remove certain types of organic contaminants that are resistant to primary and secondary treatment when such removal is required. There are some cases where an organic stream contains contaminants that are valuable enough to recover with carbon adsorption. All Volatile Organic Compounds (VOCs) can be adsorbed onto activated carbon to a greater or lesser extent. Adsorption is a surface phenomenon and defined as the increase in the concentration of a particular component at the surface or interface between two phases (Gupta et al., 2007). There are some parameters, such as pH, temperature, concentration of pollutants, contact time, particle size of the adsorbent and nature of adsorbents and the pollutants that affect the adsorption efficiency. Pre-filtration is required due to the presence of suspended particles and oils that reduce the efficiency of the adsorption. Different types of adsorbents are used in adsorption process. Most commonly used adsorbents are activated carbon, fly ashes, metal oxides, zeolites, moss, biomass, and goethites. Recently, nanomaterial, such as carbon nanotubes and composites of carbon has been utilized for the removal of heavy metals and organic pollutants. At an industrial level, pollutants are removed from wastewater by using the columns and contractors filled with the required adsorbents (Gupta et al., 2011). Adsorption on activated carbon may be used for removal many compounds from wastes such as: pesticides; 2, 4-D herbicides and carbamate insecticides, DTT, chlorobenzene, and p-chlorobenzene sulfonic acid. The important of adsorbents for the removal of heavy metals (HMs) and organics from municipal and industrial wastewater has been reported (Bansode et al., 2004).

## Bioremediation of Wastewater

Bioremediation uses microorganisms such as fungi, bacteria, plants or algae to degrade substances that are harm human health or the environment. The microbes simply use contaminants as food and convert them and release in the form of carbon dioxide and water. The energy and carbon are obtained through the metabolism of organic compounds by the microbes involved in bioremediation processes (Fulekar et al., 2009). Some of the examples of microorganisms include: *Bacillins, Berijerinckia, Flavobacterium, Pseudomonas, Rhizoctomia, Methylosinus, Mycrobacterium, Mycococcus, Nitrosomonas, Nocardia, Acinethobacter, Actinobacter, Acaligenes, Arthrobacter, Penicillium* and *Xanthofacter.* The fungal pre-treatment is used in paper and pulp industry to reduce use of pulping chemicals and this led to the reduction in pollution load (Yadav, Chaudhry, & Dhiman, 2009). The technology has been successfully demonstrated in pilot projects by the Central Pollution Control Board and the National Mission for Clean Ganga is now in the process of using bioremediation for cleaning up some parts of the river.

### *Decolourization of Effluent by Bacteria and Fungi*

Presence of colour and its causative compounds has always been undesirable in water used for either industrial or domestic needs. Before biological treatment systems, the wastewater is treated with the help of several fungi and bacteria to decolourize it. Several Basidiomycetes and Ascomycetes type fungi have been used in the decolorization of wastewaters. Filamentous fungi have lower sensitivity to variations in temperature, pH, nutrients, and aeration, and have lower nucleic acid content in the biomass (Knapp et al. 2001). A wide variety of fungi like *Aspergillus niger* (Patil et al. 2003), *Phanerochaete chrysosporium* (Thakkar et al. 2006) have been used for decolorization of distillery mill effluent. Different bacterial cultures capable of both bioremediation and decolorization of distillery spent wash have been isolated for e.g. *Bacillus thuringiensis* (Kumar and Chandra, 2006), and *Pseudomonas aeruginosa* (Mohana et al. 2007).

## Automatic Variable Filtration (AVF) Technology

Automated Variable Filtration (AVF) technology is a state of the art technology used for wastewater treatment in which upward flow of influent is cleaned by downward flow of filter media. During the treatment process itself, the filter media is cleaned by the filtered influent thus there is no requirement for any additional filter media cleaning or fresh water. The AVF process comprises two sets of media filters that can be operated in series or in parallel. The two stage series configuration is used to produce very high quality filtrate. This mode is ideal for refining secondary wastewater for reuse. The AVF process is equipped with actuated valves, sensors and programmable logic controllers to automatically.

# FUTURE PROSPECTS

Anaerobic granular sludge technology is now perceived to be the most promising and favourable wastewater treatment process, as it fulfils the desired criteria for future technology in sustainable green development. Suitable policy measures need to be formulated to encourage the reuse of treated wastewater for irrigation purposes. Developments in wastewater treatment and reuse practices from developed countries could be made available through the creation of an information network, which can serve as a forum for exchange of information, ideas and knowledge about the latest research, techniques and technological advancements in the realm of wastewater management and practices. Such a network must be broad in scope, addressing various aspects of wastewater management, including appropriate and affordable wastewater collection, treatment and disposal technologies and best practices as well as the planning and regulation issues that are fundamental to wastewater management.

## CONCLUSION

There are many parts of the world where river beds have dried up and industrial waste has killed millions of marine life. Freshwater is a vital source of survival for this biological world and if certain renewable strategies are adopted, this source can be sustained for the next billion years. In a few developed countries, environmental problems have led to the formulation of high quality standards for discharged water. To meet these standards, a combination of anaerobic and aerobic treatment is required, often coupled to nutrient removal systems. Industries can play a pivotal role in preserving fresh water resources if they install and maintain waste water treatment plants. The wastewater treatment plant reduces the stress on the water sources available. After treatment, the water can be recycled or reuse in different processes. Future treatment plants may incorporate additional anaerobic processes, or chemical and physical barriers, to remove pollutants without aerobic bacteria thus creating energy rather than using energy. However, there are limits to the ability to increase the energy efficiency of existing processes. The focus first should be on adaptive modern technologies that not only treat wastewater, but are also easily added-in to the existing infrastructure - thereby having a lower cost implication.

## REFERENCES


Abdallh, M. N., Abdelhalim, W. S., & Abdelhalim, H. S. (2016). *Industrial Wastewater Treatment of Food Industry Using Best Techniques*. Academic Press.

Amuda, O. S., & Amoo, I. A. (2007). Coagulation/flocculation process and sludge conditioning in beverage industrial wastewater treatment. *Journal of Hazardous Materials*, *141*(3), 778–783. doi:10.1016/j.jhazmat.2006.07.044 PMID:16959404

Anonymous. (2004). NATP – MM project report on 'Use of Urban and Industrial Effluent in Agriculture. CSSRI, Karnal, India.

Bachmann, C., Groskreutz, R., Awad, J., & Lozier, J. (2005). Implementing a multi-barrier approach using ultrafiltration and UV disinfection. Proceedings of the Water Environment Federation, 2005(1), 183-195. doi:10.2175/193864705783978339

Bansode, R. R., Losso, J. N., Marshall, W. E., Rao, R. M., & Portier, R. J. (2004). Pecan shell-based granular activated carbon for treatment of chemical oxygen demand (COD) in municipal wastewater. *Bioresource Technology*, *94*(2), 129–135. doi:10.1016/j.biortech.2003.12.009 PubMed

Bowker, R. P. (2000). Biological odour control by diffusion into activated sludge basins. *Water Science and Technology*, *41*(6), 127–132.

Buyukkamaci, N., & Koken, E. (2010). Economic evaluation of alternative wastewater treatment plant options for pulp and paper industry. *The Science of the Total Environment*, *408*(24), 6070–6078. doi:10.1016/j.scitotenv.2010.08.045 PubMed

Chan, Y. J., Chong, M. F., Law, C. L., & Hassell, D. G. (2009). A review on anaerobic–aerobic treatment of industrial and municipal wastewater. *Chemical Engineering Journal*, *155*(1-2), 1–18. doi:10.1016/j.cej.2009.06.041

Cheremisinoff, N. P. (2001). *Handbook of water and wastewater treatment technologies*. Butterworth-Heinemann.

CPCB. (2009). *Status of water supply, wastewater generation and treatment in Class I cities and Class II towns of India. Series: CUPS/70/2009-10*. Central Pollution Control Board.

CPCB. (2016). Final Document on revised classification of industrial sectors under red, orange, green and white categories. New Delhi: Central Pollution Control Board.

Edition, F. (2011). Guidelines for drinking-water quality. *WHO Chronicle*, *38*(4), 104–108. PMID:6485306

Falsanisi, D., Liberti, L., & Notarnicola, M. (2010). Ultrafiltration (UF) pilot plant for municipal wastewater reuse in agriculture: Impact of the operation mode on process performance. *Water (Basel)*, *2*(4), 872–885. doi:10.3390/w2040872

Frankland, A. (1869). *Franklands Bericht "über Wasserversorgung an die Royal Institution of Great Britain", Dtsch Vierteljahresschr.* öffentliche Gesundheitspfl.

Fulekar, M. H., Geetha, M., & Sharma, J. (2009). Bioremediation of Trichlorpyr Butoxyethyl Ester (TBEE) in bioreactor using adapted Pseudomonas aeruginosa in scale up process technique. *Biology and Medicine*, *1*(3), 1–6.

Ganesh, R., Rajinikanth, R., Thanikal, J. V., Ramanujam, R. A., & Torrijos, M. (2010). Anaerobic treatment of winery wastewater in fixed bed reactors. *Bioprocess and Biosystems Engineering*, *33*(5), 619–628. doi:10.100700449-009-0387-9 PMID:19876651

Garg, S., & Chaudhry, S. (2017). Treatment of Wastewater of Food Industry by Membrane Bioreactor.

Glance, A. (2016). Local and global environment. *TERI Energy & Environment Data Diary and Yearbook (TEDDY) 2015/16 (Updated Edition)*, 339.

Gupta, V. K., Agarwal, S., & Saleh, T. A. (2011). Synthesis and characterization of alumina-coated carbon nanotubes and their application for lead removal. *Journal of Hazardous Materials*, *185*(1), 17–23. doi:10.1016/j.jhazmat.2010.08.053 PMID:20888691

Gupta, V. K., Jain, R., & Varshney, S. (2007). Removal of Reactofix golden yellow 3 RFN from aqueous solution using wheat husk—an agricultural waste. *Journal of Hazardous Materials*, *142*(1-2), 443–448. doi:10.1016/j.jhazmat.2006.08.048 PMID:17010514

Judd, S., & Jefferson, B. (Eds.). (2003). *Membranes for industrial wastewater recovery and re-use*. Elsevier.

Knapp, J. S. (2001). Use of wood-rotting fungi for the decolorization of dyes and industrial. *Fungi in Bioremediation*, (23), 242.

Ko, C. H., & Fan, C. (2010). Enhanced chemical oxygen demand removal and flux reduction in pulp and paper wastewater treatment using laccase-polymerized membrane filtration. *Journal of Hazardous Materials*, *181*(1-3), 763–770. doi:10.1016/j.jhazmat.2010.05.079 PMID:20638964

Kumar, P., & Chandra, R. (2004). Detoxification of distillery effluent through Bacillus thuringiensis (MTCC 4714) enhanced phytoremediation potential of Spirodela polyrrhiza (L.) Schliden. *Bulletin of Environmental Contamination and Toxicology*, *73*(5), 903–910. doi:10.100700128-004-0512-z PMID:15669736

Lamprecht, H. O. (1988). Bau und Materialtechnik bei antiken Wasserversorgungsanlagen. *Geschichte der Wasserversorgung*, *3*, 127–155.

Latifossglu, A. A., Surucu, G., & Evirgen, M. (1997). Water Pollut. IV: Model. In *Meas., Predict., 4th Int. Conf* (pp. 733-742). Academic Press.

Lee, C. S., Robinson, J., & Chong, M. F. (2014). A review on application of flocculants in wastewater treatment. *Process Safety and Environmental Protection*, *92*(6), 489–508. doi:10.1016/j.psep.2014.04.010

Marwas, A. G. (2010). *Water & water treatment in India, Market opportunities for Swiss companies*. Academic Press.

Mateo-Sagasta, J., Raschid-Sally, L., & Thebo, A. (2015). Global wastewater and sludge production, treatment and use. In *Wastewater* (pp. 15–38). Dordrecht: Springer. doi:10.1007/978-94-017-9545-6_2

Mohana, S., Desai, C., & Madamwar, D. (2007). Biodegradation and decolourization of anaerobically treated distillery spent wash by a novel bacterial consortium. *Bioresource Technology*, *98*(2), 333–339. doi:10.1016/j.biortech.2005.12.024 PMID:16473005

Mohana, S., Shrivastava, S., Divecha, J., & Madamwar, D. (2008). Response surface methodology for optimization of medium for decolorization of textile dye Direct Black 22 by a novel bacterial consortium. *Bioresource Technology*, *99*(3), 562–569. doi:10.1016/j.biortech.2006.12.033 PMID:17804220

Patil, P. U., Kapadnis, B. P., & Dhamankar, V. S. (2003). Decolorisation of synthetic melanoidin and biogas effluent by immobilised fungal isolate of Aspergillus niger UM2. *International Sugar Journal*, *105*(1249), 8–13.

Perdigón-Melón, J. A., Carbajo, J. B., Petre, A. L., Rosal, R., & García-Calvo, E. (2010). Coagulation–Fenton coupled treatment for ecotoxicity reduction in highly polluted industrial wastewater. *Journal of Hazardous Materials*, *181*(1-3), 127–132. doi:10.1016/j.jhazmat.2010.04.104 PMID:20537462

Raza, A., Friedel, J. K., & Bodner, G. (2012). Improving water use efficiency for sustainable agriculture. In *Agroecology and strategies for climate change* (pp. 167–211). Dordrecht: Springer. doi:10.1007/978-94-007-1905-7_8

Sahu, O. P., & Chaudhari, P. K. (2013). Review on chemical treatment of industrial waste water. *Journal of Applied Science & Environmental Management*, *17*(2), 241–257.

Sato, T., Qadir, M., Yamamoto, S., Endo, T., & Zahoor, A. (2013). Global, regional, and country level need for data on wastewater generation, treatment, and use. *Agricultural Water Management*, *130*, 1–13. doi:10.1016/j.agwat.2013.08.007

Thakkar, A. P., Dhamankar, V. S., & Kapadnis, B. P. (2006). Biocatalytic decolourisation of molasses by Phanerochaete chrysosporium. *Bioresource Technology*, *97*(12), 1377–1381. doi:10.1016/j.biortech.2005.07.005 PMID:16115759

Trivedy, R. K., & Nakate, S. S. (2001). *Treatment of hospital waste and sewage in hyacinth ponds*. Jaipur, India: Low Cost Wastewater Treatment Technologies. ABD.

UN-Water. (2009). *World Water Day brochure*. Available at: http://www.unwater.org/worldwaterday/downloads/wwd09brochureenLOW.pdf

Üstün, G. E., Solmaz, S. K. A., & Birgül, A. (2007). Regeneration of industrial district wastewater using a combination of Fenton process and ion exchange—A case study. *Resources, Conservation and Recycling*, *52*(2), 425–440. doi:10.1016/j.resconrec.2007.05.006

White, C. (1972). *Handbook of chlorination*. Van Nostrand Reinhold Company.

Woo, S. H., Jeon, C. O., Yun, Y. S., Choi, H., Lee, C. S., & Lee, D. S. (2009). On-line estimation of key process variables based on kernel partial least squares in an industrial cokes wastewater treatment plant. *Journal of Hazardous Materials*, *161*(1), 538–544. doi:10.1016/j.jhazmat.2008.04.004 PMID:18486333

Woodard, C. (2005). Industrial Waste Treatment Handbook (2nd ed.). Butterworth Heinemann.

World Health Organization. (1989). *Health guidelines for the use of wastewater in agriculture and aquaculture: report of a WHO scientific group*. WHO.

World Health Organization. (2003). *Assessing Microbial Safety of Drinking Water Improving Approaches and Methods: Improving Approaches and Methods*. OECD Publishing.

Xu, T. (2005). Ion exchange membranes: State of their development and perspective. *Journal of Membrane Science*, *263*(1-2), 1–29. doi:10.1016/j.memsci.2005.05.002

Yadav, R. D., Chaudhry, S., & Dhiman, S. S. (2009). Biopulping and its potential to reduce effluent loads from bleaching of hardwood kraft pulp. *BioResources*, *5*(1), 159–171.

## KEY TERMS AND DEFINITIONS

**Biochemical Oxygen Demand (BOD):** Refers to the amount of dissolved oxygen (DO) that aerobic organisms need in order to break down organic material in water over time.

**Chemical Oxygen Demand (COD):** A measurement of the oxygen required to oxidize soluble and particulate organic matter in water.

**Coagulation:** Coagulation flocculation involves the addition of polymers that clump the small, destabilized particles together into larger aggregates so that they can be more easily separated from the water.

**Nano Filtration:** The membrane separation process known as nanofiltration is essentially a liquid phase one, because it separates a range of inorganic and organic substances from solution in a liquid mainly water. This is done by diffusion through a membrane, under pressure differentials.

**Sludge Treatment:** Biological/chemical waste water treatment reduces the solved and unresolved pollutants existing in the waste water. These are to be regained in the sewage sludge at the end of the water treatment. The sludge treatment is necessary to reduce and to ameliorate the sludges, which are produced within the biological wastewater treatment.

**Trickling Filter:** Trickling filters (TFs) are used to remove organic matter from wastewater. The TF is an aerobic treatment system that utilizes microorganisms attached to a medium to remove organic matter from wastewater. It needs moderate level of skill and technical expertise needed to manage and operate the system.

**Ultrafiltration:** Ultrafiltration (UF) technology uses a membrane barrier to exclude particles as small as 0.01 microns, including bacteria, viruses and colloids. Advantages of UF compared to conventional treatment such as clarifiers and media filters, are its high tolerance to feed water quality upsets, absolute barrier and improved water quality.

**Wastewater:** Wastewater refers to all used water from household, commercial establishments, and institutions, hospitals, industries and so on. It includes and urban runoff, agricultural, horticultural, and aquaculture effluent. It has many pollutants which are harmful for environment and water resources.

# Chapter 7
# Wastewater Pollution From the Industries

**Tabassum Ara**
*National Institute of Technology Srinagar, India*

**Rafia Bashir**
*National Institute of Technology Srinagar, India*

**Hamida Chisti**
*National Institute of Technology Srinagar, India*

**Tauseef Ahmad Rangreez**
*National Institute of Technology Srinagar, India*

## ABSTRACT

*Water is one of the most precious natural resources of the earth, without which the living beings cannot survive. Water is important for the sustenance of human civilization. Man uses water for many purposes like drinking, cleaning, washing, bathing, heating, rearing cattle, and farming. Mankind, for the bettering of themselves and society, advanced towards industries and industrial products. But this progress towards industrialization not only utilizes huge amounts of fresh water, but returns water to the environment with pollutants, which changes its natural quality. Thus, mankind is heading towards misery, instead of comfort. Effective measures need to be taken to prevent, minimize, and control water pollution before it becomes too late.*

## INTRODUCTION

Earth is estimated to have 1400 million cubic kilometers of water. Water covers about 70% of the earth's surface and it exists naturally in the earth in all the three physical states of matter. Out of the earth's percentage of water covering the surface, 97.3% is salt water in the oceans, and seas and only 2.7% is fresh water. Of the 2.7% of fresh water that is available a major portion of it (75%) is frozen in the Polar Regions. Groundwater constitutes 23% of the remaining 25% of fresh water and surface water forms the rest 2% only. The water resources are not uniformly distributed over the world. Water is one of the

DOI: 10.4018/978-1-5225-5754-8.ch007

most precious natural resources of the earth, without which the living beings cannot survive. Water is important for the sustenance of human civilization. The protoplasm of most living cells contains about 80% of water. Most of the metabolic activities of human body involve water. Man uses water for many purposes like drinking, cleaning washing, bathing, heating, rearing cattle and growing crops in the farms, air conditioning and industries etc. Every activity of man involves some use of water. The used water returns to the environment with pollutants, which changes its natural quality in the environment. Of great concern is that water is also used for disposal of pollutants. Water is considered polluted if there is any undesirable change in its physical, chemical or biological characteristics that may harmfully affect human life or industrial process that deteriorate natural resources (Khan and Ullah, 1991). Olaniran (1995) defined water pollution to be the presence of excessive amounts of a hazard (pollutants) in water in such a way that it is no long suitable for drinking, bathing, cooking or other uses. Pollution is thus the introduction of a contamination into the environment

There are natural and man-made causes of water pollution. In case of natural causes many kinds of natural salts mix with rain water and finally fall in the water bodies. The biodegraded portions of plants and animals mix with water and pollute it. Polluted water is generated through industrial and commercial waster, agricultural practices, human activities in everyday life and even models of transportation and the cause is man. Such water is very harmful for both human and aquatic lives. Billions of people in the world do not have safe drinking water and millions of deaths occur due to waterborne diseases (Onsdorff, 1996). Mostly the industrial wastes are disposed of untreated into waters in developing countries and contaminate existing water supplies (United Nations, 2009). The UN also estimates that the amount of wastewater produced annually is six times more water than exists in all the rivers of the world (UN WWAP, 2003).

It is thus important that depletion of water should neither take place through contamination nor careless use otherwise it can result in serious consequences. The three main types of pollution are land pollution, air pollution and water pollution.

Water pollution arises from various activities, among which are sewage leakages, pollution of ground water through drilling activities, oil spillage, industrial waste dumped into our waters, floods which carries waste deposits into our waters, eroded sediments building lavatories and visionaries over running water or even the sea, toxic waste disposal at sea, mineral processing plant as in coal production, herbicides and fertilizers, failing septic system, house hold chemicals, animal wastes, radioisotopes, heavy metal, combustion, deforestation, mining littering, animal wastes. Humans are generally responsible for water pollution as they carry on activities for better self which ultimately leads to water pollution. Water pollution could be categorised under the various activities that man engages in, that lead to pollution. The growth of human population, industrial and agricultural practices is the major causes of pollution. This chapter describes the pollution of water from the waste effluents of the industries.

## BACKGROUND

Although increase in industrialization leads to development, advancement in addition to the upgrading the standard of living but the biosphere becomes highly contaminated. Thus pollution of the biosphere with toxic substances has increased greatly with the evolution of the industries (Tiwari et al., 2008). Rapid industrialization causes severe degradation in pedosphere, hydrosphere and atmosphere.

Industries use pure water for various processes and during those processes poisonous gases are released into the atmosphere. Moreover, wastewater is also released which is full of various contaminants such as heavy metals and that act as a potential hazard for the whole of our environment as it is disposed of either on land or in water resources like rivers, seas, etc. (Azumi and Bichi, 2010). The release of pollutants differs from industry to industry. The waste from the pulp industry mainly contain carbohydrates, hydrochloric acid, sodium hydroxide, sodium carbonate, bisulfites, chlorine dioxide, calcium oxide, from petroleum industry oil spillage Products, textile industry contain dyes, plating industry contain nickel. Chemical industry has the main pollutants as nitrates of metals, sulphates, fluorine, acids, alkalies, etc. and leather tanning wastes contain mainly chromium, zinc, copper, suphides, carbonates, sodium and many other toxic organic compounds and inorganic compounds (Table 1) (Nouri et al., 2009). All factories produce toxic chemicals which are discharged into the water system.

The accumulation of metals in an aquatic environment has direct consequences to man and ecosystem (Alam and Mahbub, 2007). Contamination of water bodies with metals from industries is one of the major health concerns especially in the developing countries as they cannot meet the expenses for the treatment of diseases caused by such pollution. Heavy metals contaminated soils if used for the production of crops can easily enter the food chain (Principi et al., 2006) and thus affect humans and all other kinds of life.

## MAIN FOCUS OF THE CHAPTER

As we know this is the age of industries and water is the main component which is used in all type of industries. Water is used for different processes in industries. It is used as a means of production and for cooling, for washing, dilution and other several purposes. With the result waste products are brought into the water during the industrial process. Thus, industries play a major role in the pollution of water. Industrial wastewater is thus one of the important sources pollution of water and disturbs the eco-system and life on earth. Generally, almost all the industries generate wastewater that needs urgent attention. The main concerning contaminants of industrial wastewater are parasites, viruses, bacteria, inorganic and organic pollutants. Industrial wastes have been found to contaminate underground water also. The wastewater from the industries varies considerably with regard to pollutants depending upon the kind of industry. The treatment procedures for effluent from the industries must be designed specifically.

*Table 1. Water Pollutants by the Industrial Sector*

| Sector | Pollutant |
|---|---|
| Iron and steel | BOD, COD, oil, metals, acids, phenols, and cyanide |
| Textiles and leather | BOD, solids, sulfates and chromium |
| Pulp and paper | BOD, COD, solids, Chlorinated organic compounds |
| Petrochemicals and refineries | BOD, COD, mineral oils, phenols, and chromium |
| Chemicals | COD, organic chemicals, heavy metals, SS, and cyanide |
| Non-ferrous metals | Fluorine and SS |
| Microelectronics | COD and organic chemicals |
| Mining | SS, metals, acids and salts |

Broadly speaking, the wastewater can be inorganic or organic in nature. Many metals like lead, zinc, cadmium, mercury, arsenic and their compounds are inorganic pollutants. Many salts also belong to this group. Electroplating plants, nonmetallic minerals industry, coal and steel industries mostly produce inorganic industrial wastewater. Organic industrial wastewater is produced by factories which manufacture paper, textiles, chemicals, oil refining, various cleaning agents like soaps and detergents, pharmaceuticals, cosmetics, organic dye-stuffs, glue, pesticides and herbicides, synthetic detergents, synthetic detergents etc. Chlorinated pesticides are dangerous water pollutants. They affect human nerve through food chain.

## SOME INDUSTRIES CAUSING WATER POLLUTION

### Food-Related Industries

The food industry has one of the highest consumptions of water. In addition they generate a large volume of sludge during biological treatment. The wastes generated by the food-processing industries-meat, poultry, dairy, sugar, brewing, canning, etc., all are highly degradable and are dissolved oxygen (DO) repleting and water supply impairing in much the same way as domestic sewage.

Food processing can be divided mainly into five categories as:

1. Fruit and vegetables
2. Meat, poultry, and seafood
3. Beverage and bottling
4. Dairy operations
5. Sugar industries

All of these consume a large amount of water for food processing.

#### Fruits and Vegetables

In the processing of fruits and vegetables, general cleaning is done, followed by removal of leaves, skin, and seeds. Other steps include blanching, packaging, cleanup etc. Their wastewater is high in suspended solids that is, rinds, seeds, and skins from raw material, organic sugars and starches and may contain residual pesticides.

#### Meat, Poultry, and Seafood

Poultry farms and slaughter-houses always supply micro-organisms to the water bodies. The rendering bleeding, skin removal, washing, chilling, and cooling processes create blood by-products and waste streams, which are extremely high in BOD. The pathogenic organisms carried by livestock, poultry, and seafood lead to the spread of diseases. The people living by the river which is polluted by various germs and micro-organisms suffer from various diseases like cholera, diarrhea, disentry etc.

## Beverage and Fermentation Industries

Wastewater (Leung and Sinha, 1998) and solid waste is produced from raw material handling and processing, mixing, fermentation, cooling etc. Fermentation processes are higher in BOD .In bottled water plants, ozone can be used to disinfect product water without leaving any residual taste or odor. At beverage plants, ozone can be used as a disinfectant and there can be either little or no need for chemical or high temperature disinfect.

## Dairy Industry

It involves processing raw milk into products like yogurt, condensed milk, milk powder, consumer milk, butter, cheese etc. by chilling, pasteurization, and homogenization. The by-products of milk are whey, buttermilk and their derivatives. In the dairy industry the liquid wastes originate out of rinsing, milk drippings and chemicals used for cleaning containers and washings of bottles, cans and equipments. These industries release the organic pollutants high in proteins, fats and pathogens. With time, milk waste degrades to form corrosive lactic and formic acids. Generally this wastewater contains large quantities of fat, casein, lactose, oils and grease, nitrogen and phosphorus besides detergents, sanitizers etc. used for washing. The dairy industry generate on an average 2.5-3.0 litres of wastewater per litre of milk processed. Dairy effluents decompose or deplete the dissolved oxygen level of the water resulting in anaerobic conditions and release of strong foul odour. The casein decomposes further into a highly odorous black sludge. The dairy waste is found to be toxic for aquatic living life and becomes breeding place for flies and mosquitoes. Dairy effluent promotes eutrophication (Rao and Datta, 1978).

These all contribute largely towards their high biological oxygen demand (BOD), chemical oxygen demand (COD) and oil and grease much higher than the permissible limits. In addition they generate a large volume of sludge during biological treatment. Table 2 shows the different values of BODs and CODs from different Food processing industries.

## The Sugar Industries

The milling processes in sugar mills need large quantities of water and the untreated wastewater from these industries contaminate surface water. It also leaches into ground water and change its chemical composition property (Agale et al, 2013). Most chemicals used in sugar processing are toxic. Wastewater from sugar industry, if discharged without treatment, poses pollution problems in both aquatic and

*Table 2. Typical values of BOD5 and COD for different food processing industries*

| Type of Processor | BOD5(mg/l) | COD (mg/l) | BOD5/CODs |
|---|---|---|---|
| **Bakery products** | 3,200 | 7,000 | 0.46 |
| **Dairy processing** | 2,700 | 4,700 | 0.57 |
| **Jams and jellies** | 2,400 | 4,000 | 0.60 |
| **Meat packing** | 1,433 | 2,746 | 0.52 |
| **Meat specialties** | 530 | 900 | 0.59 |
| **Poultry processor** | 1,306 | 1,581 | 0.83 |

terrestrial ecosystems. They need to be treated before discharging into water bodies otherwise it results in poor quality of fresh water bodies (Moses et al, 2010).The chemical and poisonous industrial waste discharged by sugar mills is becoming serious health hazards who are suffering from various skin (The Nation Karachi, 2012) and other diseases such as asthma, eye infection etc. The sugar mills generate effluent which makes environmental problems related to water and land pollution. The wastewater generated from sugar mills infiltrate into subsoil and leaches into ground water forming contaminated pool which disturb the groundwater quality by changing its chemical composition property. The untreated wastewater discharging from these industries contaminates surface and ground water. The effluent characteristics from a typical sugar plant are presented in Table 3.

## Petroleum Industry

Petroleum is naturally occurring liquid oil found beneath the surface of the earth. Before refinement, it is called crude oil. It is formed from the remains of microorganisms living millions of years ago and is one of the three main fossil fuels, along with coal and natural gas. Petroleum is an organic compound and consists of a complex mixture of molecules called hydrocarbons, various gases, solids, and trace minerals. Petroleum refineries separate crude oil into a wide array of petroleum products through a series of physical and chemical separation techniques. These techniques include fractionation, cracking, hydrotreating, blending processes, manufacturing and transport. Through refinement processes, a variety of products can be made, for example, fuels like gasoline, jet fuel, diesel fuel and kerosene. It is also used to make synthetic chemicals, motor oil, asphalt and lubricant grease. Solvents, refrigerants, cleaning fluids, plastics, pesticides paints, detergents, antifreeze are also prepared from it. Petroleum products are highly toxic. Petroleum also contains trace amounts of dangerous sulfur and nitrogen compounds. These can further react with the environment to produce secondary poisonous chemicals.

*Table 3. Characteristics of wastewater from a typical sugar plant*

| Parameter | Concentration |
|---|---|
| BOD | 1,250 mg/L |
| COD | 2,250 mg/L |
| pH | 5–7 |
| Temperature | 25 to 35°C |
| Total suspended solids | 600 mg/L |
| Total dissolved solids | 2,000 mg/L |
| Total volatile solids | 1,300 mg/L |
| Oil and grease | 60 mg/L |
| Sulphates | 500 mg/L |
| Nitrogen | 10 mg/L |
| Phosphorus | 60 mg/L |
| Chlorides | 60 mg/L |
| Calcium | 180 mg/L |
| Magnesium | 80 mg/L |

Petroleum refineries are a major source of hazardous and toxic air pollutants such as benzene, toluene, ethylbenzene, and xylene and many more. Some of the chemicals released are suspected to be cancer-causing agents. They may also aggravate certain respiratory conditions such as childhood asthma. The combination of volatile hydrocarbons and oxides of nitrogen also contribute to ozone formation, one of the most important air pollutants.

Petroleum products pollute water at every stage of their production, storage and transport. They generate the pollutants like salty water and oil spillage Products from petroleum lead to environmental degradation. Petroleum pollution also results from oil spilled in large quantities from broken pipes of oil tankers.

These spills also occur during the transportation of crude oil by either ocean tanker or land pipeline to other countries. These spills kills sea weeds, mollusks, marine birds, crustaceans, fishes and other sea organisms that serve as food for humans. This leads to calcium deficiencies in the diet of humans.

Leaking underground storage tanks can contaminate soil. It can then leach into the underlying groundwater and move into residential areas making it dangerous to the health of people living there. In *bioremediation an* engineering field, bacteria feed on hydrocarbons and transform them into carbon dioxide and thus help in cleaning up petroleum-contaminated sites.

## Paper and Pulp Industry

Paper and pulp mills are not only air polluters but produce large amounts of inorganic pollutants such as sulphides, bleaching liquors, and organic pollutants. The organic pollutants include organic acids, fibres, cellulose bark and wood sugars. Pulp and paper Industries have high water pollution potential (Thompson *et al.*, 2001; Sumathi & Hung, 2006). The pulp and paper industries produce large quantities of toxic and intensely colored waste effluents, called black liquor which has high level of chemical oxygen demand (COD) causing severe water pollution. The production process consists two main steps -the pulping and bleaching. In pulping the raw material is wood chips. These are treated to remove lignin and improve fibers for paper making. Bleaching is the last step and aims to whiten and brighten the pulp. Whole processes require lot of energy and water (Pokhrel & Viraraghavan, 2004). The waste water discharge of this industry include high organic content, toxic pollutants, dark brown coloration, adsorbable organic halide, concentration of chemicals such as sodium hydroxide, sodium carbonate, sodium sulfide, bisulfites, chlorine dioxide, calcium oxide, hydrochloric acid etc (Sumathi & Hung, 2006). Solid wastes include lime mud, lime slaker, furnace ash, scrubber sludges, wood processing residuals, grits and wastewater treatment sludges. Effective disposal and treatment approaches are essential. Disposal of these solid wastes cause environmental problems because of high organic content, pathogens, ash and trace amount of heavy metal content (Monte *et al.*, 2009).

Physicochemical and biological treatment technologies are used extensively for the pulp and paper mills. The sedimentation, coagulation, precipitation, adsorption, chemical oxidation and membrane filtration have been carried out to examine physico-chemical approach (Pokhrel & Viraraghavan, 2004). Some fungi species are also used to remove color from the effluents. Biological treatment are preferred treatments because of wastewater consists of high organic compounds. Air emissions from pulp and paper mills are other problems. The most important gas emission is water vapours. Air emissions are of great concern because of the adverse health effects of the volatile organic compounds, sulfur dioxide, nitrogen oxides, and total reduced sulfur compounds. In workers of the paper and board industries a statistically significant excess of risk has been found for lung cancer (Lee et al, 2002; Henneberger and

*Table 4. Common Water Pollutants from Pulp and Paper Processes*

| Source | Effluent Characteristics |
|---|---|
| Water used in wood handling/debarking and chip washing | Solids, BOD, Color |
| Chip digester and liquor evaporator condensate | Concentrated BOD, can contain reduced sulfur |
| White waters from pulp screening, thickening, and cleaning | Large volume of water with suspended solids can have significant BOD |
| Bleach plant washer filtrates | BOD, color, chlorinated organic compounds |
| Paper machine water flows | Solids, often precipitated for reuse |
| Fiber and liquor spills | Solids, BOD, color |

Source: United States Environment Protection Agency (EPA)

Lax, 1998, Toren et al, 1996) and other tumour sites (Band et al, 2001; Andersson et al, 2002; Sala-Serra et al, 1996; Hogstedt, 1990 and Zambon et al 2007). due to the exposure to sulphur gases and airborne organochlorinated compound mixtures (Lee et al, 2002; Toren et al, 1996 and Andersson et al, 2001).

## Chemical Industry

The common industrial pollutants responsible for causing water pollution are, asbestos, sulfur, polychlorinated biphenyl, poisonous solvents, lead, mercury, chlorobenzene, carbon tetrachloride, toluene, and volatile organic chemicals. Chemicals like sulfur are harmful for marine life, while asbestos is known to be a potential carcinogen. Drinking water contaminated with asbestos may increase the risk for benign intestinal polyps (Figure 1).

The other chemical wastes from chemical industries are immense but the main pollutants are nitrates of metals, acids, sulphates, fluorine, silica, alkalies, phosphorus, and suspended particles. All factories that produce chemicals discharge toxic chemicals into the water system. They also produce large quantities

*Figure 1. Contaminated water from Chemical Industry*

of by-products from the preparations of pesticides, fungicides, insecticides etc. Some insecticides like DDT are particularly dangerous when allowed into bodies of water because its concentration increases along the food chain. Sulphur dioxide and nitrogen oxides cause acid rain which lowers the PH value of soil. Emission of carbon dioxide cause ocean acidification and decreases in the PH of the oceans as CO2 becomes dissolved. The problem of water pollution has become still worse due to toxic metals. Although some trace metallic elements are important for proper functioning of biological systems, their deficiency or excess could lead to a number of disorders (Ward, 1995). The toxic metals have a marked effect on the aquatic flora and fauna which enter the food chain and affect the human beings. The increase in residue levels of toxic metal content in water will result in decreased productivity (Lokhande and Kelkar, 1999) and increase in health risk in case of human beings (Kazi et al 2009; Gbaruko et al, 2008; Cai et al, 2009 and Pokhrel et al, 2009).

## The Pharmaceutical Industries

Usually the pharmaceutical industries manufacture pharmaceutical products together with other chemical products The basic raw materials in the production of pharmaceuticals varies and so does the wastes vary accordingly. The wasteproducts include solvents, poisonous substances and and many other organics. Pharmaceutical waste is generated through I. V preparation, general compounding, breakages, the partially used ampoules, needles, out-dated, unused preparations, and outdated pharmaceuticals and pharmaceutical compounds may enter the environment. Trace concentrations of human and veterinary pharmaceutical compounds as well as their metabolites have been detected in surface, ground and drinking water sources (Bruce et al, 2010 and Kolpin et al, 1999-2000).The wastewater produced by the pharmaceutical industry has a has a poor biodegradability, high or low pH value very bad quality and colour. The effluents containing organic substances can be removed by strong pretreatment of the wastewater followed by biological treatment.

## Tannery Industry

Tanning is the chemical process that converts animal hides and skin into leather and related products. In tanneries fresh and raw hides are first treated and cured as fresh and raw hides cannot be used for manufacturing leather goods. Large amounts of components, organic wastes, solids, salts, sulphides, chromium, alkalinity, lime etc., are produced in tanneries which have to be disposed of in water bodies thus polluting them. Bacteria virus and other micro-organisms combine with oxygen dissolved in water to break them down and as a result there is immediate drop in the amount of dissolved oxygen in the water and thus the Biological Oxygen Demand (BOD) reduces. Such organic wastes are oxygen demanding. The flora and fauna of the rivers experience reduction in number due to death by suffocation.

Highly polluted rivers have obnoxious smell and contain little or no flora or fauna. Industrial wastes also cause depletion of DO by occurring chemical reaction. DO is very much essential for aquatic animals. The wastewater produced by a tannery which includes the preparation of the hides is fairly acidic and has high chloride content up to 5 g Cl/L. It also contains a high concentration of COD, dichromate content, emulsified fat, and tends to form foam. So the tannery wastewater is highly harmful.

## Textile Industry

In textile industry wastes arise from the impurities in the fibre and from chemicals used in the processing. The various stages of process operations the textile industry are Grey, singein, desizing, scouring, bleaching and dyeing (Figure 2).

The industry requires acids, bases like caustic soda, inorganic chlorine compounds and oxidants like hypochlorite of sodium hydrogen peroxide and peracetic acid, chloride dyestuff, optical bleachers, finishing chemicals, starch and related synthetic polymers, salts of heavy metals like copper, zinc and iron for these processes. All these wastes are passed into an effluent tank and then drained into a drainage system. The dyeing industries produce chemicals such as zinc sulphate and copper salts which are non biodegrable. The volume of the water required for production and the wastewater generated depend upon the type of raw material used for example wool, cotton etc. Chemical pollutants from dyeing industries have been found to be animal carcinogens. When they are discharged into waterbodies they produce devastating effects on aquatic environments.

## Metal Industries

Metal industries involved in processing of chromium, lead, nickel, zinc, copper, cadmium, silver, etc., generate acids, alkaline cleaners and grease. Steel mills produce waste water from the cocking of coal, washing of blast furnace, flue (chimney), gases, etc., These waste tend to be acidic and contain cyanogens, phenols, ore, coke, limestone, alkalis, oils, and fine suspended solids. All these ultimately find their way into water bodies.

*Figure 2. Various Stages of Process Operation*

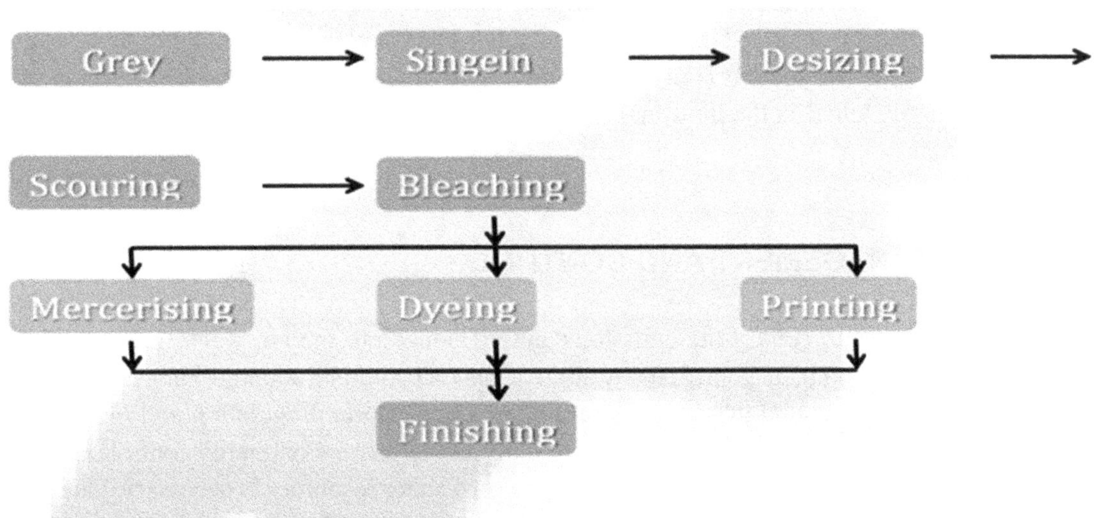

### Mining Industry

The important sources of water pollution are coal and iron ore mining in the areas where they are carried out. The pollutants arising out of of mining are hydrogen sulphide, chlorides, ferrous sulphate, sulphuric acid, ferric hydroxide, suspended solids, etc.

### Hot Water

Another source of water pollution is the discharge of hot water obtained from cooling engines in the thermal industries is thrown to the nearby water bodies. The temperature of water increases and the metabolic rate of organisms decrease. This then raises their oxygen demand. The effects of pollution are greater in shallow, enclosed or slow flowing streams.

### Plastic and Rubber Industries

Plastic manufacturing generates hydrocarbon and other organic matter Synthetic rubber is made from butadiene and styrene in a soap solution and coagulated with an acid-brine solution. The wastes containing these materials have a high BOD. Waters contaminated by it develop bad taste and odour.

### Miscellaneous Industries

There are a number of other industries that generate waste water pollution. Plants nutrients including nitrogen, phosphorus and other substances that support the growth of aquatic plant life could be in excess causing algal gloom and excessive weed growth. This makes water to have odour, taste and sometimes colour. Rottening of the plants make water to have odour, taste and colour. This changes the ecological balance of the water body. Some industrial wastewaters contain high concentration of nitrogen which may exist in the forms of ammonia, nitrate and organic nitrogen. It is important to remove nitrogen from wastewater before being discharged into water bodies. Nitrogen is commonly found in toxic water as as $NO_3$. It is not dangerous as such. It is reduced to the highly toxic NO2 by certain bacteria at subtoxic conditions commonly found in the intestinal tract. Nitrate (Iii) causes the disease known as methemoglobinemia in infants.

## POLLUTION MANAGEMENT AND CONTROL

Pollution control means to control the emissions and effluents into the air, water and land or soil. In water pollution control and management many approaches that could be adopted.They could be through prevention in the first place and the second could be practice efforts. Regulation and monitoring program could be joined in order to help in reducing or minimizing waste. For quality control operation and treatment of water resource systems proper management of water resources is necessary. Due to lack of safe drinking water sources and sanitation, people are forced to consume water directly from rivers and ponds leading to a high risk to their health by causing waterborne diseases. These diseases are cholera, diarrhea,hepatitis, typhoid, dysentery, trachoma and other parasitic infections. Water pollution due to toxic organic compounds has become a serious problem and has necessitated the legal regulations for more restrictive disposal of chemicals into the natural environment . This problem also concerns the coloured wastewaters. Since water pollution has direct consequences on human well beings, an effective teaching strategy in the formal education sector is essential for a better understanding so as to develop the right attitude towards water. Pollution prevention and waste minimization are more desirable than pollution control.

Prevention of water pollution according to Wikipedia includes use non-toxic cleaning materials, preventing throwing of chemicals, trash, into sewer drains,inspecting the septic system after every 3 – 5 years, cleaning up of oil and any other liquid spills with kitty litter and sweeping them up, avoiding use of pesticides and fertilizers that can run off into water systems, avoiding washing of paints brushes in the sink.

Pollution could be minimizing by adopting these practices (i) by recycling (ii) by reusing (iii) waste minimization (iv) by mitigating (v) by preventing (vi) by compost. Pollution control devices can also be used which include Dust Collection system e.g. bag houses, cyclones, electrostatic precipitators, scrubbers e.g. baffle spray scrubber,

Conventional wastewater treatment includes primary, secondary and tertiary treatments like physical sedimentation, chemical sedimentation and disinfection respectively. In conventional process chlorine is used for disinfection. The reduction of microorganisms in wastewater treatment plants with chlorination was found insufficient, and it may generate the formation of carcinogenic chlorinated hydrocarbons. Thus other methods are needed to replace conventional processes. Further, most of the pollutants in industrial wastewater are chemically and biologically resistant; therefore, the conventional treatment of wastewater is usually not sufficient and may be replaced by techniques like co-precipitation, adsorption on charcoal and resins, flotation, biodegradation, incineration and recycling. One of the most widely used methods to treat the organic compounds is adsorption on activated carbon. However, the process has the drawback that it only transfers the contaminant from liquid to solid phase. In this treatment the organic compounds are not degraded, and the used activated carbon has to be decontaminated or stored properly.

Nanotechnology has been considered effective in solving water problems related to quality and quantity. Carbon nanotubes (CNTs) and dendrimers) are contributing to the development of more efficient treatment processes among the advanced water systems.

Engineering designers have, for the last two decades, been incorporating membranes for removing soluble ions in the industrial water treatment systems.The membrane technology in water treatment applications by industries are expanding. Membranes are replacing ion-exchange systems, filters, clarifiers, deaerators and wastewater bioreactors.

Membrane processes like microfiltration, ultrafiltration, NF, and RO, are pressure-driven filtration processes. These have been found to be new and highly effective processes. These also act as alternative methods of removing huge amounts of organic micro-pollutants. Membrane techniques for treatment of wastewater from industries are cost-effective and shows high efficiency in removal of pollutants and meets the high environmental standards. These membranes are technically feasible and can be better alternatives for the traditional treatment systems. Separation by a membrane is achieved by creating a boundary between different bulk gas or liquid mixtures. As different solvents and solutes flow through a membrane at different rates, separation is achieved as shown in Figure 3.

*Figure 3. Membrane Filtration Process*

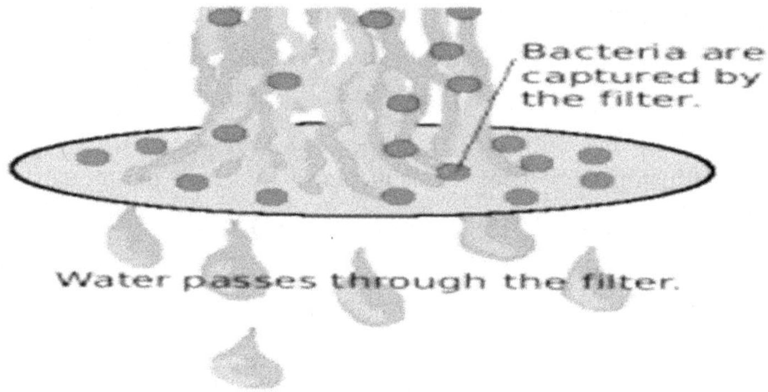


Industrial wastewater, in addition to toxic chemicals, contains valuable constituents, such as macro or micro-nutrients and organic matters which all are essential for the promotion of plant growth. Sludge is produced as a by-product of industrial wastewater treatment process and has also become a good source of organic fertilizer. From agriculture and environmental preservation point of view, the use of industrial wastewater sludge as fertilizer is the best recycling option. However, its high load of pollutants restricts its beneficial use or safely disposes to the water environment.

## CONCLUSION

The discharge of wastewater from industrial sources is a great problem in almost all the countries of the world. This situation is even worse in developing countries. Most major industries have treatment facilities for industrial effluents. But the small scale industries cannot afford investments in pollution control equipment and other chemical treatments as their profit margin is not much. So it is better to follow the preventive measures than control measures. Water pollution has direct consequences on the well being of humans so measures should be taken to prevent it. One of the preventive measures can be the effective teaching with regards to water pollution in educational institutions and awareness programmes among the masses so as to develop the right attitude towards water. Next, waste minimizations are more desirable than pollution control. This can be achieved by adopting some of the practices as recycling, reusing, waste reducing or by compost.

## REFERENCES

Alam, S. S., & Mahbub, M. N. (2007). Karyotype comparison in two varieties of Vigna mungo after staining with orcein and CMA. *Bangladesh Journal of Botany*, *36*(2), 167–170.

Andersson, E., Hagberg, S., Nilsson, T., Persson, B., Wingren, G., & Toren, K. (2001). A case-referent study of cancer mortality among sulfate mill workers in Sweden. *Occupational and Environmental Medicine*, *58*(5), 321–324. doi:10.1136/oem.58.5.321 PMID:11303081

Andersson, E., Nilsson, R., & Toren, K. (2002). Gliomas among men employed in the Swedish pulp and paper industry. *Scandinavian Journal of Work, Environment & Health*, *28*(5), 333–340. doi:10.5271jweh.683 PMID:12432987

Band, P.R., Le, N.D., Fang, R., Astrakianakis, G., Bert, J., Keefe, A., Krewski, D. (2001). Cohort cancer incidence among pulp and paper mill workers in British Columbia. *Scand J Work Environ Health*, *27*, 113-119.

Dan'Azumi, S., & Bichi, M. H. (2010). Industrial Pollution and Heavy Metals Profile of Challawa River in Kanao, Nigeria. *Journal of Applied Sciences in Environmental Sanitation*, *5*, 23–29.

Henneberger, P. K., & Lax, M. B. (1998). Lung cancer mortality in a cohort ofolder pulp and paper workers. *International Journal of Occupational and Environmental Health*, *4*(3), 147–154. doi:10.1179/oeh.1998.4.3.147 PMID:10026475

Hogstedt, C. (1990). Cancer epidemiology in the paper and pulp industry. *IARC Scientific Publications*, 382–389. PMID:2228137

Khan, B., & Ullah, Z. (1991). *Aquatic Pollution Due to industries in river Kabul at Aan Garh (Nowshera)* (M.Sc thesis report). Department of environmental planning and management, Uni. of Peshawar, NWFP, Pakistan.

Leung, A., & Sinha, D. (1998). *Brewery Industry China Hong Kong Management Case Study*. Hong Kong: Management Development Center of Hong Kong.

Monte, M. C., Fuente, E., Blanco, A., & Negro, C. (2009). Waste management from pulp and paper production in the European Union. *Waste Management (New York, N.Y.)*, *29*(1), 293–308. doi:10.1016/j.wasman.2008.02.002 PMID:18406123

Nouri, J., Khorasani, N., Lorestani, B., Karami, M., Hassani, A. H., & Yousefi, N. (2009). Accumulation of heavy metals in soil and uptake by plant species with phytoremediation potential. *Environmental Earth Sciences*, *59*(2), 315–323. doi:10.100712665-009-0028-2

Olaniran, N. S. (1995). Environment and Health: An Introduction. In Environment and Health. Academic Press.

Onsdorff, K. A. (1996). What the Weitzenhoff Court got wrong. *Journal of Water Environment Law and Practice*, *4*(1), 14–18.

Principi, P., Villa, F., Bernasconi, M., & Zanardini, E. (2006). Metal toxicity in municipal wastewater activated sludge investigated by multivariate analysis and in situ hybridization. *Water Research*, *40*(1), 99–106. doi:10.1016/j.watres.2005.10.028 PMID:16343589

Rao, M. N., & Datta, A. K. (1978). Waste water treatment (3rd ed.). Academic Press.

Sala-Serra, M., Sunyer, J., Kogevinas, M., McFarlane, D., & Anto, J. M. (1996). Cohort study on cancer mortality among workers in the pulp and paper industry in Catalonia, Spain. *American Journal of Industrial Medicine*, *30*, 87–92. doi:10.1002/(SICI)1097-0274(199607)30:1<87::AID-AJIM153.0.CO;2-4 PMID:8837689

Sumathi, S., & Hung, Y. T. (2006). Treatment of pulp and paper mill wastes. In L. K. Wang, Y. T. Hung, H. H. Lo, & C. Yapijakis (Eds.), *Waste treatment in the process industries* (pp. 453–497). Taylor & Francis.

Teschke, W. J. (2002). Mortality from lung cancer in workers exposed to sulfur dioxide in the pulp and paper industry. *Environmental Health Perspectives*, *110*(10), 991–995. doi:10.1289/ehp.02110991 PMID:12361923

The Nation Karachi. (2012). Industrial discharge playing havoc with lives, ecosystem. *The Nation*. Retrieved from http://www.nation.com.pk/karachi/21-Nov-2012/industrialdischargeplaying-havoc-with-lives-ecosystem

Thompson, G., Swain, J., Kay, M., & Forster, C. (2001). The treatment of pulp and paper milleffluent: A review. *Bioresource Technology*, *77*(3), 275–286. doi:10.1016/S0960-8524(00)00060-2 PMID:11272013

Tiwari, K. K., Dwivedi, S., Mishra, S., Srivastava, S., Tripathi, R. D., Singh, N. K., & Chakraborty, S. (2008). Phytoremediation efficiency of Portulaca tuberose sox and Portulaca oleracea L. Naturally growing in an industrial effluent irrigated area in Vadodara, Gujarat. India. *Environmental Monitoring and Assessment*, *147*(1-3), 15–22. doi:10.100710661-007-0093-5 PMID:18193484

Toren, K., Persson, B., Wingren, G. (1996). Health effects of working in pulp and paper mills: malignant diseases. *American Journal of Industrial Medicine*, *29*, 123-130.

UN WWAP. (2003). *United Nations World Water Assessment Programme. The World Water Development Report 1: Water for People - Water for Life*. Paris, France: UNESCO.

United Nations-Water. (2009). *World Water Day brochure*. Retrieved from http://www.unwater.org/worldwaterday/downloads/wwd09 brochure en LOW.pdf

Zambon, P., Ricci, P., Bovo, E., Casula, A., Gattolin, M., Fiore, A. R., Chiosi, F., & Guzzinati, S. (2007). Sarcoma risk and dioxin emissions from incinerators and industrial plants: a population-based case-control study (Italy). *Environmental Health*, *6*, 19.

## ADDITIONAL READING

Abdel-Gawad, S., & Abdel-Shafy, M. (2002). Pollution control of industrial wastewater from soap and oil industries: A case study. *Water Science and Technology*, *46*(4-5), 77–82. PMID:12361052

Neşe, T., Nüket, S., & İsmail, T. (2007). Pollutants of Textile Industry Wastewater and Assessment of its Discharge Limits by Water Quality Standards. *Turkish Journal of Fisheries and Aquatic Sciences*, *7*, 97–103.

## KEY TERMS AND DEFINITIONS

**BOD:** It is the amount of dissolved oxygen needed by aerobic biological organisms to break down organic material present in a given water sample at certain temperature over a specific time period.

**COD:** It is a measure of the capacity of water to consume oxygen during the decomposition of organic matter and the oxidation of inorganic chemicals such as Ammonia and nitrite.

**Industrial Wastewater:** The water or liquid carried waste from an industrial process.

**Nanotechnology:** It is the study and application of extremely small things and can be used across all the other science fields.

**Wastewater:** Any water that has been affected by human use by using it in the home, in a business, or as part of an industrial process.

# Chapter 8
# Water Pollutants and Their Removal Techniques

**P. Senthil Kumar**
*Sri Sivasubramaniya Nadar College of Engineering, India*

**A. Saravanan**
*Vel Tech High Tech Dr. Rangarajan Dr. Sakunthala Engineering College, India*

## ABSTRACT

*A developing number of contaminants are entering into water supplies from industrialization and human actions. On account of pharmaceuticals, individual care items, hormones, pesticides, and other substance intensifies that are discharged into the water supply, rising contaminations distinguished in water may have contrary effects to human wellbeing and aquatic environments. In perspective of the previously mentioned issues, late consideration has been centered on the improvement of more effective, low cost, vigorous strategies for wastewater treatment, without additionally focusing on nature or endangering human wellbeing by the treatment itself. The treatment methodologies include primary, secondary, and tertiary treatment using chemical and biological processes. Various strategies, for example, electrochemical, ion exchange, chemical precipitation, coagulation, membrane separation, adsorption, dialysis, flotation, and biological methods have been utilized for the expulsion of harmful contaminations from water and wastewater.*

## INTRODUCTION

Water is the most essential regular asset on the planet. It is a fundamental component in the upkeep of all types of life. Without it life can't exist and generally ventures couldn't work. The nearness of a protected and dependable wellspring of water is accordingly a fundamental essential for the foundation of a steady group. Water, as a whiz asset, requires watchful administration and protection. In spite of the fact that nature frequently has awesome capacity to recoup from ecological harm, the developing requests on water assets require the expert use of basic information about the water cycle to guarantee the support of value and amount (Vorosmarty et al. 2010). A progression of physical, concoction and

DOI: 10.4018/978-1-5225-5754-8.ch008

organic treatment techniques have been produced to expel the poisons in water and wastewater. Albeit a few of these operations also, forms are consolidated in most treatment frameworks, they are generally considered independently.

## CHARACTERISTICS OF WATER

Water quality measures can be arranged in various ways, however frequently are gathered as physical, substance and natural. To give a point of view to considering the attributes of water it will be informative to consider first the hydrologic cycle and its relationship to those attributes, and after that survey the strategies used to measure them.

### List of Water Pollutants

A growing number of contaminants are entering water suppliesfrom industrialization and human activity like heavy metals, dyes; pharmaceuticals; pesticides, fluoride, phenols, insecticides, pesticides and detergents (Speight 2017). The list of water pollutants was shown in Figure 1.

#### Pathogens

Microscopic organisms are generally found in water; it is the point at which they begin to increment in numbers that are above safe levels that water defilement happens. Two of the most well-known pathogen toxins are Coliform and E. coli microbes. Coliforms are ordinarily present in the earth in safe levels and can really be utilized to identify different pathogens in water. In any case, Water Filter Review reports that if coliforms increment in numbers, it can be hazardous for the strength of nature. The closeness of E. coli microorganisms demonstrates that water has been sullied with human or animal misuse (Chicone 2017).

*Figure 1. Water pollutants*

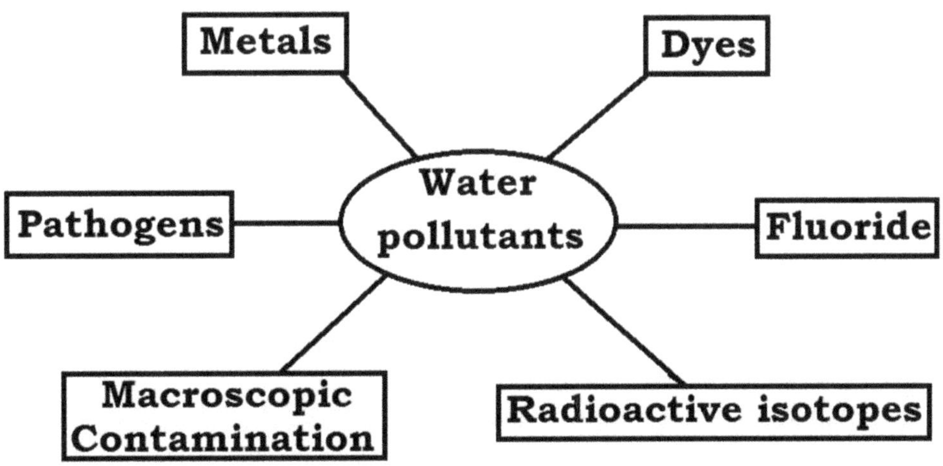

## Inorganic Material

Inorganic materials, for example, substantial metals - arsenic, mercury, copper, chromium, zinc and barium, for instance - however innocuous in little fixations, go about as toxins when they wind up in the water because of overwhelming industrialization or modern mishaps. This sort of water contamination can cause extreme medical issues and can even be deadly. Some inorganic pollutants are not especially harmful, but rather are as yet a peril to the earth since they are utilized so broadly. These incorporate manures, for example, nitrates and phosphates. Nitrates and phosphates cause algal blossoms in surface water, which causes the oxygen level of the water to decrease. This causes oxygen starvation due to the take-up of oxygen by microorganisms that brake down green growth. This is called eutrophication.

1. Metals

Metals can respond to unsafe items with different particles. They are regularly required in electron exchange responses including oxygen. This can prompt the development of dangerous oxyradicals. Metals can frame metalloids and afterward cling to natural mixes to shape lipophilic substances that are frequently exceedingly lethal and can be put away in the fat-supply of creatures and people. Metals can likewise attach to cell macromolecules in the human body. Substantial metals are the most unsafe metals. They have a thickness more noteworthy than 5 and are in this manner called substantial. Metals can't be separated into less unsafe segments, as they are non-biodegradable. The main possibility life forms have against metals is to store them in body tissues where they can't do any mischief.

2. Radioactive isotopes

The half-lives and the methods for rot of radioactive isotopes decide that they are so perilous to people. People make every single radioactive isotope in the atomic business. There are still civil arguments going ahead about whether the advantages of atomic power surpass the perils of radioactive radiation. At the point when a molecule of a radioactive substance rots, it can create four sorts of particles: alpha, beta, gamma and neutrons.

Alpha particles can just travel a short separation through air and human tissues, however they can be exceptionally harming in the event that they slam into cells in light of their extensive mass. They are emphatically charged. Beta particles are all the more entering, yet they do a great deal less harm than alpha particles. They are adversely charged. Gamma beams are profoundly entering. Their harm is like that of beta beams. Neutrons are freed through radiation and respond with different components through crash. They are the reason for atomic splitting in a reactor (Ojovan and Lee 2014).

The radioactivity of a substance is measured in bequerels, yet this does not express the measure of tissue harm the radiation causes. That is the reason the measure of radiation causing 1 kg of tissue to retain 1 joule of vitality is currently communicated in grays. Various types of radiation can do various types of harm, in light of the fact that the vitality is bestowed into tissues in various ways. This is communicated in sieverts. A measure of alpha radiation can do twenty times the harm of a similar measure of beta radiation. Radioactive issue must be held away for various timeframes, with a specific end goal to delete the threat. To what extent it must be put away relies on the half-existence of the isotopes; the time taken for half of the iotas of a radioactive isotope to rot.

3. Organic pollutants

Organic compounds are intensifying that comprise of long bonds, typically made up of carbon. Numerous natural mixes are essential textures of living beings. Particles worked of carbon and of carbon and hydrogen are non-polar and have almost no water dissolvability. They have next to zero electrical charge. The conduct of natural mixes is needy upon their atomic structure, size and shape and the nearness of useful gatherings that are imperative determinants of lethality. It is essential to know the structure of natural mixes, keeping in mind the end goal to foresee their destiny in living life forms and the earth. The natural aggravates that are risky to the earth are all man-made.

There are a wide range of sorts of organic contaminations, illustrations are:

- Hydrocarbons. These are carbon-hydrogen bonds. They can be isolated up into two classes, the first being single-fortified alkanes, twofold reinforced alkenes and triple reinforced alkynes (gasses or fluids) and the second being fragrant hydrocarbons, which contain ring structures (fluids or solids). Fragrant hydrocarbons, for example, PAH's are a great deal more receptive than any of the top of the line sorts of hydrocarbons.
- PCB's are steady and lifeless liquids that are utilized as water powered liquids, coolant/protection liquids in transformers and plasticizers in paints. There are a wide range of PCB's. None of them are water-dissolvable. In numerous nations PCB's are limited.
- Insecticides, for example, DDT's are exceptionally unsafe in light of the fact that they amass in fat tissues of lower creatures and afterward enter the evolved way of life. They have been limited for quite a long time.

4. Macroscopic contamination

Macroscopic contamination is when extensive, obvious things dirty the water. The primary normal toxin is junk - paper, plastic or food waste. It is either tossed specifically into the water or washed away by the rain into a waterway. Different sorts of plainly visible contamination incorporate nurdles (little waterborne plastic pellets); bits of wood; metals; and even clear things like wrecks. This type of contamination is the most reasonable; be that as it may, these poisons must be evacuated with a specific end goal to dodge death toll in sea-going creatures and tainting upon the substance breakdown of these articles (Jin et al. 2017).

## Organic Pollutants and Its Measurement

Various tests have been produced to quantify the natural issue substance of water. As a rule, the tests might be partitioned into those used to gaugeconvergences of natural issue more noteworthy than 1 g/ $m^3$and those used to gauge follow focuses in the scope of $10^{-12}$ to $10^{-3}$ g/m3. The previous is utilized to gauge the blends of natural mixes found in water. The last is utilized to gauge the nearness of follow amounts of anthropogenic natural mixes.

Lab tests ordinarily raced to quantify net measures of natural issue (more noteworthy than 1 g/$m^3$) incorporate tests for:

- Chemical oxygen demand (COD)
- Biochemical oxygen demand (BOD)
- Total organic carbon (TOC)
- Total oxygen demand (TOD)

To guarantee full oxidation of the different mixes found in the examples, a solid corrosive what's more, a compound impetus is included. The COD test is more reproducible and less tedious. The COD test and BOD test can be corresponded, yet the relationship at last gives a subjective esteem. The COD test measures the non-biodegradable and additionally the extreme biodegradable organics. An adjustment in the proportion of biodegradable and non-biodegradable organics influences the connection amongst COD and BOD.

The BOD test was created to decide the consumption of oxygen that would happen in a stream because of usage by living beings as they debase natural issue. That is to let's assume, it is utilized to decide the measure of biodegradable substance in a sample of wastewater. The test reproduces the conditions as close as conceivable to those that happen actually. Moreover, the BOD test relies upon the control of such ecological and nutritious factors as pH and osmotic conditions, basic supplements, temperature, and populace of life forms illustrative of common conditions (Ambec and Coria 2017).

The TOC test is another technique utilized for estimation of the natural issue exhibit in a wastewater. The test is done, by setting an example into a high-temperature heater or, on the other hand artificially oxidizing condition, in which natural carbon is oxidized to carbon dioxide. The carbon dioxide that is delivered would then be able to be measured. While the TOC test does specifically measure the grouping of natural mixes, it doesn't give a direct estimation of the rate of response or the level of biodegradability. Consequently, the TOC test has been acknowledged as an observing procedure however has not been used in the foundation of treatment directions.

## TYPES OF POLLUTANTS

Contaminants act in various ways when added to water. Non-moderate materials including most natural exacerbates, some inorganic mixes and numerous microorganisms are debased by regular self-cleansing procedure with the goal that their focuses lessen with time. The rate of rot of these materials is a component of the specific contamination, the getting water quality, temperature and other ecological elements. Numerous inorganic substances don't influence by the characteristic procedure, so that these traditionalist toxins can just have their focuses diminished by weakening. Traditionalist toxins are frequently unaffected by ordinary water and wastewater treatment forms with the goal that their essence in a specific water source may confine its utilization (Sharma and Kennedy 2017).

- The pollutants existing in the waters can be characterized into following classifications: Toxic mixes which result in the restraint or devastation of natural action in the water. The majority of these materials begin from modern releases also, would incorporate overwhelming metals from metal completing and plating operations, moth anti-agents, herbicides and pesticides and so on.
- Anything which may influence the oxygen adjust of the water, including

- Substances which devour oxygen: these might be natural issues that are biochemically oxidized or inorganic decreasing operators;
    - Substances, which thwart oxygen exchange over the air-water interface. Oils and cleansers can frame defensive movies at the interface which diminish the rate of oxygen exchange and may consequently enhance the impacts of oxygen-devouring substances;
    - Thermal contamination, which can annoy the broke down oxygen (DO) adjust since the immersion DO fixation decreases with expanding temperature.
    - Inert suspended or disintegrated solids in high focuses can cause issues.

## Pollutants Enter Into the Environment

Release of sewage water speaks to a leader worldwide wellspring of contamination. Residential and modern squanders are released unto surface water through sewage frameworks. At times modern waste is discharged straightforwardly unto surface water. The nature of sewage water that enters the surface water relies on the contaminations that are available in the sewage water and the reach out to which it is dealt with before it is acquired contact with surface water. Local sewage water for the most part comprises of paper, cleanser, pee, dung and cleansers. Modern squanders are shifted and rely on the particular procedures of the plants that they starting point from.

Heavy metals are related with mining and purifying operations, chlorophenols and fungicides with mash factories, bug sprays with mothproofing production lines, a few distinctive natural chemicals with the concoction business and radioactive substances with atomic power plants.

Amid the development time of yields nitrates and phosphates are consumed by plants, yet when the plants bite the dust they are discharged from dead plant material into the dirt and will regularly wind up in surface water.

With the exception of the consider reasons for surface water contamination, poisons can likewise enter the water condition inadvertently, for example through air affidavit. Pesticides can enter surface water effortlessly along these lines, since they are connected as beads or splashes. Poisons exhibit ashore can enter surface water through overwhelming precipitation or penetrate into the dirt and enter surface waters through groundwater.

The impacts of poisons are seen generally in little inland oceans and lakes. This is on account of the seas have a characteristic weakening framework for approaching contaminations, while lakes have no compelling outlet. Because of this, much relies on the rate of corruption and precipitation that will expel the poisons from water.

## Transportation of Pollutants Through Water

Pollutants can exist in water in various states. They can be disintegrated or they can be in suspension, which implies that they exist as beads or particles. Contaminations can likewise be broken down in beads or consumed by particles. All conditions of poisons can travel incredible separations through water in a wide range of ways. Particulate issue may tumble to the base of streams and lakes or ascend to the surface, contingent upon its thickness. This implies it for the most part stays on a similar area when the water does not stream quick. In waterways, poisons ordinarily travel extraordinary separations.

The separation they travel relies on the steadiness and physical condition of the poison and the speed of stream of the waterway. Poisons can travel most remote when they are in arrangement in a waterway that is quick streaming. The fixations on one site are then by and large low, however the poison can be distinguished on numerous a greater number of destinations than when it would not have been so effortlessly transported (Schwarzenbach et al. 2006).

In lakes and seas contaminations are transported through streams. There are numerous streams in the seas, which are wind-driven. This empowers a toxin to head out starting with one mainland then onto the next. At the point when relentless toxins aggregate in fish or ocean fowls they can't just turn into a poisonous peril to sea-going evolved ways of life, they can likewise travel awesome separations inside these creatures and wind up in the natural pecking orders of non-contaminated regions.

### Organism Responds to Water Pollutants

At the point when contamination enters the body of a living being it causes an assortment of changes. These progressions can either serve to ensure the life form against destructive impacts or not. The primary reaction of a life form to poisons is to bring a defensive component energetically. As a rule, these instruments keep up the detoxification of toxins, yet at times they create dynamic substances that can make more harm the cell than the first poison.

Another reaction is to diminish the accessibility of poisons by restricting them to another particle, to discharge or store them. Alongside defensive components a creature can likewise bring a system without hesitation that repairs harm caused by poisons. Reacts to lethality and the take-up of contaminations not just relies on the poison that enters the living beings body.

## WASTEWATER TREATMENT TECHNIQUES

Physical, chemical and biological strategies are utilized to expel contaminants from squander water. In request to accomplish distinctive levels of contaminant evacuation, singular waste-water treatment strategies are joined into an assortment of frameworks, delegated essential, optional, and tertiary waste-water treatment. More thorough treatment of waste-water incorporates the evacuation of particular contaminants and also the expulsion what's more, control of supplements. Normal frameworks are likewise utilized for the treatment of waste-water in arrive based applications. Muck coming about because of waste-water treatment operations is dealt with by different techniques all together to diminish its water and natural substance and make it reasonable for conclusive transfer and reuse (Reddy and Lee 2012). This chapter depicts the different customary and propelled innovations in current utilize and clarifies how they are connected for the viable treatment of civil waste-water. Wastewater treatment is basically a blend of settlement and organic or compound unit process. Typical wastewater treatment methodologies were shown in Figure 2.

Unit treatment procedures can be arranged into five phases:

1. **Preliminary Treatment:** The expulsion and deterioration of gross solids, the expulsion of coarseness. Oil and oil are additionally evacuated at this stage if introduce in expansive sums.
2. **Primary Treatment:** The primary significant phase of treatment following preparatory treatment, which typically includes the evacuation of settleable solids, which are isolated as slime.

*Figure 2. Typical wastewater treatment methodologies*

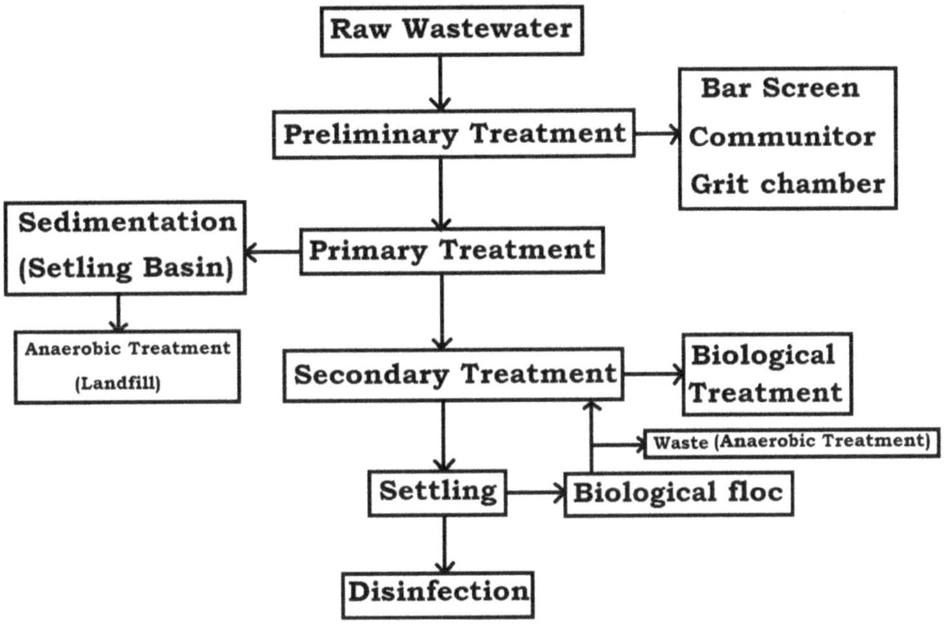


3. **Secondary (Natural) Treatment:** The broke up and colloidal natural mixes are oxidized within the sight of microorganisms.
4. **Tertiary Treatment:** Facilitate treatment of an organically treated emanating to evacuate $BOD_5$, microbes, suspended solids, particular poisonous mixes or supplements to empower the last gushing to agree to a standard more stringent some time recently release.
5. **Sludge Treatment:** The dewatering, adjustment, and transfer of ooze.

## Preliminary Treatment

Preliminary treatment of wastewater for the most part incorporates those procedures that expel trash and coarse biodegradable material from the waste stream as well as balance out the wastewater by evening out or substance expansion (Raghuvanshi et al. 2017).

### Screening

The screening of waste-water, one of the most seasoned treatment techniques, expels net poisons from the squander stream to shield downstream hardware from harm, evade impedance with plant operations and keep questionable skimming material from entering the essential settling tanks. Screening gadgets may comprise of parallel bars, poles or wires, grinding, wire work, or punctured plates, to capture expansive gliding or suspended material. The coarse screen classification incorporates physically or mechanically cleaned bar screens and junk racks. Bar screens comprise of vertical or slanted steel bars dispersed similarly over a channel through which squander water streams. They are utilized in front of mechanical gear including crude sewage pumps, coarseness chambers, and essential sedimentation tanks. Junk racks, as far as it matters for them, are built of parallel rectangular or round steel bars with clear openings. They are typically trailed by customary bar screens or comminutors. Criteria utilized as a part of the outline of coarse screens incorporate bar size, dividing, and edge from the vertical, and channel width and waste-water approach speed.

## Comminution

Comminutors are utilized to pummel vast drifting material in the waste stream. They are introduced where the treatment of screenings would be unrealistic, for the most part between the coarseness chamber and the essential settling tanks. Their utilization diminishes smells, flies and unattractiveness. A comminutor may have either turning or swaying cutters. Turning cutter comminutors either draw in a different stationary screen close by the cutters, or a joined screen and cutter turning together. An alternate kind of comminutor, known as a barminutor, includes a blend of a bar screen and turning cutters.

## Flow Equalization

Flow equalization out is a system used to enhance the adequacy of optional and propelled wastewater treatment forms by levelling out operation parameters, for example, stream, contamination levels and temperature over some undefined time frame. Varieties are damped until the point when a close consistent stream rate is accomplished, limiting the downstream impacts of these parameters. Stream evening out might be connected at various areas inside a waste-water treatment plant, e.g. close to the head end of the treatment works, preceding release into a water body, and before cutting edge squander treatment operations.For instance, supplemental methods for air circulation are regularly utilized with an evening out bowl to give:

- Better blending.
- Chemical oxidation of lessened mixes.
- Some level of natural oxidation.
- Agitation to counteract suspended solids from settling.

On the off chance that air circulation is not given, puzzles or mechanical blenders must be given to maintain a strategic distance from stratification what's more, short-circuiting in levelling bowls. The size and state of a levelling office will shift with the amount of waste and the examples of squander release. Bowls ought to be intended to give sufficient ability to oblige add up to volume of intermittent variety from the wastewater source.

## Sedimentation

Sedimentation, an essential and broadly utilized unit operation in squander water treatment, includes the gravitational settling of substantial particles suspended in a blend. This procedure is utilized for the expulsion of coarseness, particulate issue in the essential settling bowl, natural floc in the enacted muck settling bowl, and synthetic stream when the concoction coagulation process is utilized. Sedimentation happens in a settling tank, additionally alluded to as a clarifier. There are three principle plans, in particular, level stream, solids contact and slanted surface. In outlining a sedimentation bowl, itis critical to hold up under as a top priority that the framework must deliver both a cleared up gushing and a concentrated slime. Four sorts of settling happen, contingent upon molecule fixation: discrete, woolly, frustrated and pressure. It is basic for more than one kind of settling to happen amid a sedimentation operation.

## Flotation

Flotation is a unit operation used to expel strong or fluid particles from a fluid stage by presenting a fine gas, for the most part air bubbles. The gas bubbles either stick to the fluid or are caught in the molecule structure of the suspended solids, raising the light drive of the consolidated molecule and gas bubbles. Particles that have a higher thickness than the fluid would thus be able to be made to rise. In squander water treatment, flotation is utilized basically to evacuate suspended issue and to think natural muck. The boss favourable position of buoyancy over sedimentation is that little or light particles can be evacuated more totally and in a shorter time. Once the particles have been drifted to the surface, they can be skimmed out. Flotation, as of now rehearsed in civil waste-water treatment, utilizes air solely as the coasting specialist. Moreover, different substance added substances can be acquainted with upgrade the expulsion procedure.

## Granular Medium Filtration

The filtration of effluents from squander water treatment forms is a generally late practice, however has come to be generally utilized for the supplemental expulsion of suspended solids from squander water effluents of natural and substance treatment forms, notwithstanding the expulsion of synthetically accelerated phosphorus. The entire filtration operation contains two stages: filtration and cleaning or discharging. The waste-water to be separated is gone through a channel bed comprising of granular material (sand, anthracite as well as garnet), with or without included chemicals. Inside the channel bed, suspended solids contained in the squander water are expelled by methods for an intricate procedure including at least one expulsion systems such as stressing, capture attempt, impaction, sedimentation, flocculation and adsorption. The marvels that happen amid the filtration stage are essentially the same for a wide range of channels utilized for squander water filtration. The cleaning/discharging stage contrasts, contingent upon whether the channel operation is constant or semi continuous (Araneda et al. 2017).

## Preaeration

Strategies for presenting supplemental oxygen to the crude wastewater are in some cases utilized as a part of preparatory treatment. This process is known as preaeration and the goals are to:

- Improve wastewater treatability.
- Provide oil partition, smell control, also, flocculation.
- Promote uniform appropriation of suspended also, drifting solids to treatment units.
- Increase BOD expulsions in essential sedimentation.
    - This is for the most part given by either isolated air circulation or expanded confinement time in a circulated air through coarseness chamber. Arrangements for coarseness expulsion are given in just the primary part of the tank.

### pH Control

Additionally, to adjustment, the utilization of pH control as a preparatory treatment step is normally restricted to treatment of mechanical process squanders. It is important to manage pH since treatment procedures can be hurt by too much acidic or fundamental squanders. Direction of this parameter might be important to meet emanating levels determined for auxiliary treatment. Control of the pH at hoisted levels is generally required to encourage certain substantial metals and additionally lighten a smell delivering potential.

## Primary Treatment

Primary treatment for the most part alludes to a sedimentation process in front of the fundamental framework or, on the other hand auxiliary treatment. In household wastewater treatment, preparatory and essential procedures will expel around 25 percent of the natural stack and essentially the majority of the nonorganic solids. In mechanical waste treatment, preparatory or essential treatment may incorporate stream levelling, pH alteration or compound expansion that is to a great degree vital to the general treatment process.

Primary treatment for the motivations behind this manual will be constrained to sedimentation with and without substance expansion. Other unit forms are typically joined with sedimentation as a piece of "essential treatment", counting some level of preparatory treatment, slop treatment and transfer, and chlorination as a sanitization step. For some a long time, water quality criteria determined just the utilizationof essential treatment for household wastewaters. Primary treatment is not any worthier as add up to wastewater treatment venture before release to an accepting waterway and optional treatment should now be utilized to meet administrative criteria (Narvaez et al. 2017). In this way, the exchange exhibited thus on essential treatment might be used by military staff worried about:

- Alternatives that must be considered for existing treatment offices which are to be moved up to meet emanating restrictions and water quality criteria.
- Design components and choices that must be considered when arranging another wastewater treatment office.

### Primary Sedimentation

Primary sedimentation evacuates both the settleable and floatable solids. Likewise, the flocculants particles which tend to total will be permitted to settle by the expansion of chemicals (press salts, lime and alum). The material that buoys in the essential settling bowl is referred to all in all as oil. Typically, a portion of the oil settles with the slime and buoys to the surface, which can be expelled by skimming. This procedure of flocculants settling happens when the settling speed of the particles increments because of mixture with different particles. This sort of marvel is plainly seen in essential clarifiers. The open door for combination increments with increment in bed profundity, and therefore the molecule expulsion proficiency relies upon both the flood and bed profundity.

### Plain Sedimentation

Wastewater, after preparatory treatment, experiences sedimentation by gravity in a bowl or tank measured to create close tranquil conditions. In this office, settleable solids and most suspended solids settle to the base of the bowl. Mechanical authorities ought to be given to persistently clear the slop to a sump where it is evacuated for further treatment and transfer. Skimming hardware ought to be given to evacuate those floatable substances, for example, filth, oils, and oils which accumulate at the fluid surface. These skimming are joined with muck for transfer. Expulsions from local wastewaters experiencing plain sedimentation will extend from around 30 to 40 percent for BOD and in the scope of 40 to 70 percent for suspended solids. With ideal plan conditions for sedimentation, BOD and suspended solids expulsion proficiency is reliant upon wastewater attributes and the extent of organics exhibit in the solids. One of the most critical plan parameters if the flood rate, typically communicated in gal/day/sqft, which is equivalent to the stream in lady/day partitioned by the settling surface zone of the bowl in square feet. Typically, normal every day stream rates are utilized for estimating offices.

### Sedimentation With Substance Coagulation

Sedimentation utilizing substance coagulation has been inferred essentially to pre-treatment of mechanical or, on the other hand process wastewaters and expulsion of phosphorus from residential wastewaters. Concoction use as a pre-treatment venture for mechanical squanders and phosphorus expulsion is talked about later. The utilization of substance coagulating specialists to improve the evacuation of BOD and suspended solids has not been utilized broadly on residential wastewaters, since it is not normally efficient or operationally alluring. In any case, extraordinary applications may exist at a few establishments. Favourable circumstances of expanded solids division in essential sedimentation offices are:

- A diminish in natural stacking to optional treatment process units.
- A diminish in amount of optional muck delivered.
- An increment in amount of essential muck delivered which can be thickened and dewatered more promptly than secondary sludge.

Chemicals normally utilized, either independently or in blend, are the salts of iron and aluminum, lime, and manufactured natural polyelectrolytes. It is attractive to run shake concentrates to decide the ideal chemicals and measurements levels. The utilization of a given chemical(s) and gushing quality must be painstakingly adjusted against the measure of extra slime created in the sedimentation office.

## Secondary Treatment

The goal of secondary treatment is the further treatment of the profluent from essential treatment to evacuate the remaining organics and suspended solids. Much of the time, secondary treatment takes after essential treatment and includes the evacuation of biodegradable broke down and colloidal natural issue utilizing oxygen consuming organic treatment forms. Aerobic biological treatment is performed within the sight of oxygen by high-impact microorganisms that process the natural issue in the wastewater, accordingly creating more microorganisms and inorganic final results (mainly $CO_2$, $NH_3$, and $H_2O$). A few high-impact natural procedures are utilized for optional treatment contrasting fundamentally in the way in which oxygen is provided to the microorganisms and in the rate at which creatures process the natural issue.

High-rate natural procedures are described by generally little reactor volumes and high centralizations of microorganisms contrasted and low rate forms. Therefore, the development rate of new living beings is considerably more noteworthy in high-rate frameworks on account of the very much controlled condition. The microorganisms must be isolated from the treated wastewater by sedimentation to deliver elucidated optional emanating. The sedimentation tanks utilized as a part of optional treatment, regularly alluded to as auxiliary clarifiers, work in an indistinguishable fundamental way from the essential clarifiers portrayed beforehand. The natural solids expelled amid optional sedimentation, called auxiliary or organic slime, are typically consolidated with essential muck for slop handling (Bajpai 2014).

Regular high-rate forms incorporate the initiated slime forms, streaming channels or bio filters, oxidation dump, and turning natural contactors (RBC). A blend of two of these procedures in arrangement (e.g., bio filter took after by initiated muck) is now and again used to treat metropolitan wastewater containing a high centralization of natural material from modern sources.

## Activated Sludge

In the activated sludge process, the scattered development reactor is an air circulation tank or bowl containing a suspension of the wastewater and microorganisms, the blended alcohol. The substance of the air circulation tank is blended energetically by air circulation gadgets which additionally supply oxygen to the organic suspension. Air circulation gadgets normally utilized incorporate submerged diffusers that discharge packed air and mechanical surface aerators that present air by unsettling the fluid surface. Pressure driven maintenance time in the air circulation tanks typically goes from 3 to 8 hours however can be higher with high $BOD_5$ wastewaters. A part of the natural slop is reused to the air circulation bowl to keep up a high blended alcohol suspended solids (MLSS) level. The rest of expelled from the procedure and sent to ooze handling to keep up a moderately steady grouping of microorganisms in the framework. A few varieties of the essential actuated slime process, for example, broadened air circulation and oxidation trench, are in like manner utilize. Typical flow diagram for an activated sludge process was shown in Figure 3.

## Trickling Filters

A trickling filter or bio filter comprises of a bowl or tower loaded with help media, for example, stones, plastic shapes, or wooden braces. Wastewater is connected irregularly, or now and then constantly, finished the media. Microorganisms end up noticeably appended to the media and frame an organic layer or settled film. Natural issue in the wastewater diffuses into the film, where it is utilized. Oxygen is ordinarily provided to the film by the characteristic stream of air either up or down through the media, contingent upon the relative temperatures of the wastewater and surrounding air. Constrained air can likewise be provided by blowers yet this is seldom important. The thickness of the biofilm increments as new living beings develops. The sloughed material is isolated from the fluid in an optional clarifier and released to muck handling. Cleared up fluid from the auxiliary clarifier is the optional emanating and a segment is regularly reused to the bio filter to enhance water powered circulation of the wastewater over the channel.

*Figure 3. Typical flow diagram for an activated sludge process*

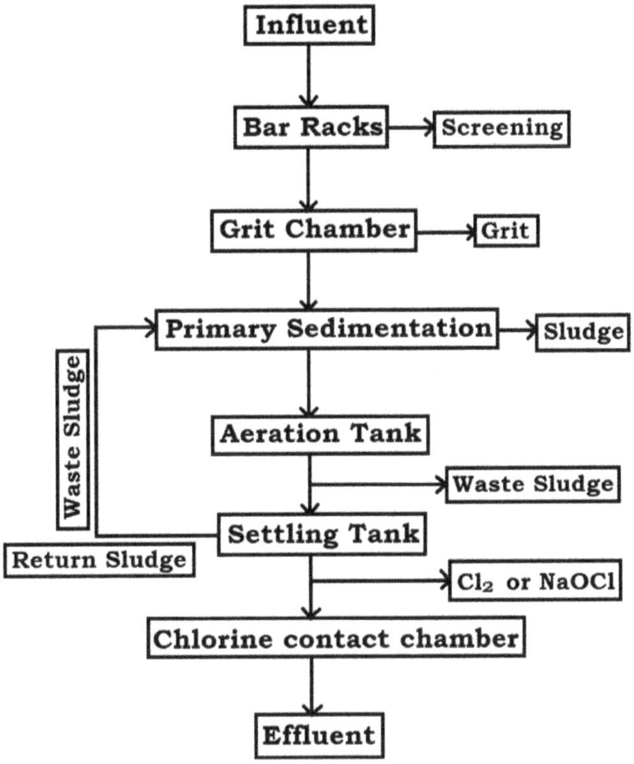


## Rotating Biological Contractors

Rotating biological contactors (RBCs) are settled film reactors like bio filters in that creatures are connected to help media. On account of the RBC, the help media are gradually pivoting plates that are halfway submerged in streaming wastewater in the reactor. Oxygen is provided to the connected biofilm from the air when the film is out of the water and from the fluid when submerged, since oxygen is exchanged to the wastewater by surface turbulence made by the plates' revolution. Sloughed bits of biofilm are evacuated in a similar way depicted for bio filters. Typical flow diagram for a rotating biological contractors was shown in Figure 4.

## Oxidation Pond

Oxidation ponds are substantial, shallow normally 1-2 m profound, where crude or somewhat treated sewage is decayed by microorganisms. The conditions are like those that win in a eutrophic lake. The lakes can be intended to keep up oxygen consuming conditions all through, yet more frequently the deterioration taking put close to the surface is oxygen consuming, while that close to the base is anaerobic. Such lakes having a blend of oxygen consuming and anaerobic conditions, are called facultative lakes. The oxygen required for high-impact decay is gotten from surface air circulation and algal photosynthesis; more profound lakes called tidal ponds, are mechanically circulated air through.

*Figure 4. Typical flow diagram for rotating biological contractors*

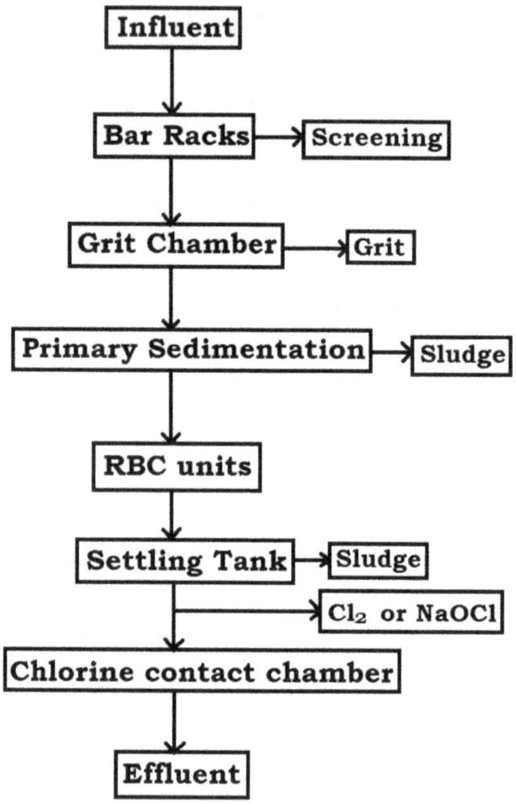


Oxidation ponds, can be intended to give finish treatment to crude sewage, yet they require great arrangement of space. These lakes have been utilized broadly in little groups where arrive imperatives are not all that basic. They are anything but difficult to construct and oversee. They oblige substantial vacillations in stream, and they can give treatment that methodologies that of customary organic frameworks yet at much lower cost. The emanating however may contain undesirable centralizations of green growth and, particularly in the winter when less oxygen is freed by photosynthesis which may create obnoxious smells. They have the inconvenience that the emanating may not meet the EPA auxiliary treatment necessity of 30 mg/L $BOD_5$ and suspended solids. Be that as it may they are straightforward what's more, successful in annihilating pathogenic living beings which make these lakes valuable in creating nations. Oxidation lakes are additionally used to supplement optional treatment and in such cases they are called cleaning lakes.

## Aerated Lagoons

A circulated air through tidal pond is a bowl in the vicinity of 1 and 4 meters inside and out in which squander water is dealt with either on a move through premise or with solids reusing. The microbiology required in this procedure is like that of the enacted ooze process. In any case, contrasts emerge in light of the fact that the substantial surface range of a tidal pond may cause more temperature impacts than are commonly experienced in traditional actuated slop forms. Squander water is oxygenated by surface, turbine or diffused air circulation. The turbulence made by air circulation is used to keep the substance of the bowl in suspension. Contingent upon the maintenance time, circulated air through tidal pond emanating contains roughly 33% to one a large portion of the approaching BOD esteem as cell mass. The majority of these solids must be evacuated in a settling bowl before definite profluent release.

## Biological Supplement Expulsion

Nitrogen and phosphorus are the important supplements of worry in squander water releases. Releases containing nitrogen and phosphorus may quicken the eutrophication of lakes and repositories and empower the development of green growth and established sea-going plants in shallow streams. Huge groupings of nitrogen may have other unfavourable impacts also: exhaustion of broke up oxygen in getting waters, poisonous quality to sea-going life, antagonistic effect on chlorine cleansing proficiency, production of a general wellbeing danger, and waste-water that is less appropriate for reuse. Nitrogen and phosphorus can be expelled by physical, substance and natural strategies. Organic expulsion of these supplements is depicted beneath.

### *Nitrification*

Nitrification is the initial phase in the evacuation of nitrogen by methods for this procedure. Natural nitrification is the work of two bacterial genera: Nitrosomonas, which oxidize smelling salts to the middle of the road item nitrite, and Nitrobacter, which change over nitrite to nitrate. Nitrifying microorganisms are delicate living beings furthermore, are to a great degree vulnerable to a wide assortment of inhibitors, for example, high groupings of alkali and nitrous corrosive, low DO levels (< 1 mg/L), pH outside the ideal range (7.5-8.6), et cetera. Nitrification can be accomplished through both suspended-development and connected development forms. In suspended-development forms, nitrification is achieved either in a similar reactor that is utilized for carbonaceous BOD expulsion, or in aisolate suspended-development reactor following a traditional initiated ooze treatment process. Smelling salts is oxidized to nitrate with either air or high-immaculateness oxygen. So also, nitrification in an appended development framework might be realized either in the same appended development reactor that is utilized for carbonaceous BOD evacuation or in a different reactor. Streaming channels, pivoting natural contactors and pressed towers can be utilized for nitrifying frameworks.

### *Denitrification*

Denitrification includes the expulsion of nitrogen as nitrate by change to nitrogen gas under anoxic conditions. In denitrifying frameworks, DO is a basic parameter. Its essence stifles the compound framework required for denitrification. The ideal pH lies in the vicinity of 7 and 8. Denitrification can be accomplished through both suspended-and joined development forms. Suspended-development denitrification takes put in a fitting stream sort of actuated slop framework. An outer carbon source is normally important for small scale life form cell union, since the nitrified gushing is low in carbonaceous issue. Some denitrification frameworks utilize the approaching waste-water for this reason. A nitrogen-gas-stripped reactor ought to go before the denitrification clarifier on the grounds that nitrogen gas prevents the settling of the blended alcohol. Appended development denitrification happens in a segment reactor containing stone or one of various manufactured media upon which the microscopic organisms develop. Occasional discharging and an outer carbon source are important in an arrangement of this kind.

### *Phosphorus Removal*

Phosphorus shows up in water as orthophosphate ($PO_4^{-3}$), polyphosphate ($P_2O_7$), and naturally bound phosphorus. Microorganisms use phosphorus amid cell union and vitality transport. Therefore, 10 to 30per penny of all influent phosphorus is evacuated amid auxiliary organic treatment. More phosphorus canbe expelled in the event that one of various uncommonly created natural phosphorus evacuation forms is utilized. These procedures depend on the introduction of microorganisms in an actuated slop framework to exchanging anaerobic what's more, vigorous conditions. This burdens the miniaturized scale creatures, with the goal that their take-up of phosphorus surpasses typical levels.

The A/O process is a solitary ooze suspended-development framework that consolidates oxygen consuming and anaerobic areas in grouping. Settled slop is come back to the influent end of the reactor and blended with the approaching waste-water. In the PhoStrip procedure, a segment of the arrival enacted slop from the auxiliary natural treatment process is redirected to an anaerobic phosphorus stripping tank. There, phosphorus is discharged into the supernatant, which is consequently treated with lime or some other coagulant. The phosphorus-poor actuated slop is come back to the air circulation tank.

## Tertiary Treatment

### Disinfection

Disinfection typically includes the infusion of a chlorine arrangement at the head end of a chlorine contact bowl. The chlorine measurements rely on the quality of the wastewater and different components, yet doses of 5 to 15 mg/L are normal. Ozone and ultra violet (uv) illumination can likewise be utilized for sanitization however these strategies for purification are not in like manner utilize. Chlorine contact bowls are normally rectangular channels, with confuses to forestall short-circuiting, intended to give a contact time of around 30 minutes. Be that as it may, to meet propelled wastewater treatment prerequisites, a chlorine contact time of as long as 120 minutes is at times required for particular water system employments of recovered wastewater. The bactericidal impacts of chlorine and different disinfectants are reliant upon pH, contact time, natural substance, and profluent temperature (Munoz et al. 2008).

## ADMINISTRATION OF TREATED EFFLUENT

After treatment, squander water is either reused or released into the earth. Exceptionally treated squander water emanating from civil waste-water treatment plants can be reused as a solid wellspring of water for rural water system, scene water system, mechanical reusing and reuse, groundwater revive, recreational utilizations, non-consumable urban reuse or even consumable reuse. If not reused, treated waste-water is regularly released into a water body and weakened. Natural controls, rules and strategies guarantee worthy release of waste-water emanating (Jones et al. 2017).

## Emanating Reclamation and Reuse

Emanating recovery and reuse has gotten much consideration of late, inferable from developing interest for water what's more, unsustainable rates of utilization of normal water assets. A noteworthy worry in reuse applications is the nature of the recycled water, which is the primary factor managing the choice of the waste-water treatment process arrangement. This area depicts the different profluent reuse applications with accentuation on profluent quality issues (Hamoda et al. 2004).

### Water System

Treated waste-water emanating can be utilized for the water system of yields or arranged zones. The fundamental thought related with this emanating application strategy is the nature of the treated water and its reasonableness for plant development. A few constituents in recycled water that are of specific importance in wording of agrarian water system incorporate hoisted groupings of disintegrated solids, poisonous chemicals, remaining chlorine and supplements. Another exceptionally critical thought is general wellbeing and security risks coming about from the potential nearness of bacterial pathogens, intestinal parasites, protozoa and infections. Concerns shift with the proposed water system utilize and the level of human contact. Potential requirements related with the utilization of recovered waste-water for water system incorporate the attractiveness of yields and open acknowledgment, surface furthermore, groundwater contamination without sufficient administration, and high client costs, eminently the cost of directing emanating to inundated land (Englande et al. 2003).

### Industrial Use

Recycled water is perfect for enterprises utilizing forms that don't require water of consumable quality. Mechanical employments of recycled water incorporate evaporative cooling water, kettle nourish water, process water, and water system and upkeep of the grounds and scene around the plant. Each kind of reuse is related with various requirements on its immaterialness; the utilization of recycled water in cooling towers, for instance, makes issues of scaling, erosion, natural development, fouling and frothing. These issues are too experienced when new water is utilized, however less much of the time. Recycled water utilized as heater sustains water must be relaxed and demineralized, while process water quality is reliant on the necessities of the producing process included (Ding 2017).

### Recreational Employments

Recycled water is generally utilized for recreational purposes, including scene upkeep, stylish impoundments, recreational lakes for swimming, angling, and sculling, fancy wellsprings, snow makingwhat's more, angle cultivating. The required treatment level for recycled water is directed by the proposed utilize: the more prominent the potential for human contact, the higher the treatment level required. For instance, non-confined recreational water utilize requires the treatment of auxiliary profluent by coagulation, filtration, and cleansing to accomplish an aggregate coliform tally of less than 3 for every 100 millilitres.

## Groundwater Revive

Groundwater revive utilizing recovered waste-water serves to moderate water table decrease, ensure groundwater in waterfront aquifers against salt-water interruption, and store recycled water for some time later. Groundwater revive techniques incorporate surface spreading in bowls and by coordinate infusion into aquifers. Surface spreading uses flooding, edge and wrinkle, developed wetlands, and invasion bowls. This application technique enhances the nature of the recycled water extensively as it permeates progressively through soil, unsaturated zone and aquifer. Coordinate infusion includes the pumping of recycled water straightforwardly into an aquifer. Disadvantages of this technique incorporate high emanating treatment cost and the high cost of the fundamental infusing offices. The significant burden of groundwater revives utilizing recycled water is the expanded danger of groundwater tainting.

## REFERENCES

Ambec, S., & Coria, J. (2017). Policy spillovers in the regulation of multiple pollutants. *Journal of Environmental Economics and Management*. doi:10.1016/j.jeem.2017.05.011

Araneda, M., Pavez, J., Luza, B., & Jeison, D. (2017). Use of activated sludge biomass as an agent for advanced primary separation. *Journal of Environmental Management*, *192*, 156–162. doi:10.1016/j.jenvman.2017.01.030 PMID:28160643

Bajpai P (2014). 14 – Water Reuse, Waste water Treatment and Closed-Cycle Operation. *Recycling and Deinking of Recovered Paper,* 251-269.

Chicone, C. (2017). Chapter 3 – An Environmental Pollutant. An Invitation to Applied Mathematics, 31-37.

Ding, G. K. C. (2017). Waste water Treatment and Reuse – The Future Scope of Water Supply. Encyclopedia of Sustainable Technologies, 43-52.

Englande, A. J., Krenkel, P., & Shamas, J. (2003). Waste water Treatment & Water Reclamation. Encyclopedia of Physical Science and Technology, 639-670.

Hamoda, M. F., Ghusain, I. A. I., & Mutairi, N. Z. A. (2004). Sand Filtration of wastewater for tertiary treatment and water reuse. *Desalination*, *164*(3), 203–211. doi:10.1016/S0011-9164(04)00189-4

Jin, Y., Wu, S., Zeng, Z., & Fu, Z. (2017). Effects of environmental pollutants on gut microbiota. *Environmental Pollution*, *222*, 1–9. doi:10.1016/j.envpol.2016.11.045 PMID:28086130

Jones, D. L., Freeman, C., & Rodriguez, A. R. S. (2017). Waste Water Treatment. *Encyclopedia of applied. Plant Science*, *3*, 352–362.

Munoz, J. I., Roca, J. A. M., & Clar, A. I. (2008). Study of different alternatives of tertiary treatments for wastewater reclamation to optimize the water quality for irrigation reuse. *Desalination*, *222*(1-3), 222–229. doi:10.1016/j.desal.2007.01.157

Narvaez, O. M. R., Hernandez, J. M. P., & Goonetilleke, A. (2017). Treatment technologies for emerging contaminants in water: A review. *Chemical Engineering Journal*, *323*, 361–380. doi:10.1016/j.cej.2017.04.106

Ojovan, M.I., & Lee, W.E. (2014). 14 – Treatment of Radioactive Waste. *An Introduction to Nuclear Waste Immobilisation*, 171-203.

Raghuvanshi, S., Bhakar, V., Sowmya, C., & Sangwan, K. S. (2017). Waste water treatment Plant life-cycle Assessment: Treatment Process to Reuse of water. *Procedia CIRP*, *61*, 761–766. doi:10.1016/j.procir.2016.11.170

Reddy, D. H. K., & Lee, S. M. (2012). Water pollution and treatment technologies. *Journal of Environmental & Analytical Toxicology*, 2.

Schwarzenbach, R. P., Escher, B. I., & Fenner, K. (2006). The challenge of micropollutants in aquatic systems. *Science*, *313*(5790), 1072–1077. doi:10.1126cience.1127291 PMID:16931750

Sharma, S. K., & Kennedy, M. D. (2017). Soil aquifer treatment for wastewater treatment and reuse. *International Biodeterioration & Biodegradation*, *119*, 671–677. doi:10.1016/j.ibiod.2016.09.013

Speight, J. G. (2017). Chapter 4 – Sources and types of organic pollutants. Environmental Organic Chemistry for Engineers, 153-201.

Vorosmarty, C. J., McIntyre, P. B., Gessner, M. O., Dudgeon, D., Prusevich, A., Green, P., ... Davies, P. M. (2010). Global threats to human water security and river biodiversity. *Nature*, *467*(7315), 555–561. doi:10.1038/nature09440 PMID:20882010

# Chapter 9
# Agriculture Pollution

**P. Senthil Kumar**
*Sri Sivasubramaniya Nadar College of Engineering, India*

**P. R. Yaashikaa**
*Sri Sivasubramaniya Nadar College of Engineering, India*

## ABSTRACT

*Vast industrialization and population expansion forced an increase in the utilization of natural resources for human survival. Agriculture is a critical process in which livestock and growth of crops have to be balanced equally. Humans completely depend on agriculture for food and nutrition. Presently, the agricultural sector faces destructive activities due to chemical fertilizers and pesticides which cause environmental pollution. Agricultural pollution signifies biotic and abiotic side effects of cultivating practices that outcome in infection and deterioration of the environment and encompassing biological communities, as well as causing health effects to humans. Eutrophication, loss of biodiversity, contamination of soil, air, and water, depletion of fertility of soil are a few causes of agricultural pollution. Management of these pollutants includes replacement with biological formulations, run-off water treatment, minimizing use of nutrients, and livestock management. Thus, there is a need for securing the agriculture sector for crop productivity and handling country's economy.*

## INTRODUCTION

Agricultural practice is the development and reproducing of creatures, plants and growths for nutrition, fibre, biofuel, therapeutic plants and different items used to maintain and upgrade human life. Agriculture was the key advancement in the ascent of inactive human progress, whereby cultivating of trained species made sustenance surpluses that sustained the improvement of development. The investigation of farming is known as agricultural science. The historical backdrop of agriculture goes back a large number of years, and its improvement has been driven and characterized by extraordinary atmospheres, societies, and innovations. Modern agriculture in light of extensive scale monoculture cultivating has turned into the overwhelming agrarian procedure. For quite a long time, agriculture has been related with the generation of fundamental nutritional crops. At present, agribusiness well beyond cultivating incorporates dairy, natural product development, poultry, honey bee keeping, mushroom, discretionary,

DOI: 10.4018/978-1-5225-5754-8.ch009

and so forth. Today, preparing, advertising, and circulation of harvests and animals products and so forth are altogether recognized as a major aspect of current agriculture. Thus, agriculture could be termed as the creation, handling, advancement and appropriation farming products. Farming assumes a basic part in the whole existence of a given economy. Agriculture is the foundation of the economic arrangement of a given nation. In spite of providing nourishment and crude material, agriculture additionally gives business chances to substantial rate of the population. The advancement of farming exceeding subsistence levels made it workable for societies to then have other individuals in the group represent considerable authority in other skills. It is the main biggest business giving area and biggest uncomposed division on the earth. Over 70% of the world representatives are reliant on this division both specifically and indirectly. Unfortunately it is the totally disregarded part in India.

Present day agriculture, plant reproducing, agrochemicals, for example, pesticides and composts, and mechanical improvements have resulted in expanded yields from development, yet in the meantime have caused natural harm and negative human wellbeing impacts. Particularly reproduction and present day techniques in animal husbandry have correspondingly expanded the yield of meat, yet have raised specific notices about animal husbandry and the health impacts of the anti-toxins, development hormones, and different chemicals normally utilized as a part of industrial large scale meat generation. Genetically modified organisms are an expanding segment of farming, despite the fact that they are restricted in a few nations. Farming food generation and water administration are progressively getting to be worldwide issue. Critical contamination of land and water resources, including the consumption of aquifers, has been seen in late decades, and the impacts of an Earth-wide global warming and of agricultural practise on an unnatural weather change are as yet not completely recognised (Jones et al., 2017).

Though agriculture remains as a major source of country's nutrition and economy, the pollution caused by agricultural practise leads to various health and environmental issues. The pollutants released into atmosphere and their root of behaviour is of highly important.

## AGRICULTURE POLLUTION

Agriculture pollution is defined as the situation that causes destruction, contamination and damage to the agricultural land, water resources and finally to the ecosystem resulting in health effects to humans and other living creatures. Agricultural pollution may arise from various sources such as point source pollution (from a single point) or non-point source pollution (from landscape level). The resource majorly influenced due to this type of pollution is soil followed by water resources and air. The run-off water with contaminants from agricultural land leads to eutrophication finally death of aquatic organisms (Chen et al., 2017). As ecological conditions impact to farming practices, agricultural practices additionally have consequences for condition. Specifically; agriculture influences to worldwide streaming of greenhouse gasses. The fundamental explanation behind the deforestation is to get agricultural land. Because of population expansion, greenhouse gasses are produced to a greater extent. These greenhouse gasses reflect the second real negative effect after the negative impacts of nursery gasses which made by the utilizing of the petroleum products. Woodlands gather carbon 20 or 40 times more than rural grounds and the greater part of the carbon are discharged into environment when forest land has been destructed to convert to urban land (Abbasi et al., 2014). The factors that connect the environment and agriculture can be correlated using PSR Loop as follows (Figure 1).

*Figure 1. PSR loop*

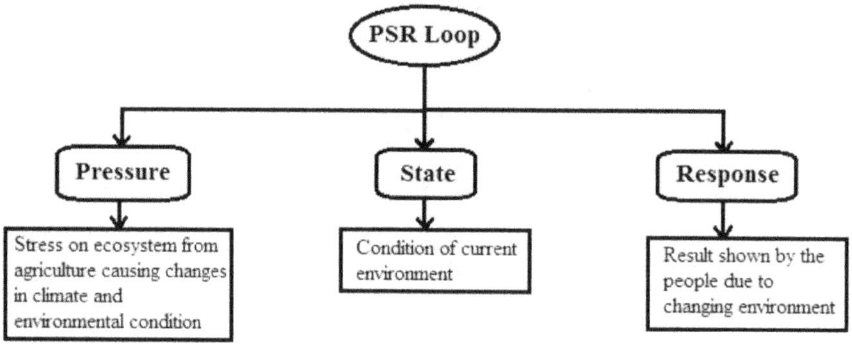


## CAUSES OF AGRICULTURE POLLUTION

Agriculture is a critical process in which the crop growth and development and livestock have to be properly balanced. Current cultivating and agricultural practices add to the damage, destruction and contamination of environmental condition and the neighbouring biological communities. Composts, pesticides, herbicides, animal fertilizer and other agro-chemicals are rich in chemical supplements and dangerous substances like toxins which are frequently the main reason of agricultural pollution. The abundance utilization of these chemical manures can influence water quality when it rains, and the affected soils and water is washed into the adjoining waters or into ground waters (Emmanuel et al., 2016). The chemicals can likewise be consumed by plants which are then consumed by animals and people, unsafe and cause various diseases (Figure 2).

### Air Pollution

The presence of pollutants caused either due to unwanted solid, liquid or gaseous materials in the air are termed as air pollution. Agricultural pollution can be associated to air pollution as follows:

- Agricultural sources that directly contaminate air and eco-system.
- Non-agricultural sources causing air pollution thus affecting crop and land quality.

*Figure 2. Causes of agricultural pollution*

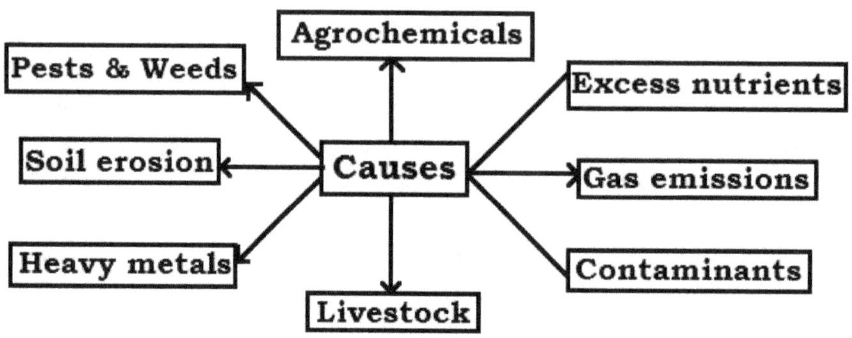

The quality of crops produced is greatly influenced by the pollutants in the agricultural field and the severity depends on the quantity of pollutants present. Some of the pollutants affecting agricultural land and crops are soil erosion, sedimentation, green house gases, ozone, toxic chemicals and air-borne pollutants (Agarwal 2005).

## Greenhouse Gases

Greenhouse gases absorb the infrared radiations from the sunlight and emit them again into the atmosphere thus maintaining the Earth's temperature through a process termed as greenhouse effect. The imbalances due to pollutant sources and these gases, the centralization of these gases in the atmosphere is increasing thereby indirectly raising Earth's temperature. This temperature rise is a major threat to life in earth. These gases affect agriculture leading to production of more greenhouse gases. More than 25% of these gases are produced through agriculture practice only (Preston and Leng 1989). Most important greenhouse gases are chlorofluorocarbon (CFC), water vapour, methane, hydrofluorocarbon (HFC), nitrous oxide, carbon dioxide, etc. Chlorofluorocarbon (CFC) is defined as the mixture of compounds such as chlorine, fluorine and carbon. CFCs are usually gases or liquids at room temperature which evaporate quickly. These gases are stable, non-toxic, non-reactive and non-flammable. They are used as propellants in aerosol sprays, in refrigerators and air conditioners, as cleaning solvents and also in electrical appliances. These gases are released into the atmosphere through leakage in refrigerators or in aerosols. These gases don't have any direct effect on ecosystem but when present as volatile organic compounds they increase the ground level of ozone in the atmosphere causing destruction to plants and animals. CFCs cause adverse effects to humans also. When present in excess level it raises the atmospheric temperature damaging ozone layer resulting in direct rays of sun's harmful UV rays into earth. Inhalation of excess amount of these gases leads to damage in lungs, liver, heart, kidney and central nervous system. Harmful UV radiations cause skin cancer.

### *Carbon Dioxide*

Carbon dioxide is an important gas for all plants since they inhale carbon dioxide for their growth and development by a process termed as carbon fertilization. This gas when present beyond limit causes improper growth cycle and physiological structure in plants. This is the major gas responsible for nearly 70% of greenhouse effect. Industries, manure and through burning of fossil fuels are the main sources of release of carbon dioxide into atmosphere. Apart from burning of fossil fuels in discharging carbon dioxide, deforestation can likewise influences the level of carbon dioxide (Ainsworth and McGrath 2010). Trees decrease the measure of carbon dioxide from the climate through the procedure of photosynthesis, so less trees implies more carbon dioxide left in the environment. Organic substances when present in the soil increase the nutrient level and its decomposition results in $CO_2$ formation. It reaches the stomata of plants and causes necrotic lesions in plants like cucumber and tomato. The height of carbon dioxide causes increment in the grouping of non-structural sugars inside the leaves, which can be a reason of diminished nitrogen in plant tissues. The capacity of stomata to lead gasses additionally diminishes. In grain crops, the quality is gravely influenced as well as protein and mineral amount is likewise diminished by a specific rate. Farmers are attempting to develop more harvests, and for this reason they utilize composts to upgrade the yield of products. Accordingly more carbon dioxide is discharged by plants during breath, which at last has some uncommon impacts on the atmosphere.

### Nitrogen Oxide

Nitrogen oxide is a gas composed of gas pollutants such as nitrogen dioxide, nitric oxide and nitrous oxide resulting from burning of fossil fuels and other combustion processes. These gases are produced as result of interaction between oxygen and nitrogen in the atmosphere at high temperatures. Microorganisms enhance the formation of nitrous oxide if nitrogen is present in excess quantities unconsumed by plants. Cultivation of nitrogen-fixing crops and also chemical fertilizers with nitrogen enhances the production of nitrous oxide. It also acts as agent for acid rain affecting crops such as carrot, tomato, tobacco, etc. The effect can be analysed by gray and green colour patches on the surface of the leaves leading to lesions. Plants require nitrogen for protein synthesis. Studies reveal that for efficient growth and development of crops concentration of these oxides must be low. Continuous exposure to nitrogen can cause severe lung and respiratory problems in humans (Aneja 2009).

### Ozone ($O_3$)

Ozone is an allotrope of carbon powerful oxidant blue colour pungent smelling gas. It is formed from dioxygen by the action of UV rays and electrical discharges. This gas is usually present in atmosphere at very low concentration. It is also formed by the reaction of nitrogen oxide, carbon dioxide and other volatile substances (Roy et al., 2009). Burning of fossil fuels is also responsible for ozone formation. It enters into plants through stomata and causes damage to crops like tomato, cucumber, radish, potato, onion, etc. The effect of ozone in plants can be identified by formation of yellow colouring of leaves resulting in necrosis, lesions on upper surface and fleck formation. It decreases photosynthetic process and enhances senescence. It also causes destruction of respiratory and mucous tissues in animals. In humans it causes heart attack, asthma, respiratory issues and sometimes leads to death also.

## Other Gases

Apart from greenhouse gases, there are few other gases namely sulphur dioxide and fluoride responsible for causing air pollution. Sulphur dioxide at high concentrations can cause necrosis in plants even during short duration. It is the major pollutant with combination of sulphur and oxygen. This gas is directly emitted into the air. Burning fossil fuels and heating process are the major source of these gas emissions. The main crop being affected due to this pollutant is soybean because of its high solubility and hydration property it directly enters into the stomata and causes injuries (Matsubara et al., 2009). Sulphur dioxide and oxygen react to produce sulphur trioxide and on further reaction with water vapour produces sulphuric acid or acid rain causing damage to plants. Fluoride is another important pollutant released in the form of hydrogen fluoride from mainly fertilizer producing industries. It reduces photosynthesis in plants, changes the physiological structure and causes necrosis (Gautam et al., 2010).

## Soil Erosion and Sedimentation

In-depth cultivating practices incredibly add to soil disintegration and sedimentation. Soil is composed of many layers and it is just the highest layer that can reinforce cultivating. Because of inefficient cultivating operations, this soil is left open for disintegration and prompts declining quality every year. Regardless of whether disintegrated by water or wind energy, this soil must be kept some place or the other. The subsequent sedimentation causes soil develops in regions, for example, waterways, streams, neighbouring fields, etc. Thus the procedure of agricultural contamination keeps the common development of water, oceanic creatures and supplements to other fertile lands. Through the cultivating operations, a large number of fertile soils are separated, debased, and dissolved by means of water overflows which wind up collecting as sediments in waterways, streams, lakes, seas or other land areas. Along with this, it influences water quality in different areas, for example, by making it contaminating or polluting it with the agrochemical in the soils. Sedimentation may likewise confine light penetration in water in this manner influencing aquatic living things. Sedimentation additionally adds to development of the farming contaminations in water bodies and other land regions. Continuous farming, erosion and sedimentation, deforestation, increased salt concentration and excessive nutrient removal from soil are the main reasons of soil degradation (Zapata and Nguyen 2009).

## INFLUENCE OF AGRICULTURE ON AIR

Air quality is defined as the measure of pollutants present in the air. The effect of agricultural activity has great impact on quality of air. The mode of agricultural practice fills air with hazardous pollutants. On inhalation of this contaminated air by humans and animals leads to various health diseases and respiratory problems.

### Burning of Agricultural Waste

Detach of agricultural waste into environment is a major drawback in agricultural sector. The wastes released during agricultural process are burnt in an open environment to destroy unwanted weeds, pests and also for production of high quality of crops in the upcoming seasons. This process of burning releases enormous amount of particulate matter, smoke and chemical toxic substances and gases like carbon dioxide, carbon monoxide and sulphur dioxide into atmosphere causing various health effects (Figure 3). The livelihood in surrounding areas is difficult due to inhalation of this polluted air. This combustion process usually occurs at low temperature only. This process is performed for crop management and farmers must be provided with certain guidelines for waste disposal process to avoid land and air pollution (Venkataraman et al., 2006).

### Particulate Matter

Particulate matter is a complex mixture of solid and liquid particles present in the air mostly toxic. Dust, smoke, sulphate, elemental carbon, solid compounds, fine solid particles and liquid droplets are few hazardous components present in the particulate matter. Particulate matter is considered as the deadliest form causing air pollution that directly affects the lungs and blood streams (Lemke et al., 2004). It may be caused due to erosion, soil tillage or during crop burning with the release of oxides of nitrogen and sulphur gases. Particulate matter can be classified into two categories based on their size

#### Coarse Particles

These particles vary in size between 2.5 – 10 μm in diameter. They are usually found in roadways and from industries producing dusts and smoke.

*Figure 3. Impact of burning agriculture waste*
*(Source: Rinkesh, n.d.)*

### Fine Particles

Fine particles vary in diameter about 2.5 µm. These particles are found mostly in smoke and are emitted directly during forest fires, power plants, etc.

## Fertilizers and Pesticides

Fertilizers are nutrient supplements added in the agricultural field for enhancement of agro productivity. It may be chemical or mineral fertilizers based on the requirement. The application of fertilizers has increased in recent years due to demand of food and nutrition owing to population expansion. Nitrogen, Phosphorous and Potassium are the three major elements required for plant growth. In addition to chemical fertilizers, pesticides are also added to prevent the attack of insects and pests in the field. These fertilizers and pesticides are applied by spraying technique. These products get mixed with the water and further consumption of this ground water leads to infection. Due to their negative impact on environment, these chemical fertilizers and pesticides have been band in certain developing nations. Fertilizers also contribute about 2% to greenhouse gas production. It also results in eutrophication leading to death of aquatic creatures with increased content of nitrates and phosphates in water (Khan et al., 2017).

## Gas Production

Methane ($CH_4$) is the main gas constituent produced in agricultural field. Paddy fields with flooded water are the major source of methane production. The methanogenic microbes especially bacteria termed methanogens get activated under favourable conditions and produce methane gas through process called methanogenesis (Tian et al., 2017). Decay and decomposition of organic matter in agricultural field result in emission of carbon dioxide, acetate and hydrogen gas. The bacteria feed on those organic matter and releases methane gas into atmosphere thus polluting air.

## Water Pollution

Concentrated point contamination loads released into the earth are promptly noticeable, effectively checked, cause critical effects and can be reduced by direct administrative and mechanical measures. Effluents from urban focuses and from real industries fall into this class. Interestingly, contamination loads radiating from agriculture, animal husbandry and even small-scale sectors, for the most part have no characterized or identifiable purpose of outfall and enter nature in little amounts at an extensive number of focuses. Expanding utilization of chemical fertilizers, pesticides and other agricultural chemicals, and expanding amounts of nutritional supplements and pesticides in agricultural spill over waters resulted in pollution of water sources. The fundamental water quality issues caused by diffuse agricultural contamination in developed nations showed increments in concentration of nitrates and pesticides, all of which are dangerous and hazardous. Alternate issues are increments in phosphates, total salts and sodium in water. The relationship of both aquatic also, agricultural frameworks are very difficult, and the work that they make has multidimensional viewpoints (Smith et al., 2017). The most critical collaboration in this relationship is between catchment region and the getting water. A few components like phosphorous, nitrogen, and a couple of different minerals are known as rare components, which have been to a great extent drained from scenes due to unreasonable deforestation and broad agricultural practice. The undisturbed normal lands are more equipped for holding soil particles and minerals. Besides, actually chose species have proficient root framework and microorganisms equipped for holding soil particles and minerals by forming defensive soil outside layers. Clearly, such land areas will be influenced by natural calamities like volcano, fall of trees or hurricanes; however such regular occasions are unavoidable. In addition, the delayed consequences like introduction of soil and its disintegration can be vanished by consecutive progression of species that in the long run restabilizes the soil surface.

It is notable that agriculture is the single biggest user of freshwater assets, utilizing a worldwide normal of 70% of all surface water supplies. Aside from water lost through evapotranspiration, agricultural water is reused back to surface water and additionally groundwater. In any case, farming is both reason and sufferer of water contamination. It is a reason through its release of toxins and sedimentation to surface as well as groundwater, through loss of soil by poor farming practices, and through salinization and water logging of flooded land. It is a victim through utilization of wastewater and contaminated surface and groundwater which pollute crops and transmit infection to farmers as well as consumers. Agriculture exists inside a beneficial interaction of land and water (Braden and Shortle 2013).

### Agriculture and Water Pollution

Agriculture aggravates common soil and supplement preservation technique. It replaces the origin of wood debris; aborts the predators like wolves and bears to secure the household stock, and may totally modify the complex organic and physiological framework keeping in mind the end goal to advance water system and waste framework. Besides, it might cause tremendous modification in prey-predator relationship by favouring the creation of particular fish species because of supplement advancement. It likewise presents unknown species likes biocides, which can't survive longer because of underdevelopment of protective components. Such species have invested less developmental energy and subsequently can't be kept in common natural aquatic ecosystems. So, agriculture has no constructive outcome on environmental working and biodiversity of oceanic living spaces. Land areas chosen for agriculture activity constitute a major threat to biodiversity. However farming is a fundamental need for individual for nutrition, agriculture can never be disregarded. Certain exercises can be improved like environmental control, water purifying and its storage, arrangement of bug control offices, advancing characteristic touching, and wood production. Therefore, eco-proficiency has advanced to a wild subject in both ecological and agriculture teaches these days (Williams et al., 2004).

The real favourable circumstances of farming incorporate forestation, crop productivity, and generation of natural material for food and fuel for household animals. Agriculture can be done in differing potencies, and that is the fact behind the contamination of freshwater and marine water biological communities by agricultural adjustment of common earthly environment. It influences the water science and general water quality with extreme enhancement of food chain, biocide permeation, disintegration of soil with suspended burdens and alteration of water cycles. Contamination is not just the consideration of some unsafe things into ecosystem that execute life form; in any case, it is characterized as the collection of man-made influencing effects that harm biological community. For example, agricultural interruptions into aquatic systems enhance the production of disease causing vectors such as the naucorid insects which causes infectious diseases and severe ulcer.

## Impacts of Agriculture on Water Quality

Determining the effects of agriculture on water quality is quite complicated and difficult to evaluate or estimate. The agricultural activities influencing water quality are excessive application of chemical fertilizers and pesticides, poultry feeding, etc that results in increase in concentration of chemical oxygen demand (COD) and other chemicals like nitrogen, ammonia, phosphorous, etc led into aquatic resources.

Agriculture raises different issues like water logging, disintegration, salinization, and desertification. Likewise, it might raise the issues of water quality destruction by agrochemicals, salts, and dangerous leachates. Lately, the salinization of water bodies is a more boundless fact and conceivably has more concern than that of soil salinization. Major trace elements like molybdenum, selenium, arsenic, mercury, etc have more observation and its occurrence in agricultural waste results in contamination serious toxic and hazard to the planted crops.

Abundant use of nitrogen-and phosphorus-based chemical fertilizers with respect to agricultural lands has prompted improvement of nitrogen and phosphorous in surface water, groundwater, and soil. This procedure of spillage of these supplements into water bodies at last causes eutrophication. Farmers employ these nitrogen-and phosphorus-based manures to build the production or yield of the products. These two supplements are required for the development of the plant; they utilize the required measures of these supplements, and overabundance supplements are normally connected with draining and unnecessary runoff exchange from land to water bodies. The harm done by such supplement misfortunes to the water bodies would be specific to particular area and would rely upon different components like unsteady association of water frameworks, kind of soil and science of environment. Influencing parameters like biological oxygen demand (BOD), chemical oxygen demand (COD), dissolved oxygen (DO) and presence of chemical constituents like phosphate, nitrogen, ammonia, etc affects water quality to a greater extent (Wu et al., 2013).

Polluted water leads to infection of many waterborne diseases. Human population is greatly affected due to these infections. The surface runoff water and nonpoint pollution sources contribute greatly to spread of disease causing pathogens. Contaminated water used for irrigation and its direct contact with the food crops are the major sources of crop contamination by microbes. Diseases such as cholera, typhoid, amoebiasis, etc are caused to humans consuming those crops. The types of crops that easily get infection to these pollutants are cabbage, carrot, onions, etc. Hormone system in fish, humans and animals are also affected causing imbalances. Severe condition leads to toxicity in reproductive system.

## SOIL POLLUTION

Soil pollution is defined as the surface accumulation of toxic chemicals, salts, compounds, radioactive materials and other disease causing organisms that pose serious threats to humans and other life on earth. The thin layer of soil is composed of organic and inorganic constituents that forms layer over the rock surface. The organic top layer is formed from decomposition of plant and animal remains. The solid rock particle forms the inorganic portion of the soil. Soil can be polluted or contaminated using different methods such as

- Application of fertilizers and pesticides
- Discharge from a landfill
- Discharge of industrial waste into soil
- Breakage of storage tanks
- Escape of soil waste
- Land and forest clearance
- Overfeeding by grazing animals

The presence of toxic chemicals in these wastes deposits into fertile lands and converts them into infertile or barren unfit for any purpose. Some of the chemicals that cover the land are heavy metals including mercury, chromium, lead, etc, solvents, hydrocarbons and pesticide compounds. Soil can be polluted by the following methods:

- Agricultural pollution of surface and underground soil
- Industrial pollution of surface soil and soil quality
- Man-made pollution by deposition of waste into fertile lands
- Nuclear pollution by radioactive materials

Soil contamination is caused by the involvement of man-made chemicals or other modification in the normal soil condition. This sort of pollution ordinarily emerges from the crack of underground storage joins, use of pesticides, and permeation of polluted surface water to subsurface level, oil and fuel dumping, draining of wastes from landfills or direct release of industrial wastes to the soil. This event is associated with the level of industrialization and forces of substance usage. An immense amount of chemicals is connected to soils, which brings about the expansion level of heavy metals, for example, cadmium, mercury, arsenic, and lead. The utilization and assortment of pesticides have expanded definitely around the world with expanded utilization of nutrition, in respect to expanded yield production. Pesticides have brought about their abuse, therefore posturing severe contamination and ill-effects. Pesticides can be any substance or blend of various substances that are deliberate for degradation or repulsion of any pest. Pesticide application is mandatory because of requirement of increased food productivity to fulfil human demands. Pesticides and its products produced after decomposition can get away into ecosystem, soil, or streams, at last prompting the aggregation of dangerous substances. Pesticides, for example, DDT (dichlorodiphenyltrichloroethane), chlorinated hydrocarbons and organophosphates are consumed by the soil particles where they gather and cause contamination of root crops like potato developed in soil. Subsequently, the utilization of such polluted crops causes the pesticide molecule to enter human system causing infectious diseases. Soil and crop pollution with chemicals and heavy metals remains as major drawback in agricultural sector. Heavy metals, for example, cadmium (Cd), zinc (Zn), lead (Pb), and copper (Cu) are exceptionally normal in upper layers of soil particularly in rice fields. Heavy metal gathering in fields and vegetation brings about dangerous impacts on human health. Copper is broadly utilized in animal feed, and there is an expanded risk of contamination in the soil by the expansion of such excrement in crops with copper (Xu et al., 2013). Manures containing more state of sodium and potassium reduce the soil pH and lessen the efficiency of field crops. Phosphate composts are an imperative reason for cadmium metal aggregation when compared with different manures. Other sources of heavy metal deposition apart from chemical fertilizers are different agrochemicals, for example, pesticides, animal fertilizer, and utilization of contaminated water for water system.

## AGRICULTURE IMPACTS ON HUMAN HEALTH

Alterations have been acquired in the rural region for enhancing health and safety measures for farmers and other individuals working in the agricultural field. These progressions incorporate change in innovation being utilized and knowledge about fitness among people. Food Quality Protection Act (1996) and Worker Protection Standard (1992) are the examples of the administrative methodologies taken to avoid word related and ecological health risks that can be caused because of agricultural practices. The present conditions demonstrate that there is a requirement for research and consciousness of farming security. Steps should be taken to teach agriculturists and different people identified with agricultural industry. Cultivating has been announced a profoundly unpredictable industry as indicated by an investigation done by the Bureau of Labour Statistics. Agricultural specialists may result in wounds and ailments because of physical effort, contact with farm animals, utilization of apparatus, and a high rate of introduction to poisonous materials. Studies on health effects must consider a few altering factors in agrarian exposures bringing about physical ailments including work compel age and ethnicity, designing controls, and utilization of individual defensive equipment. There has likewise been an expansion on fundamental level administrators that work off the farm, which includes extra introduction issues. It is known that agriculture is presently at a low level in labour strength and as the quantities of farmers diminish, there will be an expansion in the measure of the labour force. Aging of the farmer population may result in increased illness such as respiratory and muscular problems (Lu et al., 2015). Communication obstruction exists which can block following fitness data on marks and preparing in appropriate work operations. Temporary workers rather than cultivate owners now contract vast quantities of farm specialists, raising new health and security concerns. These progressions may expand health and security issues in the agricultural working environment.

## Dust Exposure

Exposure to dusts may be organic, inorganic or grain dust. These particles enter human respiratory system and cause breathing problems and lung related problems. Most respiratory problems arise due to organic dusts only spreading through animal feeding and grain processing methods. Organic dust exposure leads to farmer's lung disease, asthma, bronchitis and organic dust toxicity syndrome (ODTS). Inorganic dusts are released as a result of ploughing, harvesting and tilling processes. It is estimated that of total dust exposure, nearly 43% of the dusts are inorganic. The illness caused by inorganic dusts is quite less compared to organic. They cause minor respiratory problems. Grain dusts result from pesticides, pollens, microbes, animal and bird faeces, insects and waste vegetable products. The size of respirable dust varies up to 100 μm. Respirable dusts are those particles possessing diameter of 4 μm. Harvesting, Processing, Production, Storage are few method of release of grain dusts (Morman and Plumlee 2013).

## Microorganisms

Microbes play a major role in spreading infectious diseases to farmers and other workers in agricultural field. The food crops infected by the bacteria are consumed by humans that cause diseases. Animals and other small insects that feed on land are affected to greater extent. The pathogenic microbes enter into human physiological system through inhalation working in the field. For example, tuberculosis causing bacteria *Mycobacterium bovis* if present in the farm animals can easily be transferred to humans. There is more possibility of transmitting Swine Influenza disease from pigs to humans. Query fever (Q fever) caused by goats, sheep and cattle damage human's organs such as brain, lungs, heart and liver.

## Pollutants

Formation of pollutants such as hazardous gases, allergens, inflammations, endotoxins and chemicals constitutes major part to human illness. The toxic gases such as methane, nitrogen oxides, carbon dioxide, hydrogen sulphide, carbon monoxide, etc are produced through various sources in agricultural field. Longer exposure of these gases results in various infectious non-curable diseases. For example, exposure to nitrogen oxide beyond the limit leads to chemical pneumonitis termed as silo filler's disease. Ammonia, methane and hydrogen sulphide are produced by deposition of swine waste in lands. Another gas hydrogen sulphide causes irritation in respiratory tract at low levels itself. Farmers, workers or common people entering into lands with increased concentration of any one these gases experiences nausea, loses consciousness and finally collapses. People who managed to survive will suffer from pneumonia causes by microbes.

## Cancer

Though cancers are not caused because of agricultural activity, studies reveal that there are few cancers that affect farmers only. The reason behind this is long exposure to chemical pesticides and direct inhalation of those toxic chemicals at the time of spraying. Leukemia (cancer in white blood cells), skin, lip, brain and multiple myeloma are some of cancers that attack mostly farmers. There is chance of this disease to whole farmer's family if agricultural work is shared by them. Serious threat may rise to humans in the surrounding region through indirect exposure. Toxic chemicals may spread through air as dust particles.

## ECOSYSTEM AND AGRICULTURE

The effect of agriculture on ecosystem is the impact that diverse cultivating techniques have on the biological communities around them, and how those impacts can be followed back to those practices. The natural effect of farming fluctuates in view of the wide assortment of agricultural practices utilized the world over. Eventually, the ecological effect relies upon the production practices of the framework utilized by agriculturists. The association between discharges into nature and the cultivating framework is secondary, as it likewise relies upon other atmosphere factors, for example, precipitation and temperature. Environmental change and agriculture are interrelated procedures, both of which happen on an overall scale. An unnatural weather change is anticipated to affect conditions influencing farming, including temperature, precipitation and frosty keep running off. These conditions decide the conveying limit of the biosphere to deliver enough nourishment for the humans and animals. Rising carbon dioxide levels would likewise have impacts, both unfavourable and valuable, on crop yields. In the meantime, agriculture has been appeared to deliver specific consequences for environmental change, fundamentally through the generation and arrival of greenhouse gasses, for example, carbon dioxide, methane, sulphur and nitrous oxide. Deforestation causes the loss of environment for a huge number of animal groups, and is additionally a driver of environmental change. Trees go about as a carbon sink: that is, they assimilate carbon dioxide, an undesirable ozone harming substance, out of the environment. Evacuating trees discharges carbon dioxide into the environment and deserts fewer trees to ingest the expanding measure of carbon dioxide noticeable all around. Deforestation magnifies environmental change. Among some of these issues is the exhaustion of underground aquifers through over drafting. Soil can be over-flooded as a result of poor circulation consistency or administration of waste water, chemicals, and may prompt water contamination. Over-water system can cause profound seepage from rising water tables that can prompt issues of water system saltiness requiring water table control by some type of subsurface land waste. If the soil is under watered, it gives poor soil saltiness control which prompts expanded soil saltiness with resulting development of poisonous salts on soil surface in regions with high dissipation. Soil decomposition is the decrease in soil quality that can be a consequence of many components, particularly from agriculture. Soils hold most of the world's biodiversity, and fertile soils are basic for food and nourishment generation and a sufficient water supply. Common traits of soil corruption can be salting, water logging, pesticide pollution, and decrease in soil structure quality, loss of richness, changes in soil corrosiveness, alkalinity, saltiness, and disintegration. Soil corruption additionally have huge affects on organic decomposition, which influences the microbial group of the soil and can modify supplement cycling, control of insects, pests, infectious diseases and chemical modification in the properties of the dirt. Figure 4 depicts the various impacts of traditional and modern agriculture impacts on environment.

## AGRICULTURE POLLUTION MANAGEMENT

Cultivating operations can add to supplement contamination when not legitimately managed. Composts and animal manures, which are both rich in nitrogen and phosphorus, are the essential choice of supplement contamination from agricultural sources. Abundance supplements can affect water quality when it rains or when water and soil containing nitrogen and phosphorus wash into close-by waters or drain into ground waters. Fertile soils and domesticated animals can be important sources of vaporous, nitrogen-based gases like ammonia and nitrogen oxides. Ammonia can be destructive to aquatic creatures

*Figure 4. Impact of agriculture on ecosystem*

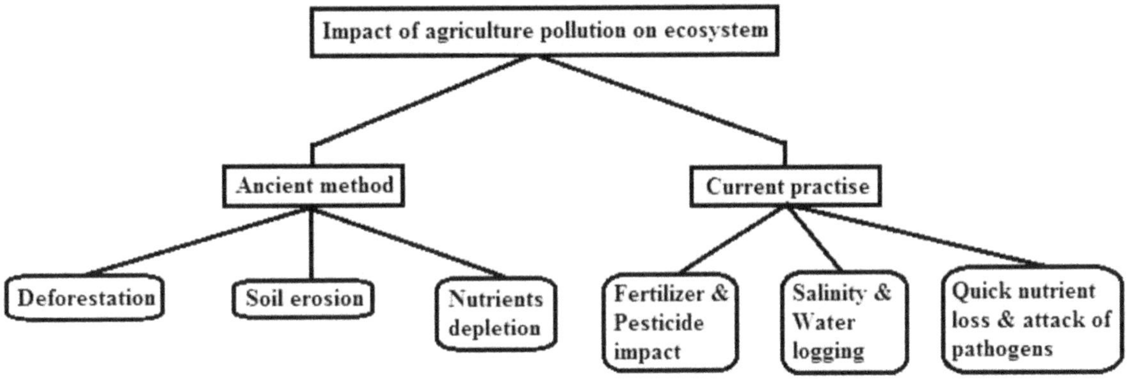


if extensive quantities are stored to surface waters. Nitrous oxide is a powerful greenhouse gas. Farmers are encouraged to find ways to relieve the impacts of spill over, pesticide harming, and disintegration, governments should start an educational awareness to spread consciousness of the threats of abusing manures and pesticides, and giving agriculturists moves they can make to lessen supplement loaded overflow, for example, split utilization of composts. The legislature ought to likewise support agricultural practices that reduce pesticide and compost overflow.

## Biofertilizers and Biopesticides

Replacement of chemical fertilizers and pesticides with bioformulations paves way for safety measure of nutrient supplement to crops for their growth and development. The utilization of organic manures has been drilled from long time past days without the scientific information and advantage of the composts. As the effect of chemical fertilizers has expanded, the utilization of inorganic composts among the farmers increased. The current pattern in biotechnology is use of microorganism in the field of agriculture for improvement of agro profitability. Microorganisms assume an important part in fixation of atmospheric nitrogen, solubilising and mobilising phosphorus, generation of anti-infectious agents, etc. Out of 16 supplements required by plants for their development, the three macronutrients, for example, nitrogen, phosphorous and potassium assumes specific part in plant development and improvement (Khan et al., 2009). The microorganisms are an important part of soil and can be sorted as Nitrogen-fixers (*Rhizobium, Azospirillum, Azobacter* and so on), Phosphate-solubilisers (*Pseudomonas, Bacillus, Agrobacterium, Flavobacterium*, etc) and Potassium-solubilizers (*Enterobacter hormaechei*, etc). The term biofertilizer or called 'microbial inoculants' can be the most part characterized as a preparation or substance containing live or dormant cells of effective strains of nitrogen fixing, phosphate and potassium solubilizing or cellulytic microorganisms utilized for incorporation in seed, soil or treating the soil regions with the goal of expanding the quantities of such microorganisms and quicken certain microbial procedure to increase the degree of the accessibility of supplements which can be acquired by plants. Microbes are likewise significant in the treatment of strong waste and sewage. They clear ecosystem by decomposition of a few toxins like pesticides, hydrocarbons, metals, dyes and paints. They also help in the improved retrieval of oil and metals from poor quality minerals. Similarly, biopesticidal formulations are those containing microbial populations against pests that attack crops in the field. Pest management is an important process in agriculture and studies reveal that effective means is through bioformulations only (Suman and Dikshit 2010).

## Formulation of Biofertilizers

Biofertilizers are formulated using suitable carrier materials. Environmental, domestic or industrial wastes can be used as carrier materials for production of biocontrol agents. Through this process environmental wastes can be minimized leaving a green ecosystem (Figure 5). The carrier materials used should possess few properties like no toxicity, good absorbing capacity, easily available, good pH and easily sterilizable (Jambhulkar & Sharma, 2014). Commercially used carrier material is talc powder.

## Agricultural Waste Management

Agricultural pollution can be controlled by managing the agricultural waste also. The waste may be on the form of solid, semi-solid or liquid depending on the type of waste released. Agricultural waste pollution can be classified as point and non-point sources pollution. The non-point source pollution includes excess nutrient and sediment run-off and pesticides sprayed. Farm animal wastes and manure are mainly considered as point source pollution. The total amount of solids present in the waste determines the method of handling the waste. The parameters influencing the amount of total solids includes the type of animal, nutritional intake of animal, climatic conditions and the quantity and type of water consumed

*Figure 5. Production of biofertilizer*

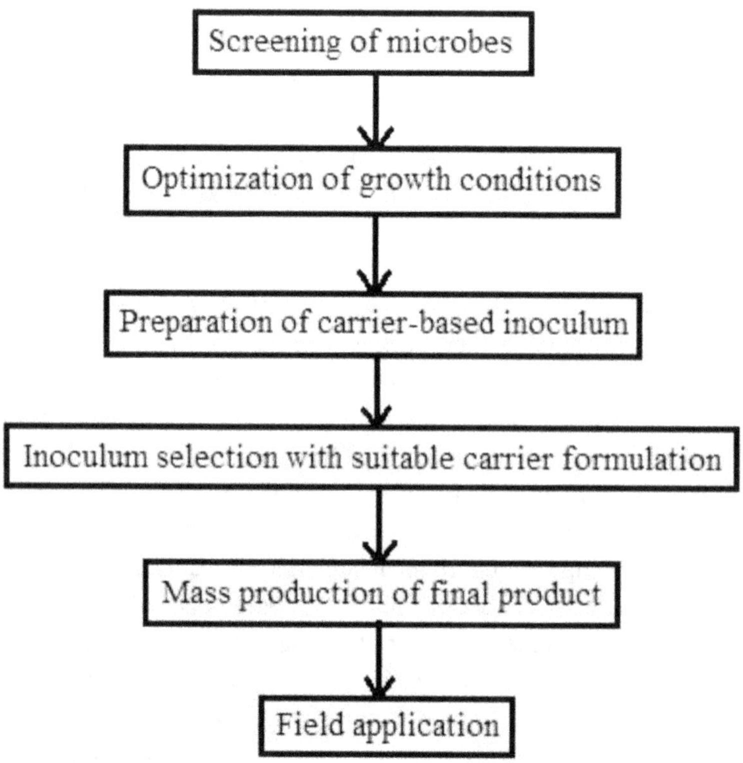

by the animal (Sakai et al., 2011). Agricultural waste management can be processed using six functions namely Production, Collection, Storage, Treatment, Transfer and Re-Utilization of the treated waste.

## Altering Agricultural Practices

Total productivity in agriculture is still increasing though only less land is left for cultivation. Increased production results in requirement of large number of workers in the field. Due to various factors, the man power involved in food production decreased. The main factors include more income from industrial sector, invention of machines in agricultural field and less interest for people in agricultural sector. Livestock requirement and crop production tend to increase due to population expansion. Agricultural pollution can be minimized only when ancient method of agriculture comes to practice again. The rise in agricultural pollution is mainly due to modern methods of cultivating crops with enhanced involvement of chemicals in the form of fertilizers and pesticides thus destructing soil quality and fertility. Organic manures such as cow dung, animal and fruit wastes and domestic wastes can be used as fertilizers which contains enormous amount of natural nutrients inbuilt in it (Loyon 2017).

## Government Plans

Government has announced few steps for short and long time management of agricultural pollution.

- Creating awareness and educating farmers regarding the impact of pollution on ecosystem, solutions for solving these problems and the economic part involved in implicating the solutions.
- Research and development program regarding the management of agricultural waste and pollution caused by it and solutions for treating those wastes.
- Awareness regarding the importance of biological formulations and consequences of using chemicals in agricultural land.
- Education regarding the quantity of water used for agriculture and also minimising run-off into water bodies.
- Farmers should know that disposal and treatment of wastes is an important step in this sector.

Proper management of soil and crop can minimize the release of sediment run-off, nutrients and other chemicals onto surface waters. Studies reveal that the amount of nutrients released by the crop is greater than those that are replaced. To enhance productivity, farmers incorporate more nutrient supplements than required by the crops. Hence the remaining excess nutrients are released into surface and ground water causing pollution. So education must be given to farmers regarding the application of quantity and quality of fertilizers and pesticides (Ntaliani et al., 2008).

## Agriculture and Biodiversity

The eco-agriculture is the blend of agriculture manageability and biodiversity protection. Speculations should be done to create the techniques and innovations that can upgrade yield and diminishing their expenses, while monitoring the biodiversity. It is additionally basic to give chances to the farmers to expanding eco-agriculture relatively. Generally financial and social issues confronted by agriculturists will influence the biodiversity protection. There ought to be an adequate number of group level associations for the eco-farming. The methodologies ought to be made to gain satisfactory advantages from eco-agriculture. Another essential perspective is to deliver and share learning about eco-farming. The needs have to be set for setting up a comprehension of collaborations among protection and creation zones. For an eco-agriculture framework to be best, completely archived data is required. The connecting of biodiversity preservation with manageable agriculture is critical to bring the eco-farming at a worldwide level. The properties of soil and water are interlinked with each other in a way that on the off chance that one thing is in poor condition, at that point the other must be influenced. Fertile soil keeps the water clean, and comparably clean water keeps the soil in a high-quality condition. The water quality is evaluated by the amount of unsafe chemicals and dissolved solids exhibit in the water. Again, the soil and water quality can be kept up by the removal of overabundance pesticide and manure utilize. The modern effluents likewise must be legitimately arranged and not poured in the water utilized for agriculture reason.

### Recycling of Manures

Farm animal's excreta can be used as manures. The advantages like supply of supplements to crops, maintaining soil quality and fertility, also, moisture retaining capacity of the soil can be acquired by applying fertilizer. Animal excreta in semi-liquid form are called slurry. Slurry is additionally utilized like fertilizer to build the soil quality. Utilizing nitrogen in excreta enables nitrogen cycling in the earth. The concentration of nitrogen i excreta differs in depending on the species. This may be because of various methods involved in process of excreta production and storage in different animals. Adequate quantity of excreta gets deposited in the land when the animal feeds while the remaining is collected for manure. Nitrogen may also be released in the form of ammonia gas. In order to a prevent slurry from foaming and fouling, aeration is provided causing loss of nitrogen. The excreta deposited gets mixed with the run-off mostly during rainy season and remains available to the crops for longer period. It is clear that in a few conditions weakened slurry connected to grass might be more impervious than concentrated slurry. The farming contamination can be reduced by utilizing fertilizer in the field. The fertilizer is delivered from compost and other farming products. It has major advantages like depletion of soil disintegration, clarification of soil surface, and minimized utilization of chemical composts (Mulbry et al., 2005). The main function is to provide nutrient supplements and control plant diseases caused by pests. Manure controls soil erosion by imparting proper composition to soil. The volume and concentration of manures used varies depending on the nature of land.

## CRITICAL AND CONSTRUCTIVE ASSESSMENT

A standout amongst the most noteworthy wellsprings of pollution from agriculture fields: overflow from nitrogen composts that causes eutrophication. In any case, this is just a single of numerous natural difficulties related with agriculturally determined changes in the nitrogen cycle. A large number of these different difficulties are related with types of nitrogen that are gotten from manures. Urea is an essential part of numerous nitrogen composts. Once connected, urea is immediately changed over to ammonium ($NH^{4+}$). At the point when the dirt is clammy, ammonium is changed over to nitrates ($NO^{3-}$). On the off chance that the dirt needs adequate oxygen, denitrification prevails, and the nitrates are changed over to nitrous oxide ($N_2O$). Nitrous oxide is an ozone harming substance that is the fourth biggest wellspring of radiative constraining. The environmental grouping of nitrous oxide has expanded altogether finished the most recent century. A lot of this expansion is related with the expanding utilization of nitrogen manures.

Pollution of water is harder to decide on than different types of contamination because of its nature, and despite of the fact that there have been numerous investigations on surface water contamination and connections with agricultural practices, its impact on the earth and on individuals' work systems and results are lesser known. The chapter has prompted improvement and the measures prescribed incorporate specialized preparing; advancement of rules for best administration hones; enactment and control of utilization of chemicals; and observing consistence with directions and suggested hones.

## CONCLUSION

The success of agricultural sector depends on agriculture policy, pollution and protection. These three are interrelated with each other. Agricultural pollution can be reduced by following proper agricultural policy and protection act laid by Government. Agriculture contributes to a country's food and nutrition as well as economy. Numerous sorts of staple products, grains, and natural products are being delivered from this area, which are making significant offer in the export. With course of time, agricultural sector faces destruction because of the pollutants released from practices involved. Farming contamination influences air, water, and soil, as well as issues identified with wellbeing and biodiversity have additionally been seen using compost, pesticides, natural issue, and ozone depleting substance discharges. There will be a disturbing circumstance when agricultural contamination will limit the farming yield itself. There is an expanding open concern with respect to farming contamination and its effect on the earth. There is a need to expand the agricultural generation to defeat the expanding interest of nourishment. Presently, farmers are utilizing new strategies to build the harvest efficiency and quality, however regardless of this; this industry is not following guidelines and controls that have been followed in different sectors. So, there ought to be an essential concentration to fortify the administrative projects to keep the farming contamination and its uncommon impacts on the earth. Finally, agricultural pollution can be minimized by managing excess nutrient run-off, contaminated water run-off into surface and ground water, tillage conservation, managing livestock and its waste and following guidelines laid by government policies.

## REFERENCES

Abbasi, A. (2014). Agricultural Pollution: An Emerging Issue. In *Improvement of Crops in the Era of Climatic Changes*. Springer.

Agrawal, M. (2005). Effects of air pollution on agriculture: An issue of national concern. *National Academy Science Letters (India)*, *28*, 93–106.

Ainsworth, E. A., & McGrath, J. M. (2010). Direct effects of rising atmospheric carbon dioxide and ozone on crop yields. *Climate Change and Food Security*, *37*, 109–130. doi:10.1007/978-90-481-2953-9_7

Aneja, V. P., Schlesinger, W. H., & Erisman, J. M. (2009). Effects of agriculture upon the air quality and climate: Research, policy, and regulations. *Environmental Science & Technology*, *43*(12), 4234–4240. doi:10.1021/es8024403 PMID:19603628

Braden, J. B., & Shortle, J. S. (2013). Agricultural Sources of Water Pollution. *Reference Module in Earth System and Environmental Sciences, 3*, 81–85.

Chen, Y., Wen, X., Wang, B., & Nie, P. (2017). Agricultural Pollution and Regulation: How to subsidize agriculture? *Journal of Cleaner Production, 164*, 258–264. doi:10.1016/j.jclepro.2017.06.216

Emmanuel, D., Sekyere, E. O., & Owusu, V. (2016). Impact of agriculture extensive service on adoption of chemical fertilizers: Implications for rice productivity and development in Ghana. *NJAS Wageningen Journal of Life Sciences, 79*, 41–49. doi:10.1016/j.njas.2016.10.002

Gautam, R., Bhardwaj, N., & Saini, Y. (2010). Fluoride accumulation by vegetables and crops grown in Nawa Tehsil of Nagaur district (Rajasthan, India). *Journal of Phytology, 2*, 80–85.

Jambhulkar, P. P., & Sharma, P. (2014). Development of Bioformulation and Delivery System of *Pseudomonas fluorescens* Against Bacterial Leaf Blight of Rice (*Xanthomonas oryzae pv.oryzae*). *Journal of Environmental Biology, 35*, 843–849. PMID:25204057

Jones, J. W., Antle, J. M., Basso, B., Boote, K. J., Conant, R. T., Foster, I., ... Wheeler, T. R. (2017). Brief history of agricultural system modelling. *Agricultural Systems, 155*, 240–254. doi:10.1016/j.agsy.2016.05.014 PMID:28701816

Khan, M. N., Mobin, M., & Abbas, Z. K. (2017). *Fertilizers and their contaminants in Soil, Surface and Groundwater*. Reference Module in Earth System and Environmental Sciences. doi:10.1016/B978-0-12-409548-9.09888-2

Khan, W., Rayirath, U. P., Subramanian, S., Jithesh, M. N., Rayorath, P., Hodges, D. M., ... Prithiviraj, B. (2009). Seaweed extracts as biostimulants of plant growth and development. *Journal of Plant Growth Regulation, 28*(4), 386–399. doi:10.100700344-009-9103-x

Lemke, R., McConkey, B., & Janzen, H. (2004). Impacts of agriculture on air quality (with emphasis on $CO_2$ and $N_2O$). *Agriculture and Agri-Food Canada,* 23–36.

Loyon, L. (2017). Overview of manure treatment in France. *Waste Management (New York, N.Y.), 61*, 516–520. doi:10.1016/j.wasman.2016.11.040 PMID:27955906

Lu, Y., Song, S., Wang, R., Liu, Z., Meng, J., Sweetman, A. J., ... Wang, T. (2015). Impacts of soil and water pollution on food safety and health risk in China. *Environment International, 77*, 5–15. doi:10.1016/j.envint.2014.12.010 PMID:25603422

Matsubara, M., Morimoto, S., Sase, H., Ohizumi, T., Sumida, H., Nakata, M., & Ueda, H. (2009). Long-term declining trends of river water pH in Central Japan. *Water, Air, and Soil Pollution, 200*(1-4), 253–265. doi:10.100711270-008-9909-3

Morman, S. A., & Plumlee, G. S. (2013). The role of airborne mineral dusts in human disease. *Aeolian Research, 9*, 203–212. doi:10.1016/j.aeolia.2012.12.001

Mulbry, W., Westhead, E. K., Pizarro, C., & Sikora, L. (2005). Recycling of Manure nutrients: Use of algal biomass from dairy manure treatment as a slow release fertilizer. *Bioresource Technology, 96*(4), 451–458. doi:10.1016/j.biortech.2004.05.026 PMID:15491826

Ntaliani, M., Costopoulou, C., & Karetsos, S. (2008). Mobile Government: A challenge for agriculture. *Government Information Quarterly*, *25*(4), 699–716. doi:10.1016/j.giq.2007.04.010

Preston, T. R., & Leng, R. A. (1989). The greenhouse effect and its implications for world agriculture, the need for environmentally friendly development. *Livestock Research for Rural Development*.

Rinkesh. (n.d.). *Agricultural Pollution*. Retrieved from https://www.conserve-energy-future.com/causes-and-effects-of-agricultural-pollution.php

Roy, S., Beig, G., & Ghude, S. (2009). Exposure-plant response of ambient ozone over the tropical Indian region. *Atmospheric Chemistry and Physics*, *9*, 4141–4157.

Sakai, S., Yoshida, H., & Hirai, Y. (2011). International comparative study of 3R and waste management policy developments. *Journal of Material Sciences and Waste Management,* 86–102.

Smith, L., Inman, A., & Zhang, H. (2017). Mitigation of diffuse water pollution in England and China, and the scope for policy transfer. *Land Use Policy*, *61*, 208–219. doi:10.1016/j.landusepol.2016.09.028

Suman, G., & Dikshit, A. K. (2010). Biopesticides: An ecofriendly approach for pest control. *Journal of Biopesticides*, *3*(1), 186–188.

Tian, Z., Niu, Y., & Fan, D. (2017). Maintaining rice production while mitigating methane and nitrous oxide emissions from paddy field in China: Evaluating tradeoffs by using coupled agricultural system models. *Agricultural Systems*. doi:10.1016/j.agsy.2017.04.006

Venkataraman, C., Habib, G., & Kadamba, D. (2006). Emissions from open biomass burning in India, Integrating the inventory approach with high-resolution moderate resolution imaging spectroradiometer (MODIS) active and land cover data. *Global Biogeochemical Cycles*, 20.

Williams, P., Whitfield, M., Biggs, J., Bray, S., Fox, G., Nicolet, P., & Sear, D. (2004). Comparative diversity of rivers, streams, ditches and ponds in an agricultural landscape in southern England. *Biological Conservation*, *115*(2), 329–341. doi:10.1016/S0006-3207(03)00153-8

Wu, M., Tang, X., Li, Q., Yang, W., Jin, F., Tang, M., & Scholz, M. (2013). Review of ecological engineering solutions for rural non-point source water pollution control in Hubei Province, China. *Water, Air, and Soil Pollution*, *224*(5), 1561–1579. doi:10.100711270-013-1561-x

Xu, X., Zhao, Y., Zhao, X., Wang, Y., & Deng, W. (2013). Sources of heavy metal pollution in agricultural soils of a rapidly industrializing area in the Yangtze Delta of China. *Ecotoxicology and Environmental Safety*, *108*, 161–167. doi:10.1016/j.ecoenv.2014.07.001 PMID:25063882

Zapata, F., & Nguyen, M. L. (2009). Chapter 7 Soil Erosion and Sedimentation Studies using Environmental Radionuclides. Radioactivity in the Environment, 16, 295-322.

## KEY TERMS AND DEFINITIONS

**Agricultural Pollution:** Indicates contamination from point and non-point sources resulting in environmental contamination causing degradation and hazardous to human and other life.

**Biofertilizers:** Are biological formulations containing microbes such as bacteria, fungi, or algae which enables nutrient uptake by plants from soil.

**Eutrophication:** The condition of excessive nutrients run-off into water bodies causing dense growth of plants on surface and lethal to aquatic life.

**Greenhouse Gases:** Gases in the atmosphere which absorbs and emits harmful radiations within the thermal infrared range.

**Manure Recycling:** Refers to the process of conversion of animal wastes into nutrient supplements essential for growth and development of crops.

**Methanogenesis:** The process of production of methane gas by special group of bacterial population called methanogens.

**Particulate Matter:** The total sum of all toxic solid and liquid particles present in the air.

**Waste Management:** The process of minimizing the toxic concentration of waste released from agriculture through recycling.

# Chapter 10
# Pesticidal Pollution

**Rizwana Mobin**
*Government College for Women, Cluster University Srinagar, India*

**Hamida-Tun-Nisa Chisti**
*National Institute of Technology Srinagar, India*

**Tauseef Ahmad Rangreez**
*National Institute of Technology Srinagar, India*

**Rafia Bashir**
*National Institute of Technology Srinagar, India*

**Altaf Ahmad Najar**
*Cluster University Srinagar, India*

## ABSTRACT

*The development and application of pesticides has contributed in a long way in making the "Green Revolution" possible. These newer pesticides have synergetic effect over the control of pests that otherwise have negative impact on the quality and quantity of food. The toxicity, persistence, and environmental pathway are some important criteria that determine the impacts on ecology and environment. The generalization of the impact of pesticides on different organisms is difficult as these are of broad spectrum chemical nature. However, the major problem that arises due to widespread use of pesticides is the contamination of water by pesticide runoff. The chemically contaminated water in turn leads to the much aggravated problems of bio-concentration and bio-magnification of these chemicals. While the bio-concentration describes the transfer of a chemical from surrounding into the tissue/body of organism, the bio-magnification is related to the increased concentration of such a chemical along a food chain.*

## INTRODUCTION

Water covers about two third part of the earth's surface and only one third part of earth's surface is covered by land (Dessouky & Ettouney, 2002). Population density on earth's surface is growing day by day which increases the pressure on its water sources. As a result of this, the quality of water is reduced and poorer water quality means unhygienic and polluted water is available for domestic use. Industrialization

DOI: 10.4018/978-1-5225-5754-8.ch010

revolution in the 19$^{th}$ century became starting point for the problem of pollution. As such, pollution is a human created problem, when there was less population density on earth's surface, no one could have thought about the pollution which became a serious problem over time.

Water pollution is the contamination of water bodies and these contaminations occur when human activity exceeded the limits. Direct or indirect addition of contaminants or pollutants, without removing its harmful components to water bodies causes environmental degradation. Such contaminated water leads to severe health problems (Zhao, Li, Ren, Chen, & Wang, 2011). Waste from industries (chemical waste), nuclear power plant (radio-active waste) and human activities add the contamination to surface water as well as ground water bodies (Zahra, Hashmi, Malik, & Ahmed, 2014). About 2 million people still die each year from water, sanitation and hygiene-related issues (Leonard et al., 2003). About 315,000 children under the age of five die from diarrhoea each year, often as a result of unsafe water. Childhood diarrhoea is the second-leading cause of childhood mortality (Ashbolt, 2004). Water sources are also polluted by agricultural waste and several products used in agricultural field such as pesticides, fertilizers etc (Mozaz, Alda, & Barcelo, 2004).

Entry of pesticides into surface and ground water from agricultural field causes water pollution especially drinking water contamination. This chapter basically deals with the agricultural pollution. Agricultural pollution is caused by pesticides, fertilizers, metals like cadmium, fluoride, radioactive elements, other metals, organic contaminants and heavy metal like lead, mercury, arsenic etc. Agricultural pollution is one of the major sources of water pollution, in which pesticides play major role. Pesticides are used to control the pest. Pesticides get leached through the soil to the ground water and pollute the ground water (Fawell, & Nieuwenhuijsen, 2003).

This chapter deals with the development of pesticides, toxicity of pesticides, their effect on environment as well as on human health, monitoring of pesticides, how to reduce the uses of pesticides and their alternatives.

## PESTICIDES

The term 'pesticide' is a composite term which includes all the chemicals used for killing of pest, or to control the pest. This term (pesticide) is collectively used for herbicides, insecticides, fungicides, nematocides and rodenticides. The application and development of pesticides for the control of insecticides and herbicides which destroy the quality and quantity of food contributed to the "Green Revolution". Uses of pesticide are a standard method for controlling of pest, and have changed the society since 1950. Unfortunately, in addition to advantages, pesticides are attached with several disadvantages. Long term use of pesticides disturbs the predator-prey relationship and causes loss of biodiversity. Pesticides can also affect the human health. The use of chemicals in the field of agriculture are restricted while agriculture is one the few activities where chemicals are released intentionally in the environment.

Today, the use of pesticides in agricultural field constitutes the larger area where industrial chemicals are used. According to American Chemical Society database 13 million chemicals were in use in 1993 and 5,00,000 new chemicals were added annually (Ongley, 1996). The use of pesticides has great impact on quality of water which leads to several environmental and human health problems. In a recent survey, 50 pesticides are commonly used in the agricultural area of western provinces of Canada (Ongley, 1996). In Asia and Africa no pesticides are used in traditional agricultural process. Excessive and inappropriate use of pesticides has inverse impact on human health, environment and water quality. According to the US-EPA's National Pesticide Survey (1992) 10.4% of community wells and 4.2% of rural wells in the United States contained detectible amount of one or more pesticides (Wilson & Tisdell, 2001). In south-western Ontario (Canada), about 35% of the groundwater wells in agricultural field were found positive for pesticides on at least one occasion (Lampman, 1995). Factors associated with the impact of pesticides on water quality:

- Active ingredient in the pesticide formulation.
- Contaminants that exist as impurities in the active ingredient.
- Additives that are mixed with the active ingredient (wetting agents, diluents or solvents, extenders, adhesives, buffers, preservatives and emulsifiers).
- Degradate that is formed during chemical, microbial or photochemical degradation of the active ingredient.

Silviculture is another field where pesticides are extensively used. Over very large area insecticides are sprayed by aircraft to control the pests especially insects. In tropical and subtropical environment, irrigated agriculture requires modification of hydrological regime which provides favourable condition for the breeding of insects, responsible for several vector-borne diseases (Ahmad, Getso, Ahmad, & Abdullahi, 2014). In order to control the various vector-borne diseases, pesticide along with additional insecticides such as DDT are also applied which have serious and widespread ecological consequences (Ongley, 1997). Several environmental management methods have been developed and tested in many projects in order to control the breeding of disease vectors. Figure 1 shows type of pesticides on the basis of type of pest.

## Historical Perspective of Pesticides

The battle of humans with pests is not new but it has historical evidence. There are several examples present how pests impacted on human society. One of the important incidence is the Black Plague in

*Figure 1. Classification of pesticides*

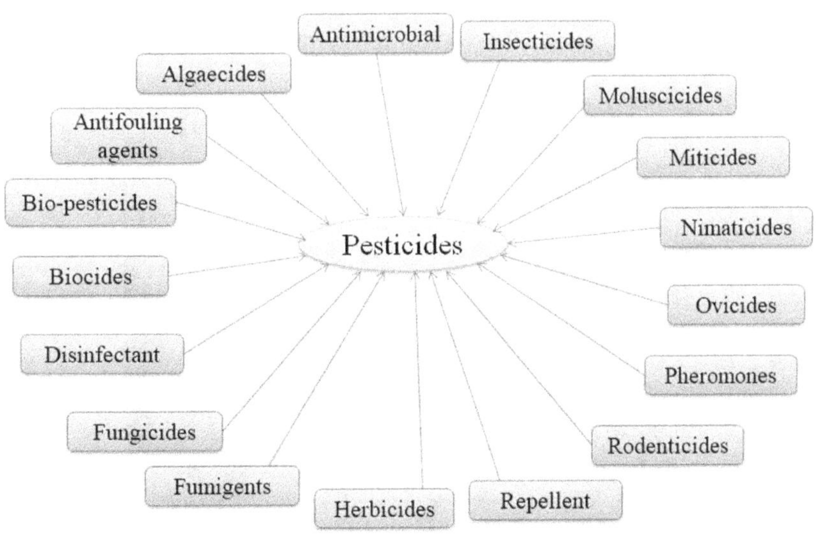

Europe in the fourteenth century, where millions of people died from a mysterious disease. Another historic example of pest affliction was in 19th century where potato crops destroyed in Ireland. This is called 'late blight' (Figure 2) essentially eliminated potatoes, the staple food crop in Ireland at that time. Million of Europeans starved to death during the Great Irish Famine of 1845-1847 (Fishel, 2013). Even today, 'late blight' is one of the major pathogens.

About 2000 years before the birth of Christ, Romans used sulphur fumes as an insecticide. Uses of sulphur as insecticides were also reported by Homer in 1000 B.C. in Odessey (Mckinlay, Dassyne, Djamgoz, Plant, & Voulvoulis, 2012). Today a number of products containing sulphur have been used as pesticide. Sulphur is mainly used for wide variety of crops in order to control the disease. Several other chemicals were developed with pesticidal activity over time. There are many plant derived insecticide identified such as 'hellebore' used to control body lice, nicotine used for controlling aphids and pyrethrins used to control a wide variety of insects. Lead arsenate was first used in 1892 as an orchard spray (Fishel, 2013).

In 1850s a veinyard grower from Bordeaux Region of France faced a problem that people pilfering grapes from his vines. He applied a mixture of copper and lime to his vineyard to make grapes unattractive in order to protect the grapes from thieves. As a result of this the grapes were protected from the thieves and it was also observed that the applied area had no any incidence of disease (Lucar, Campbell, & Lucas, 1992). This copper-lime mixture is now known as Bordeaux mixture, a commonly used fungicide, even today. This was the beginning of modern fungicide use.

Till the end of 1940s, pest controlling chemicals have been derived from plant sources and inorganic compounds. During World War II, synthetic chemical (DDT) played an important role in saving Allied soldiers from insect-transmitted diseases (Roberts, 2016). DDT has been used as insecticide to control all insect problems. Countless number of pesticides has been developed to control the pest in modern days. This is the beginning of new era in pest control.

In recent years, some drawbacks of pesticides have been recognised in which pest resistance to pesticides is one of the most disturbing. Worldwide many insects have become resistant to one or more pesticides. It was found that after few years of introduction of DDT, resistance was observed in the housefly in Sweden. Pesticide resistance also observed in hundreds of weeds biotype and several plant pathogens (Madsen & Streibig, 2003). Table 1 represent the chronological development of pesticides.

*Figure 2. Late blight disease of potato*

*Table 1. Chronological development of pesticides all over the world*

| Year | Development |
|---|---|
| 2000 BC | Sulphur was used as pesticides by pre-Roman civilizations |
| 1200 BC | Biblical armies and ash was used as non-selective herbicide |
| 100 BC | Hellebore used by Romans for control of rats, mice and insects |
| 300 | Predatory ant was used by Chinese in citrus for control of insects |
| 900 | Arsenic for the control of garden insects used by Chinese |
| 1649 | In South America rotenone was used to paralyze fish |
| 1690 | Extract of tobacco (nicotine) used as insecticide |
| 1787 | Soap used as insecticide |
| 1848 | Rotenone used by Asian as insecticide |
| 1850s | In France, mixture of lime and copper was used on grape plants |
| 1860s | Colorado potato beetle controlled by Paris green as insecticide |
| 1873 | DDT was first of all prepared in laboratory |
| 1882 | Bordeaux mixture was introduced by France |
| 1883 | Pressure sprayer was developed by John Bean |
| 1886 | In California hydrogen cyanide was used as fumigant |
| 1892 | In Massachusetts, for the control of gypsy moth lead arsenate was developed |
| 1894-1900 | Different pesticide spray devices such as steam, mechanical and horse driven were developed |
| 1907-1911 | Industrial production of lead arsenate begins |
| 1910 | Federal Insecticide Act passes |
| 1921 | Application of pesticides by airplane |
| 1927 | U.S. Food and Drug Administrator decide the tolerance limit for arsenic |
| 1932 | First use of methyl orange as fumigant in France |
| 1936 | Pentachlorophenol used as wood preservatives |
| 1942 | DDT made available for U.S. military |
| 1944 | Warfarin was used against rodent control |
| 1946 | Germany introduce organophosphate insecticide |
| 1950s-1960s | Several research have been done for development and commercialization of different pesticides |
| 1965 | Registration of atrazine as herbicide |
| 1970 | U.S. Environmental Protection Agency fromed |
| 1972 | EPA cancelled the use of DDT |
| 1973 | Photo-stable synthetic pyrethroid insecticide, permethrin developed |
| 1978 | First list of restricted use pesticides released by EPA |
| 1980s | Uses of chlorinated hydrocarbon pesticides was cancelled by EPA |
| 1990s-2000s | Merges of pesticide industry |

## Environmental Pathways of Pesticides

Pesticides are the chemical substances used to kill pest. 95% of herbicide and 98% of insecticides sprayed over entire agricultural fields reaches at different site rather than their target species (Miller, 2004). Pesticides spread in the environment by two ways: 1) wind and 2) run-off.

### Pesticide in Air

Pesticides can enter into environment by wind, suspended pesticide particles in air are carried away by wind from one place to other thereby contaminating the other places too (Cornell, 2007). Pesticides sprayed on the crops volatilizes and are blown by wind to some another place, where these pose potential hazards to human and animal lives (NPS, 2006). Low humidity and high temperature enhance the rate of evaporation of pesticides spray. Pesticides drift is lesser in case of ground spraying as compared to aerial spraying (Palmer, Bromley, & Brandenburg, 2007). Pesticides sprayed in the field, release volatile organic compounds, react with other chemicals and form pollutants which destroy the ozone layer (UC IPM, 2007).

### Pesticide in Water

Pesticides reach to the water sources through different routes: 1) sprayed pesticides leach or percolate through the soil and reach to the ground water, 2) also carried to water sources by soil erosion and 3) soluble pesticides dissolve in rain water and reach to water sources. Pesticides properties that contaminate the water are: water solubility, the distance from an application site to a body of water, weather, soil type, presence of a growing crop and the method used to apply the pesticide (Pedersen, 1997). According to the US Geological Survey 90% wells of US were found to be polluted (Gillion et al., 2006). Pesticide residues also found in rain and groundwater (Kellogg, Nehring, Grube, Goss, & Plotkin, 2000). Studies by the UK government showed that pesticide concentrations exceeded the allowable limits for drinking water in some samples of river water and groundwater (Bingham, 2007). Figure 3 represents the environmental pathways of pesticide.

### Pesticide in Soil

Pesticide reach to the soil through three possible ways: 1) with water, 2) attach to the soil particles and 3) metabolised by the microorganism present in soil. Chemicals used in pesticides are persistent soil contaminants, whose impact may endure for decades and adversely affect soil conservation (EPA, 2008). Biodiversity of the soil decreases by the use of pesticides (Johnston, 1986). By reducing or preventing the use of chemicals, higher soil quality can be retained ((Johnston, 1986), by adding more organic matter in the soil water retention capacity of soil can be increased (Kellogg, Nehring, Grube, Goss, & Plotkin, 2000). This helps in improving the farming yield in drought years. Organic farming increases the yield about 20-40% than conventional farming (Lotter, Seidel, & Liebhardt, 2003). Presence of organic matter in the soil increases the amount of pesticide, because organic matters bind to and break down the pesticides (Kellogg, Nehring, Grube, Goss, & Plotkin, 2000). Persistence of pesticides in soil depends

*Figure 3. Environmental pathway of pesticide.*

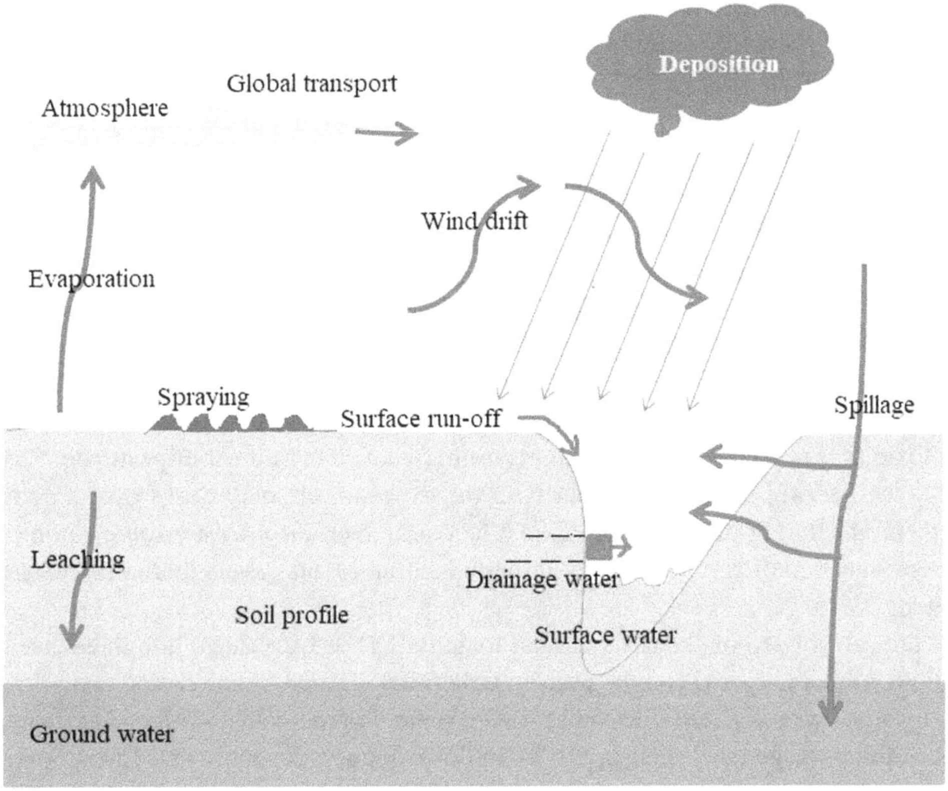


on the degradation and sorption. Degradation involves the breaking of organic matter and interaction among microorganism in the soil. Bioaccumulation of pesticide is affected by sorption, which depends on the presence of organic matter in soil.

## Pesticide in Organism

Pesticides enter into human body or other organism through food chain or by drinking pesticide contaminated water. Pesticides accumulate in the tissues of organisms and this accumulation of pesticide is known as bioaccumulation. Bioaccumulation results in higher concentrations of a chemical and increases with the food chain (Alexander, 1999). These pesticides are persistent in environment, do not break easily and retain their form after being ingested and stored in the body (Augustijn-Beckers, Hornsby, & Wauchope, 1994). In United States of America, pesticides with these characteristics have been banned, though such pesticides that were used in past are still present in the environment.

## Hazardous Effects/Toxicity of Pesticides

Basically, pesticides are used to kill the pests; all the pesticides are toxic, poisonous and have potential to harm the humans, animals, other organism and environment (USEPA, 2007). Therefore, it is important to know about the potential health effects and toxicity of pesticides for those who are in regular contact with pesticides. Pesticides have ability to cause severe health problem, illness or injury. This ability in pesticides is due to the presence of active chemical ingredient in pesticides which is responsible for killing of pests (Stoate et al., 2001; Berny, 2007; Power, 2010). Test animal is subjected to various dosages of active chemical ingredient and formulated sample for the determination of toxicity of pesticide. Acute and chronic are two types of toxicity.

Acute toxicity results from short duration or single exposure of animal or a person. There are four routes of exposure: skin (dermal), lungs (inhalation), mouth (oral) and eyes. It can be determined by examining dermal, inhalation and oral toxicity of the test specimen. Acute toxicity is measured in terms of quantity or concentration, usually expressed as LD (lethal dose to kill 50% of the animals in test population) or LC (lethal concentration to kill the 50% of animal in a test population). The lower the value of LD or LC of a pesticide, the greater its toxicity to humans and animals (Damalas & Ilias, 2011). The higher LD value of pesticide is least toxic to humans or animals.

Chronic toxicity is related with long-term exposure to active ingredient of pesticide. Chronic toxicity represents the harmful effects that occur with the exposure of small doses over a period of time. Birth defects, production of tumors, neurotoxic effect and blood disorder are some chronic toxicity of pesticide. Determination of chronic toxicity through laboratory analysis is more difficult as compared to acute toxicity.

Pesticide are categorised on the basis of their toxicity (LD or LC values) into three categories. Category I pesticides are highly toxic on the basis of either oral, dermal, or inhalation toxicity and must be represented by signal words DANGER and POISON printed in red with a skull and crossbones symbol on the front of the package label (Fishel, 2015). In this category, the acute oral LD is 50 mg/kg (Penn State Extension, 2017).

Category II pesticides are considered as moderately toxic and must be represented with single word WARNING and AVISO on the product label. In this category, the acute oral LD ranges from 50 to 500 mg/kg.

Slightly toxic or relatively nontoxic pesticides are categorised as III and IV, represented by single word CAUTION on the pesticide label. Acute oral LD values in this group are greater than 500 mg/kg. Table 2 represents limits for oral, dermal, inhalation and eye LD or LC value for these four categories of pesticides. Acute oral and dermal $LD_{50}$ values for some pesticide are listed in Table 3 with their use category either in general use or restricted.

## Effect on Human Health

About 20,000 people annually visit emergency rooms as reported by US Consumer Product Safety Commission on exposure to pesticides. Pesticide has adverse effect on human health and well being on long term exposure and chronic condition. Digestion, inhalation and dermal exposure are the three main ways through which humans are exposed to pesticide (Pesticides in the Environment: Exposure Pathways, 2003).

Pesticides enter into the digestive track by swallowing and are absorbed directly or transformed and absorbed by blood. Pesticides reach the digestive track by eating contaminated food such as vegetables and fruits with pesticide residuals. Contaminated drinking water may also expose humans to different amounts of pesticides (Van der Werf, 1996; Wilson & Tisdell, 2001; Pimentel, 2005; Maroni, Fanetti, & Metruccio, 2006; Soares & Porto, 2009).

*Table 2. LD and LC values of four categories of pesticides for oral, dermal, inhalation and eye (Penn State Extension, 2017).*

| Toxicity category | Route of exposure | | | | |
|---|---|---|---|---|---|
| | Oral $LD_{50}$ (mg/kg) | Dermal $LD_{50}$ (mg/kg) | Inhalation $LD_{50}$ (mg/L) | Skin Effect | Eye Effect |
| Cat. I | Up to 50 | Up to 200 | Up to 0.2 | Corrosive | Corrosive for corneal opacity and irreversible |
| Cat. II | 50-500 | 200-2000 | 0.2-2.0 | Severe irritation for 72 hours | Corneal opacity reversible within 7 days |
| Cat. III | 500-5000 | 2000-20,000 | 2-20 | Moderate irritation for 72 hours | No corneal opacity |
| Cat. IV | >5000 | >20,000 | >20 | Mild or slight irritation | No irritation |

*Table 3. Oral $LD_{50}$ and Dermal $LD_{50}$ and use category of some pesticides (Penn State Extension, 2017).*

| Pesticide (Trade Name) | Use category | Oral-$LD_{50}$ (mg/kg) | Dermal-$LD_{50}$ (mg/kg) |
|---|---|---|---|
| **Sulphur** | General | >5000 | >5000 |
| **Phaser** | General | 160 | 359 |
| **Di syston** | General | 2-12 | 3.6-15.9 |
| **Terra clean** | General | 330 | 1410 |
| **Mycoshield** | General | >5000 | >2000 |
| **Actigard** | General | >5000 | >2000 |
| **Vitavax** | General | 3820 | >4000 |
| **Telone** | Restricted | 127 | 423 |
| **Elevate** | General | >5000 | >5000 |
| **Degree** | Restricted | 2148 | 4166 |
| **Blazer** | General | 2025 | >2000 |
| **AAtrex** | Restricted | 1869 | >3100 |
| **Python** | General | >5000 | >2000 |
| **Resource** | General | 3200 | >2000 |
| **U 46 M-Fluid** | General | 900-1160 | >4000 |
| **Cobra** | General | >5000 | >2000 |
| **Amber** | General | >5050 | >2000 |
| **Princep** | General | >5000 | >3100 |
| **Butyrac** | General | >2000 | >10,000 |
| **Poast** | General | 2676-3125 | >5000 |
| **Agri-Mek** | Restricted | 300 | >1800 |
| **Neemix** | General | >5000 | >2000 |
| **Brigade** | Restricted | 262 | >2000 |
| **Ammo** | Restricted | 250 | 2000 |
| **B-Nine** | General | >5000 | >5000 |
| **Sumagic** | General | 2020 | >2000 |
| **Thin-it** | General | 1690 | 2000 |
| **Propel** | General | 3543 | >5000 |
| **Atrimmec** | General | 31,000 | >1000 |
| **PrimoMaxx** | General | >5050 | 2020 |

Through breathing pesticides enter into the lungs from where they are absorbed by the blood stream. If proper care is not taken, large amount of pesticides can be inhaled directly from where pesticides are applied. Pesticides also pollute the outdoor air which is inhaled by applicator, other people and by various other organisms. Major route for dermal exposure of pesticides are throat, nose and eye where mucous membrane is present. Serious exposure route for pesticides is direct skin contact.

Exposure of pesticides causes acute and delayed health effects in humans (USEPA, 2007). Adverse health effect of pesticide on human ranges from simple skin and eye irritation to severe problems such as cancer, effect on the nervous system, hormonal imbalance, reproductive problems and foetal death (USEPA, 2017; Sanborn et al., 2007). The American Academy of Pediatrics recommends limiting exposure of children to pesticides and using safer alternatives (AAP, 2012). According to world health organisation and UN Environment Programme, 3 million workers associated with agricultural field in developing countries are exposed to pesticides poisoning every year (Miller, 2004). About 99% of pesticide related deaths occur in developing countries. Studies showed that about 25 million workers in developing countries suffer mild pesticides poisoning every year (Jeyaratnam, 1990).

## Effect on Environment

Around 98% of insecticide and 95% of herbicide sprayed reaches to their non-target destination, such as air, water and soil (Miller, 2004). Pesticides are one of the major causes of water pollution and persistent pesticides cause soil contamination. Suspended pesticides present in air are blown away by wind from one place to other. Today worst situation is created as pests develop resistance to pesticides, it means that more stronger pesticides have to been used or amount of dose should be increased to kill the pest, as a result of this more ambient pollution problem raises. Pesticide pollution also includes reduced biodiversity (Wells, 2004), destruction of bird habitats (Palmer, Bromley, & Brandenburg, 2007) and also threatens the endangered species (Miller, 2004). Pesticides can harm the environment by two mechanisms: 1) Bioconcentration and 2) Biomagnification.

Some chlorinated hydrocarbons containing pesticides are lipophilic such as DDT, insoluble, not excreted and get deposited into fatty tissue such as human fatty tissue or edible fish tissue. This type of concentration of pesticides in fatty tissue, not excreted and remains for indefinite period of time is known as bioconcentration.

In biomagnification, concentration of insoluble pesticides increases at each level of food chain. In marine animals, pesticide concentrations are higher in carnivorous fishes and concentration increases in the fish-eating birds and higher concentration is found in mammals which are at the top of the ecological pyramid (Castro, Peter, & Huber, 2010). In warmer region pesticides are evaporated, can be transported by wind over thousands of kilometres to the region of lower temperature where pesticides condense and come back to the ground with rain or snow (Quinn & Amie, 2012). This process of transportation of pesticides from warmer to colder region is known as "Global Distillation".

A desirable quality in novel pesticides is their bio-degradability and reduced bio-magnification; which can be attributed to their chemical structure (Sims & Cupples, 1999). For example, the halogen slows down degradation of pesticide in an aerobic environment (Sims & Sommers, 1986). The adsorption of pesticide to soil also reduces its movement and at the same time reduces its bioavailability to microbial degraders (Wolt, Smith, Sims, & Duebelbeis, 1996). Table 4 shows the harmful effect of pesticides on environment, human and other organism.

## PESTICIDE MONITORING IN ENVIRONMENT

Different techniques based on the medium to be tested have been used for monitoring of pesticides. A very poor monitoring data are available in all over the world especially in developing countries. Developing countries have many difficulties in carrying analysis due to:

- Problems of inadequate facilities,
- Impure reagents and
- Financial constraints.

With immunoassay procedures presence and absence of specific pesticides may reduce costs, increase reliability and are available for triazines, acid amides, carbamates, 2,4-D/phenoxy acid, paraquot and aldrin (Rickert, 1993).

### Monitoring in Water

Water sources such as wells, lakes, aquifers, streams etc have been monitored regularly to determine the presence of pesticides and their degradation products. In late 1970's and early 1980's the Environmental

*Table 4. Harmful effect of pesticides on environment, human and other organism.*

| Pesticides | Effects |
|---|---|
| **DDT/DDE (Organochlorine)** | Carcinogen, endocrine disruptor, egg shell thinning in raptorial bird, thyroid disruption in rodents, birds, amphibians and fish, inhibition of acetylcholine esterase activity (Rattner, 2009; Turusov, Rakitsky, &Tomatis, 2002). |
| **Diclofol, Dieldrin, Toxaphene** | Mortality in reptiles, juvenile population decreases (Rain & Guillette, 1998). |
| **Parathion** | Fungal infection susceptibility (Galloway & Depledge, 2001). |
| **Chlordane, Carbamates** | Vertebrate immune system affected (Galloway & Depledge, 2001). |
| **Triazine** | Earthworms infected with monocystid gregarines (Kohler & Triebskorn, 2013). |
| **Organophosphate** | Oxidative damage, immunotoxicity, animal infection, thyroid disruption in rodents, birds, fish, amphibians, reduce reproduction, impaired development, impaired metabolic function (Rattner, 2009; Fleischli, Franson, Thomas, Finley, & Riley, 2004; Galloway & Handy, 2003). |
| **Anticholinesterase** | Bird poisoning (Fleischli, Franson, Thomas, Finley, & Riley, 2004). |
| **Phenoxy herbicide 2,4-D, Atrazine** | Vertebrate immune system affected (Galloway & Depledge, 2001). |
| **Carbamate** | Impaired metabolic function, reduce reproduction, impaired development, thyroid disruption (Story & Cox, 2003). |
| **Pyrethroid, Triazine, Thoicarbamate** | Thyroid disruption in fish, rodents, amphibians, birds (Rattner, 2009). |
| **Triazole** | Impaired growth, reproduction rate decline, impaired metabolic function (Rattner, 2009). |
| **Nicotinoid** | Respiratory, neurological, cardiovascular, immunological toxicity in humans and rodents (Lin, Lin, Liao, Guo, & Chen, 2013). |
| **Herbicides** | Food availability reduces, adverse effect on soil invertebrates and butterflies (Freemark, 1995). |

Protection Agency (EPA) recommended the regular water monitoring due to the presence of pesticides and their degradation products in groundwater and surface water bodies (Espana & Moran, 2010). Since then, presence of pesticides in water bodies has been monitored regularly.

Determination of lipophilic pesticide residues in fish, exposure of fish to lipophobic pesticides by the analysis of liver or bile is restricted. Hence, determination of pesticide residues used in the world is very difficult. Chlorinated pesticides like DDT have harmful impact on environment; hence these compounds are banned in many parts of the world for agricultural purpose (Rickert, 1993). Some pesticides used in previous year are hydrophobic carcinogenic containing PAHs and PCBs are poorly monitored in water sample (Richard & Sittig's, 2015). In routine analysis, the detection limit of some pesticides is found to be very high for the determination of presence/absence for the protection of human health. US Geological Survey Pesticide Monitoring Network in 1984 set the detection limit of 0.05mg/L for DDT, 0.001mg/L for aquatic life and 0.0002mg/L for human health. These detection limits are much lower than routine detection limit.

### Monitoring in Soil

As pesticides are sprayed in agricultural field in order to kill the pests and properties of soil has important role in distributing pesticides throughout the landscape. Pesticides move with water and reach the soil. Soil erosion by wind causes movement of pesticides from one place to other and is responsible for airborne pesticides. Due to these reasons high amount of pesticides are present in soil hence, its regular monitoring is very important.

### Monitoring Air

Movement of pesticides with wind from one place to another and aerial spraying of pesticides on crops are the main sources of pesticide entry into the atmosphere (GWA, 2010). By analysing air, studies show that the air contains pesticides residue which are used in past as well as those used today. As air is contaminated with several other contaminants other than the pesticides, hence, it is not possible to monitor air regularly. Rain water causes wet deposition of pesticides, so monitoring rain water shows the level of pesticides in air as well as amount of pesticides deposited on water bodies and land (Szekacs, Mortl, & Darvas, 2015).

## MEASURES TO REDUCE USE OF PESTICIDES

Pesticides emerged as a miracle in the field of agriculture in preventing and controlling of pest that harms the crops. But soon its harmful effects on environment and human health appeared. Insects, developed the resistance against the pesticides with their regular uses, also natural predators of target pests are killed with the use of pesticides (Cambridge University Press, 2010). Regular uses of pesticides have several harmful effects on environment, water quality, soil properties and as well as human and other organism. Hence, use of pesticides have to been reduced, several strategies has been employed to reduce the use of pesticides.

Integrated pest management (IPM) plan have been widely used in this field. IPM gives better understanding of pests' life cycle and their environmental impact. The main goal of IPM is to reduce or eliminate the problems before they arise. IPM plan gives idea about how to reduce the hazardous effect of pesticides on environment and humans in more economical way. Biological control and habitat modification have been used under IPM plan to control the problems arising from the pesticides. The problems which can be easily solved then only pesticides are used. Several measures have been taken to reduce the use of pesticides under IPM plan such as:

- Pesticides levels in the environment should be minimized
- Minimizes the hazardous effect on environment and organism
- Uses of natural pesticides should be promoted
- Provide better understanding of effective pest control methods
- Establish programs that will encourage voluntary participation

For successful implementation of IPM plan following step should be taken:

- Determination of actual threshold
- Identification of pest and their proper monitoring
- Prevention of pests
- Evaluation of problem

## Determination of Actual Threshold

First step involves the determination of threshold value of pesticides at which pests become actual threat to the environment (Spiegel & Maystre, 1998). On the basis of this threshold value following action should be taken to control the pest.

## Identification of Pest and Their Proper Monitoring

Identification of pests is important because all insects or organisms present may not be harmful, many of them may be beneficial in order to maintain the balance in insect community. Hence, continuous monitoring helps in the determination of target pest for which suitable pesticides should be applied without harming the others.

## Prevention of Pest

Favourable environment for pest growth should be modified to control the pest growth. This can be achieved by carrying crop rotation, planting pest resistant crops and pest free rootstocks. Plantation of only single species plant provide favourable environment for pest breeding ( Stenberg, Heil, Ahman, & Bjorkman, 2015). In case of favourable environmental condition such as sunlight, moisture content enhances the plant growth and naturally defends the pest breeding.

## Evaluation of Problem

After taking care of all the steps if there is still a problem, then proper action should be taken. Natural control method is the best option and control method which is least harmful should be employed first.

For this, natural predators should be replenished. Uses of pheromones is another method; pheromones disturb the natural mating cycle of pests. Further evaluation shows the presence of pests then pesticides have been used. Bio-pesticides are more suitable than conventional pesticides because bio-pesticides are less toxic than conventional pesticides (Hoddle & Parra, 2013). Bio-pesticides are easily degradable, can be applied in smaller quantity and having lower toxic exposure limit, decreases the pollution and environmental degradation.

## Degradation of Pesticides by Natural Means

Pesticide degradation occurs also by biological process other than chemical and photochemical reactions. There are two ways of biological process by which pesticides degrade: (i) microbiological processes occurring in soil and water bodies and (ii) degradation of pesticides by metabolic activity of organism as pesticides enter through the food chain (Javaid, Ashiq, & Tahir, 2016). By means of these two biological processes toxicity of pesticides has been reduced.

### Degradation of Pesticides in Soil

By the process of mineralisation pesticides degrade easily and quickly in soil. In this process pesticides convert into simpler compound such as $H_2O$, $CO_2$, $NH_3$ etc. Mineralisation process occurs through the hydrolysis, photolysis, microbiological process and metabolism. After mineralisation pesticides are the sources of carbon and other nutrients for soil. Some pesticides are persistent and degrade slowly (Stephenson & Solomon, 1993).

### Process of Metabolism

Pesticides enter into the human body through the food chain. Metabolic activity of humans is an important process by which humans can protect themselves from the toxic effect of pesticides. By means of metabolism pesticides are converted into less toxic form and either excreted or stored (Matsumura & Krishna Murti, 1982). Enzymes play important role in the metabolic process, secretion oxygenases enzymes in liver shows the presence of foreign substances.

## Pesticides Management in European Countries

Survey by Netherlands National Institute of Public Health and Environmental Protection (Van Doesburg, Noordijk, &Van den Anker, 1993) showed that presence of pesticides in groundwater is of concern in all European states as pesticides level in groundwater exceeded its EC standard limit (0.5 mg/L) (Ongley, 1996). Regular analysis and data calculation shows that EC standard for some of the pesticides exceeded in almost 65% of all agricultural land. To reduce the use of pesticides and their impact on human health and environment European countries adopted a number of measures which includes (FAO, 1991):

- Reduction in use of pesticides.
- Bans on certain active ingredients.
- Revised pesticide registration criteria.
- Training and licensing of individuals that apply pesticides.

- Reduction of dose and improved scheduling of pesticide application to more effectively meet crop needs and to reduce preventative spraying.
- Testing and approval of spraying apparatus.
- Limitations on aerial spraying.
- Environmental tax on pesticides.
- Promote the use of mechanical and biological alternatives to pesticides.

### Pesticides Management by Danish Government

Danish Government implements the plan for sustainable agriculture in order to prevent the use of pesticides in 1986. Reduced uses of pesticides have two purposes (i) protect the environment and (ii) save the human health. It was observed that, use of pesticides has been reduced to about 30% by 1993 (DEPA, 1993). Danish Government initiates the following points to reduce the use of pesticides are as follows:

- Danish Government provides funds to promote the conversion of traditional agricultural to organic agriculture where pesticides are not used.
- Prevention of use of pesticides within 10 m of drinking water sources and other water sources.
- Prevention of use of pesticides around the private garden and other area where cultivation is done without the use of pesticides.

### Pesticides Management by Swedes

The Swedes Government adopted certain measures which are given below and by employing these measures Swedes got success in reducing the use of pesticides (Ongley, 1996).

- Setting of targets with achievable goals and using multiple measures of reduction.
- Lead role played by the Environment Ministry and Chemicals Inspectorate.
- Active support of farmer's organizations which realize the economic and environmental advantages of reduced pesticide usage.
- A strong research and development base that provides credible support for new pesticide initiatives.
- Certification of new machinery and routine testing of farm sprayers at government-regulated test centres.
- Re-evaluation and re-registration of pesticides which has resulted in 338 products being removed from the market.

## EFFECTIVE ALTERNATIVE FOR PESTICIDES

Due to increasing risk of human health and environment reduction in the use of pesticides has been very important. For this purpose various alternatives have been used now a days which are given below:

- Use of biological pest control (Sanborn, 2006).
- Different method of cultivation such as polyculture, crop rotation, plantion in that area where pest does not live etc (Sanborn, 2006).

- Genetic engineering (Sanborn, 2006).
- Interfere the cultivation with insect breeding (Sanborn, 2006; Biology pages 2007; Skylab, 2007).
- Application of compost (McSorley & Gallaher, 1996).
- Spraying of hot water over crops has been used by US farmers to reduce the use of pesticides (Sanborn, 2006).
- Use of predators or parasite of pest (Sanborn, 2006).
- Use of biological pesticides based on entomopathogenic fungi, bacteria and viruses which cause disease in the pest (Sanborn, 2006).
- Push pull strategy has been employed. First of all "push-pull" was started in 1987 as an approach for integrated pest management (IPM). This strategy includes the use of mixture of behaviour-modifying stimuli that manipulate the distribution and abundance of insects. "Push" means the insects has been repelled away from source that is being protected. "Pull" means that certain stimuli (semi-chemical stimuli, pheromones, food additives, visual stimuli, genetically altered plants, etc.) have been used which attracts pests to trap crops where the pest has to be killed (Cook, Khan, & Pickett, 2007).

## Use of Pesticides in India

In India pesticides were first used for the malaria (DDT) and locust control (BHC) (Gupta, 2004). First pesticide (BHC) producing industry was developed in 1952 at Rishra, Kolkata. After that Hindustan Insecticides Ltd. set two production units of DDT. Small plant of pesticide formulation was set up in Bhopal by Union Carbide in 1969. Today, about 125 large and medium scale manufacturing units and more than 500 pesticide formulation units are present in India. Dusting powder contribute about 85%, water soluble pesticides contribute 12% and emulsification concentrate contribute about 2% of the total pesticides available in the market (Abhilash & Nandita, 2009). Total consumption of pesticides is about 2 million tons per year worldwide. 45% of total consumption has been consumed by Europe, 24% by USA and 25% of total pesticides has been consumed by the rest of the world (Abhilash & Nandita, 2009). In India, about 40% of total pesticides belong to organo-chlorine class of pesticide (Gupta, 2004). Lindane, DDT and malathione are the most commonly used pesticides in India and these contribute about 70% of total pesticide consumption. In India, pesticide has been used in different sectors such as agriculture, public health, personal care, building material and domestic area. In agriculture pesticide has been used for the control of pests, rodents and weeds etc. In public health sector pesticide has been used for the control of malaria, dengue fever, locust and filariasis. As scabicides and pediculicdes pesticide have been used for skin care; pesticides are also incorporated into paints, glues, plastic protection used as building materials. For domestic use, pesticides have been used as garden, household spray and also used for the control of ecto-parasites (Gupta, 2004). Andhra Pradesh and Punjab utilise the major portion of the pesticides (GOI, 2008-2012) all over India.

# CONCLUSION

The development of pesticides has significantly contributed towards the increased production of food crops. The pesticides have a positive effect over the control of pest menace. However, the uncontrolled use and over dependence on pesticides has lead to several health hazardous issues viz., respiratory, neurological, cardio-vascular and reproductive disorders. Keeping in view the hazardous effects of pesticides, there is immediate need for regulating the use of pesticides. Strict laws must be in place and also the implementation of these laws is a must to ensure the protection of human health and well being. The development of novel eco-friendly pesticides which overcome the problem of bio-concentration and bio-magnification while retaining the anti pest characteristics is need of the hour.

## REFERENCES

Abhilash, P. C., & Nandita, S. (2009). Pesticides use and application: An Indian scenario. *Journal of Hazardous Materials*, *165*(1-3), 1–12. doi:10.1016/j.jhazmat.2008.10.061 PMID:19081675

Ahmad, M. M., Getso, B. U., Ahmad, U. A., & Abdullahi, I. I. (2014). Impact of climate change on the distribution of tropical parasitic and other infectious diseases. *Toxicology and Food Technology*, *8*, 19–26.

Alexander, (1999). Bioaccumulation, bioconcentration, biomagnification. *Environmental Geology*. doi:10.1007/1-4020-4494-1_31

American Academy of Pediatrics. (2012). Policy statement: Pesticide exposure in children. pediatrics. *The Official Journal of the American Academy of Pediatrics*, *131*, 1757–1763.

Ashbolt, N. J. (2004). Microbial contamination of drinking water and disease outcomes in developing regions. *Toxicology*, *198*(1-3), 229–238. doi:10.1016/j.tox.2004.01.030 PMID:15138046

Augustijn-Beckers, P. W. M., Hornsby, A. G., & Wauchope, R. D. (1994). SCS/ARS/CES Pesticide Properties Database for Environmental Decisionmaking II. *Additional Properties Reviews of Environmental Contamination and Toxicology*, 137.

Berny, P. (2007). Pesticides and the intoxication of wild animals. *Journal of Veterinary Pharmacology and Therapeutics*, *30*(2), 93–100. doi:10.1111/j.1365-2885.2007.00836.x PMID:17348893

Bingham, S. (2007). *Pesticides in rivers and groundwater*. Environment Agency.

Castro, P., & Huber, M. E. (2010). *Marine Biology* (8th ed.). New York: McGraw-Hill Companies Inc.

Cook, S. M., Khan, Z. R., & Pickett, J. A. (2007). The use of push-pull strategies in integrated pest management. *Annual Review of Entomology*, *52*(1), 375–400. doi:10.1146/annurev.ento.52.110405.091407 PMID:16968206

Cornell University. (2007). *Pesticides in the environment, Pesticide fact sheets and tutorial*. Pesticide Safety Education Program.

Damalas, C. A., & Ilias, G. (2011). Pesticide Exposure, Safety Issues, and Risk Assessment Indicators. *International Journal of Environmental Research and Public Health*, *8*(5), 1402–1419. doi:10.3390/ijerph8051402 PMID:21655127

Danish Environmental Protection Agency. (1993). *Oversigt over revurderingen 1988, 1989, 1990, 1991 of 1992* [Summary of the reassessment of already approved pesticides in 1988, 1989, 1990, 1991 and 1992 based on stricter standards]. Author.

El-Dessouky, H., & Ettouney, H. (2002). Teaching desalination-a multidiscipline engineering science. *Heat Transfer Engineering*, *23*(5), 1–3. doi:10.1080/01457630290090590

Espana, P. S., & Moran, B. S. (2010). Pesticide degradation in water. In H. S. Rathore & L. M. L. Nollet (Eds.), *Pesticides: Evaluation of environmental pollution*. CRC Press.

FAO, Network on Erosion-Induced Loss in Soil Productivity. (1991). *Report of a Workshop, Bogor, Indonesia*. Land and Water Development Division, FAO.

Fawell, J., & Nieuwenhuijsen, M. J. (2003). Contaminants in drinking water. *British Medical Bulletin*, *68*(1), 199–208. doi:10.1093/bmb/ldg027 PMID:14757718

Fishel, F. M. (2013). Pest Management and Pesticides: A Historical Perspective. University of Florida/IFAS Extension, PI 219.

Fishel, F. M. (2015). *Pesticides labelling*. Signal Words.

Fleischli, M. A., Franson, J. C., Thomas, N. J., Finley, D. L., & Riley, W. (2004). Avian Mortality Events in the United States Caused by Anticholinesterase Pesticides: A Retrospective Summary of National Wildlife Health Center Records from 1980 to 2000. *Archives of Environmental Contamination and Toxicology*, 46. PMID:15253053

Freemark, K. (1995). Impacts of agricultural herbicide use on terrestrial wildlife in temperate landscapes: A review with special reference to North America. *Agriculture, Ecosystems & Environment*, *52*(2-3), 67–91. doi:10.1016/0167-8809(94)00534-L

Galloway, T., & Handy, R. (2003). Immunotoxicity of Organophosphorous Pesticides. *Ecotoxicology (London, England)*, *12*(1/4), 345–363. doi:10.1023/A:1022579416322 PMID:12739880

Galloway, T. S., & Depledge, M. H. (2001). Immunotoxicity in Invertebrates: Measurement and Ecotoxicological Relevance. *Ecotoxicology (London, England)*, *10*(1), 5–23. doi:10.1023/A:1008939520263 PMID:11227817

Gillion, R. J., Barbash, J. E., Crawford, G. G., Hamilton, P. A., Martin, J. D., Nakagaki, N., … Wolock, D. M. (2006). *Overview of Findings and Implications. Pesticides in the Nation's Streams and Ground Water (1992–2001)*. Academic Press.

Government of India. (2008). *Eleventh Five-Year Plan: 2008–2012*. New Delhi: Planning Commission of India.

Government of Western Australia, Department of Water. (2010). *Aerial spraying of crops with pesticides*. Water quality protection note 104. Author.

Gupta, P. K. (2004). Pesticide exposure-Indian scene. *Toxicology*, *198*(1-3), 83–90. doi:10.1016/j.tox.2004.01.021 PMID:15138033

Hajek, E. A. (2010). *Natural Enemies: An Introduction to Biological control*. Cambridge University Press.

Hoddle, M. S., & Parra, J. R. P. (2013). Potential Lepidoteran Pests Associated with Avocado Fruit in Parts of Home Range of Persea Americana. In J. E. Pena (Ed.), Potential Invasive parts of Agricultural Crop. Academic Press. doi:10.1079/9781845938291.0086

Javaid, M. K., Ashiq, M., & Tahir, M. (2016). Potential of Biological Agents in Decontamination of Agricultural Soil. *Scientifica*, *2016*, 1–9. doi:10.1155/2016/1598325 PMID:27293964

Jeyaratnam, J. (1990). Acute pesticide poisoning: A major global health problem. *World Health Statistics Quarterly*, *43*, 139–144. PMID:2238694

Johnston, A. E. (1986). Soil organic-matter, effects on soils and crops. *Soil Use and Management*, *2*(3), 97–105. doi:10.1111/j.1475-2743.1986.tb00690.x

Kellogg, R. L., Nehring, R., Grube, A., Goss, D. W., & Plotkin, S. (2000). *Environmental indicators of pesticide leaching and runoff from farm fields Archived June 18, 2002, at the Way back Machine*. United States Department of Agriculture Natural Resources Conservation Service.

Kohler, H. R., & Triebskorn, R. (2013). Wildlife Ecotoxicology of Pesticides: Can We Track Effects to the Population Level and Beyond? *Science*, *341*(6147), 759–765. doi:10.1126cience.1237591 PMID:23950533

Lampman, C. (1995). The relationship between experience and attitudes concerning epilepsy. *Journal of Applied Social Psychology*, *25*(7), 619–631. doi:10.1111/j.1559-1816.1995.tb01602.x

Leonard, P., Hearty, S., Brennan, J., Dunne, L., Quinn, J., Chakraborty, T., & Kennedy, R. O. (2003). Advances in biosensors for detection of pathogens in food and water. *Enzyme and Microbial Technology*, *32*(1), 3–13. doi:10.1016/S0141-0229(02)00232-6

Lin, P. C., Lin, H. J., Liao, Y. Y., Guo, H. R., & Chen, K. T. (2013). Acute poisoning with neonicotinoid insecticides: A case report and literature review. *Basic & Clinical Pharmacology & Toxicology*, *112*(4), 282–286. doi:10.1111/bcpt.12027 PMID:23078648

Lotter, D. W., Seidel, R., & Liebhardt, W. (2003). The performance of organic and conventional cropping systems in an extreme climate year. *American Journal of Alternative Agriculture*, *18*(03), 146–154. doi:10.1079/AJAA200345

Lucar, G. B., Campbell, C. L., & Lucas, L. T. (1992). *Introduction to plant disease identification and management* (2nd ed.). Chapman & Hall.

Madsen, K. H., & Streibig, I. C. (2003). Benefits and risks of the use of herbicide resistant crops. In R. Labrada (Ed.), Weed management for developing countries. Academic Press.

Maroni, M., Fanetti, A. C., & Metruccio, F. (2006). Risk assessment and management of occupational exposure to pesticides in agriculture. *La Medicina del Lavoro*, *9*, 430–437. PMID:17017381

Matsumura, F., & Krishna Murti, C.R. (1982). *Biodegradation of Pesticides*. Academic Press.

Mckinlay, R., Dassyne, J., Djamgoz, M. B. A., Plant, J. A., & Voulvoulis, N. (2012). Agricultural pesticides and chemical fertilizers In Pollutants, Human health and the environment: A risk based approach (pp. 147-179). Wiley Blackwell.

McSorley, R., & Gallaher, R. N. (1996). Effect of Yard Waste Compost on Nematode Densities and Maize Yield. *Journal of Nematology*, *2*, 655–660. PMID:19277191

Miller, G. T. (2004). Sustaining the Earth (6th ed.). Thompson Learning, Inc.

Mozaz, S. R., De Alda, M. J. L., & Barcelo, D. (2004). Monitoring of estrogens, pesticides and bisphenol A in natural waters and drinking water treatment plants by solid phase extraction-liquid chromatography-mass spectrometry. *Journal of Chromatography. A*, *1045*(1-2), 85–92. doi:10.1016/j.chroma.2004.06.040 PMID:15378882

National Park Service. (2006). US Department of the Interior, Sequoia & Kings Canyon National Park: Air quality-Airborne synthetic chemicals. Author.

Ongley, E. D. (1997). Pesticides, as water contaminants. In *Combating Agricultural Contamination of Water Resources*. Academic Press.

Ongley, E. D. (1996). *Control of water pollution from agriculture*. Academic Press.

Palmer, W. E., Bromley, P. T., & Brandenburg, R. L. (2007). *Wildlife & pesticides -Peanuts*. North Carolina Cooperative Extension Service.

Pedersen, T. L. (1997). *Pesticide residues in drinking water*. Academic Press.

Penn State Extension. (2017). *Toxicity of Pesticides*. Author.

*Pesticides in the Environment: Exposure Pathways*. (2003). Academic Press.

Pimentel, D. (2005). Environmental and economic costs of the application of pesticides primarily in the United States. *Environment, Development and Sustainability*, *7*(2), 229–252. doi:10.100710668-005-7314-2

Power, A. G. (2010). Ecosystem services and agriculture: Tradeoffs and synergies. *Philosophical Transaction of Royal Society B*, *365*(1554), 2959–2971. doi:10.1098/rstb.2010.0143 PMID:20713396

Quinn, L. & Amie. (2012). The impacts of agriculture and temperature on the physiological stress response in fish. Uleth. *University of Lethbridge*.

Rain, D. A., & Guillette, L. J. Jr. (1998). Reptiles as models of contaminant-induced endocrine disruption. *Animal Reproduction Science*, *53*(1-4), 77–86. doi:10.1016/S0378-4320(98)00128-6 PMID:9835368

Rattner, B. A. (2009). History of wildlife toxicology. *Ecotoxicology (London, England)*, *18*(7), 773–783. doi:10.100710646-009-0354-x PMID:19533341

Richard, P., & Sittig's, P. (2015). *Handbook of Pesticides and Agricultural Chemicals* (2nd ed.). William Andrew Publishing.

Rickert, D. (1993). Water quality assessment to determine the nature and extent of water pollution by agriculture and related activities. In *Prevention of Water Pollution by Agriculture and Related Activities. Proceedings of the FAO Expert Consultation*. Santiago, Chile: FAO.

Roberts, M. T. (2016). Food law in the United States. Cambridge University Press.

Sanborn, M., Kerr, K. J., Sanin, L. H., Cole, D. C., Bassil, K. L., & Vakil, C. (2007). Non-cancer health effects of pesticides: Systematic review and implications for family doctors. *Canadian Family Physician Medecin de Famille Canadien*, *53*, 1712–1720. PMID:17934035

Sims, G. K., & Cupples, A. M. (1999). Factors controlling degradation of pesticides in soil. *Pesticide Science*, *55*(5), 598–601. doi:10.1002/(SICI)1096-9063(199905)55:5<598::AID-PS9623.0.CO;2-N

Sims, G. K., & Sommers, L. E. (1986). Biodegradation of pyridine derivatives in soil suspensions. *Environmental Toxicology and Chemistry*, *5*(6), 503–509. doi:10.1002/etc.5620050601

Skylab. (2007). *Classroom in Space: Part III - Science Demonstrations, Chapter 17: Life Sciences*. Author.

Soares, W. L., & Porto, M. F. D. (2009). Estimating the social cost of pesticide use: An assessment from acute poisoning in Brazil. *Ecological Economics*, *68*(10), 2721–2728. doi:10.1016/j.ecolecon.2009.05.008

*Sources of common contaminants and their health effects*. (2008). Emergency Response Program, EPA.

Spiegel, J., & Maystre, L. Y. (1998). International Labour Office. Environmental pollution control and prevention. In *Encyclopedia of Occupational Health and Safety* (4th ed.). Geneva: International Labour Office.

Stenberg, A. J., Heil, M., Ahman, I., & Bjorkman, C. (2015). Optimizing Crops for Biocontrol of Pests and Disease. *Trends in Plant Science*, *20*, 698–712.

Stephenson, G. A., & Solomon, K. R. (1993). *Pesticides and the Environment. Department of Environmental Biology*. Guelph, Ontario, Canada: University of Guelph.

Stoate, C., Boatman, N. D., Borralho, R. J., Carvalho, C. R., de Snoo, G. R., & Eden, P. (2001). Ecological impacts of arable intensification in Europe. *Journal of Environmental Management*, *63*(4), 337–365. doi:10.1006/jema.2001.0473 PMID:11826719

Story, P., & Cox, M. (2001). Review of the effects of organophosphorus and carbamate insecticides on vertebrates. Are there implications for locust management in Australia? *Wildlife Research*, *28*(2), 179. doi:10.1071/WR99060

Szekacs, A., Mortl, M., & Darvas, B. (2015). Monitoring Pesticide Residues in Surface and Ground Water in Hungary: Surveys in 1990-2015. *Journal of Chemistry*, *2015*, 1–15. doi:10.1155/2015/717948

*The biological control of pests*. (2007). Biology Pages.

Turusov, V., Rakitsky, V., & Tomatis, L. (2002). Dichlorodiphenyltrichloroethane (DDT): Ubiquity, persistence, and risks. *Environmental Health Perspectives*, *110*(2), 125–128. doi:10.1289/ehp.02110125 PMID:11836138

U.S. Environmental Protection Agency. (2007). *Pesticides: Health and Safety*. National Assessment of the Worker Protection Workshop.

UC IPM Online (2007). *Statewide IPM Program, Agriculture and Natural Resources*. University of California. Retrieved from http://www.ipm.ucdavis.edu

United States Environmental Protection Agency (2017). *Human health issues related to pesticides*. Author.

Van der Werf, H. M. G. (1996). Assessing the impact of pesticides on the environment. *Agriculture, Ecosystems & Environment*, *60*(2-3), 81–96. doi:10.1016/S0167-8809(96)01096-1

Van Doesburg, M.J., Noordijk, H., & Van den Anker, I.M. (1993). *Evaluation of the smog situation in the summer period of 1992*. RIVM Report 722101005.

Wells, M. (2007). *Vanishing bees threaten U.S. crops*. BBC News.

Wilson, C., & Tisdell, C. (2001). Why farmers continue to use pesticides despite environmental, health and sustainability costs. *Ecological Economics*, *39*(3), 449–462. doi:10.1016/S0921-8009(01)00238-5

Wolt, J. D., Smith, J. K., Sims, J. K., & Duebelbeis, D. O. (1996). Products and kinetics of cloramsulam-methylaerobic soil metabolism. *Journal of Agricultural and Food Chemistry*, *44*(1), 324–332. doi:10.1021/jf9503570

Zahra, A., Hashmi, M. Z., Malik, R. N., & Ahmed, Z. (2014). Enrichment and geo-accumulation of heavy metals and risk assessment of sediments of the Kurang Nallah – feeding tributary of the Rawal Lake reservoir, Pakistan. *The Science of the Total Environment*, *470-471*, 925–933. doi:10.1016/j.scitotenv.2013.10.017 PMID:24239813

Zhao, G., Li, J., Ren, X., Chen, C., & Wang, X. (2011). Few-layered graphene oxide nanosheets as superior sorbents for heavy metal ion pollution management. *Environmental Science & Technology*, *45*(24), 10454–10462. doi:10.1021/es203439v PMID:22070750

# Chapter 11
# Sources of Groundwater Pollution

**Abderrezak Khelfi**
*National Center of Toxicology, Algeria*

## ABSTRACT

*In many regions in the world, groundwater represents an important source of fresh water. It is now established that several contaminants enter groundwater from a number of sources and pathways. These sources are both natural and anthropogenic. Contamination of groundwater resources by a variety of anthropogenic pollutants from both point and nonpoint sources represents a key global environmental problem. The most frequently identified contaminant sources are industrial manufacturing, agricultural activities, municipal landfills, and wastes. Frequently detected contaminants included nitrates, volatile organic compounds, arsenic, and fluorides. Other contaminant species include solvents, fuel hydrocarbons, heavy metals, pesticides, disinfectants, detergents, and radionuclides. In this chapter, the main sources and pathways for contaminants in groundwater are reviewed. It identifies challenges that need to be met to minimize risk to drinking water and ecosystems. Particular attention is paid to the occurrence of known and potential endocrine disrupting substances in groundwater.*

## 1. INTRODUCTION

Water is an essential component of life. Almost all organisms including human beings are dominantly made up of water and rely on water to regulate temperature, carry dissolved materials inside and between cells, remove waste products, etc. Human beings need to consume large quantities of fresh and clean water. Of the earth's total amount of fresh water, around 75% exists in the frozen state in polar ice caps and glaciers, while less than 2% is in surface waters and a relatively small amount is contained in unsaturated soil. That leaves nearly 22% of our planet's fresh water that permeates porous rocks and sediments, at some depth below the ground surface, called groundwater. The importance of this water as part of the biosphere is now widely recognized.

DOI: 10.4018/978-1-5225-5754-8.ch011

In many regions, groundwater constitutes an important source of fresh water for domestic, agricultural or industrial use. It provides also the most reliable perennial source of freshwater on earth. Groundwater resources have been threatened not only by excessive withdrawal but also by contamination resulting from past and present industrial, agricultural and commercial activities. Most natural groundwaters contain at least some amount of dissolved substances that we think of as contaminants. These contaminants result from the deliberate use of chemicals in industry and agriculture, waste disposal, unintended transport accidents, leaking storage facilities and other human activities.

Because it lacks visibility, groundwater is difficult to understand and easy to overlook. Ignorance has led to its abuse and abuse to endangering critical supplies, both in terms of quantity and quality. The reduction of groundwater supply from overuse and degradation of groundwater quality by contamination in many countries have become major issues bearing on public health and socioeconomic development. There is a growing concern about the impacts of groundwater contamination on hydrological, biological and ecological systems.

Furthermore, because of the relatively slow rates of recharge and fluid migration through the subsurface environment, and because of the vast potential for exposure of subsurface materials to sources of pollution, groundwater contamination represents a critical resource problem to the generation to become.

The term contamination is used for addition of any solute into the hydrological system as a result of man's activity while the term pollution is restricted to a situation when the contamination attains levels that are considered to be objectionable (Singhal and Gupta, 2010).

Examples of typical groundwater contaminants include industrial chemicals (solvents, fuels, heavy metals, etc), agricultural chemicals (fertilizers, pesticides, etc) and municipal wastes (fecal coliform, nutrients, $NO_3^-$, disinfectants, detergents, etc). Today, more than 1,000 organic chemicals have been identified in groundwater. Some contaminants adsorb onto the surface of aquifer solids, moving very little from their source, while others migrate freely with the flowing pore water, sometimes ending up many kilometers from their source. Chemical reactions along the way can cause a contaminant to disappear or worse, appear from apparently nowhere.

It is now established that these compounds enter the environment from a number of sources and pathways. These sources are both natural and human-induced. Based on the contamination properties, groundwater contamination sources can also be classified as point sources and nonpoint sources. Point sources are those that release contaminants from a discrete geographic location, including leaking underground storage tanks, ruptured transfer pipes, septic systems, injection wells, landfills and spillage due to traffic accidents. Such sources threaten groundwater resources directly at the point of entry, where transport processes may subsequently facilitate migration.

Nonpoint sources of contamination are more extensive in area and diffuse in nature. For example, agricultural activities, large industrial areas and urban runoff are potential nonpoint sources. These nonpoint sources threaten regional water resources on a larger scale. The main anthropogenic sources of groundwater contaminants are various industries that emit hazardous chemicals. In fact, in the last century, the rise of industrial manufacturing complexes and synthetic chemicals have led to the release of complex mixtures of contaminants. Another major source of groundwater contaminants is associated with agricultural activities. In agricultural practice, pesticides, fertilizers and sometimes animal wastes, human wastes or sewage sludge are applied to the cropland. Excessive use of fertilizers makes $NO_3^-$ the most ubiquitous contaminant in groundwater. Animal and human wastes contribute to the elevation of pharmaceutical agents, $NO_3^-$, phosphorous, pathogens and viruses in groundwater. Other sources of groundwater aquifer contamination includes, mining activities, oil and gas exploration, septic and underground storage tanks, hospital effluents, deep-well injection, accidental spills and overexploitation of groundwater, especially in coastal areas where it is responsible for contaminating fresh water aquifers due to sea-water intrusion.

Groundwater contamination has been increasingly threatening the world's already scarce freshwater resources. The problem is worsening due to continual contamination by negligence and illegal activities as well as ineffective remedial systems implemented early on. In the European context, groundwater quality is currently regulated under the Water Framework Directive and the Groundwater Daughter Directive (EC, 2000; 2006) and drinking water under the Drinking Water Directive (EC, 1998).

Because of the vast array of possible compounds, many studies have selected contaminants according to priority lists established taking into account potential sources. While individual compounds were generally found in concentrations that are considered too low to cause acute effects (i.e. typically <100 ng $L^{-1}$), it is clear that the toxic effects due to long-term exposure to a combination of low concentrations of range of emerging contaminants are not well understood. Some of these contaminants can have human or ecological health effects and there is a need for better understanding of their fate in environmental systems.

This chapter reviews the occurrence of a range of groundwater contaminants including: pharmaceuticals, personal care compounds, veterinary medicines, steroids, metabolic regulators, surfactants, well as a large range of other wastewater related compounds. It focuses on the main sources, pathways and fate of these contaminants found in public groundwater supplies. Particular attention is paid to the occurrence of known and potential endocrine disrupting substances in groundwater.

## 2. SOURCES OF GROUNDWATER POLLUTION

A groundwater contaminant is defined by most regulatory agencies as any physical, chemical, biological or radiological substance or matter in groundwater. The contaminants can be introduced in groundwater by naturally occurring activities, such as natural leaching of the soil and mixing with other groundwater sources having different chemistry. They are also introduced by planned human activities, such as waste disposal, mining and agricultural operations, etc.

This section discusses the principal sources and causes of groundwater contamination from different sources with regard to their occurrence and effects on groundwater quality.

### 2.1. Natural Sources/Geogenic Sources

Natural sources may be exemplified by trace elements contained in regional geologic materials (e.g., release of As and other heavy metals). The concentration of some toxic substances beyond acceptable limits for human consumption can be acceptable due to the dissolution of aquifer materials (Table 1).

Increased concentration of fluoride, As and other heavy metals in groundwater is usually due to this type of contamination.

The fluoride problem has received attention in Ethiopia and India and in several other countries. The hot springs in volcanic areas in Ethiopia have high fluoride concentration (90 mg $L^{-1}$). In India, high fluoride (2-10 mg $L^{-1}$) water was reported mainly from granites and pegmatites which have F-bearing minerals, e.g. fluorite ($CaF_2$) and fluor-apatite (Singhal and Gupta, 2010).

*Table 1. Major, minor and trace elements in natural groundwater*

| Major Elements | Minor Elements | Trace Elements | |
|---|---|---|---|
| (1.0–1,000 mg/L) | (0.01–10 mg/L) | (0.0001–0.1 mg/L) | (<0.001 mg/L) |
| Bicarbonate<br>Calcium<br>Chloride<br>Magnesium<br>Silica<br>Sodium<br>Sulfate | Carbonate<br>Fluoride<br>Iron<br>Nitrate<br>Potassium | Antimony<br>Aluminum<br>Arsenic<br>Barium<br>Bromide<br>Cadmium<br>Chromium<br>Cobalt<br>Copper<br>Iodide<br>Lead<br>Lithium<br>Manganese<br>Molybdenum<br>Nickel<br>Phosphate<br>Selenium<br>Uranium<br>Vanadium<br>Zinc | Beryllium<br>Bismuth<br>Cesium<br>Gold<br>Indium<br>Platinum<br>Radium<br>Silver<br>Thallium<br>Thorium<br>Tin<br>Tungsten<br>Zirconium |

Excessive concentration of fluoride in water was also reported in other countries like Thailand, China and several African countries.

Arsenic in drinking water has received much attention due to its detrimental health effects. The upper tolerated limit for As in drinking water according to World Health Organization (WHO) is 0.01 mg $L^{-1}$. High concentrations of As in groundwater was reported from several countries such as Bangladesh, India, China, Argentina, Chile, Mongolia, Taiwan and the USA.

People consuming high levels of As are prone to develop skin, bladder and lung cancers. Dissolved As may exist in groundwater either as arsenite in trivalent form ($As^{3+}$) or as arsenate in pentavalent form ($As^{5+}$). Both forms are capable of adsorbing to various subsurface materials, such as ferric oxides and clay particles. An increase in soil pH may cause both arsenite and arsenate desorption and excretion in groundwater (Kresic, 2007). Several other mechanisms, including oxidation of As-bearing sulfide minerals (e.g. pyrite, arsenopyrite), reduction of As bearing iron and manganese hydroxides and release of sorbed As from acid mine drainage can be responsible for higher concentration of As in groundwater (Figure 1).

In some areas in the USA, high As concentrations (>50 μg $L^{-1}$) in groundwater are associated with sedimentary rocks derived from volcanic rocks, geothermal systems and gold deposits (Naymik, 1993). In some Asian countries (Bangladesh, West Bengal and India), high As concentrations (0.3-1.1 mg $L^{-1}$, highest being 5 mg $L^{-1}$) in groundwater of shallow alluvial aquifers posed serious health problems to millions of people. In Bangladesh, large volumes of groundwater used for public water supply are polluted with naturally occurring As, which adversely affects the health of millions of people. Consumption of this water has led to widespread death and disease. A large part of total population of West Bengal is suffering from As related diseases including skin cancer (Chandrasekharam, 2005). The high As concentrations were attributed to the oxidation of pyrite present in clay minerals. However, recent opinion attributes it to the release of adsorbed As from the Fe-hydroxide and Mn-hydroxide surfaces under reducing conditions (Nath *et al.*, 2006).

*Figure 1. Arsenic transformations in mining and groundwater environments*

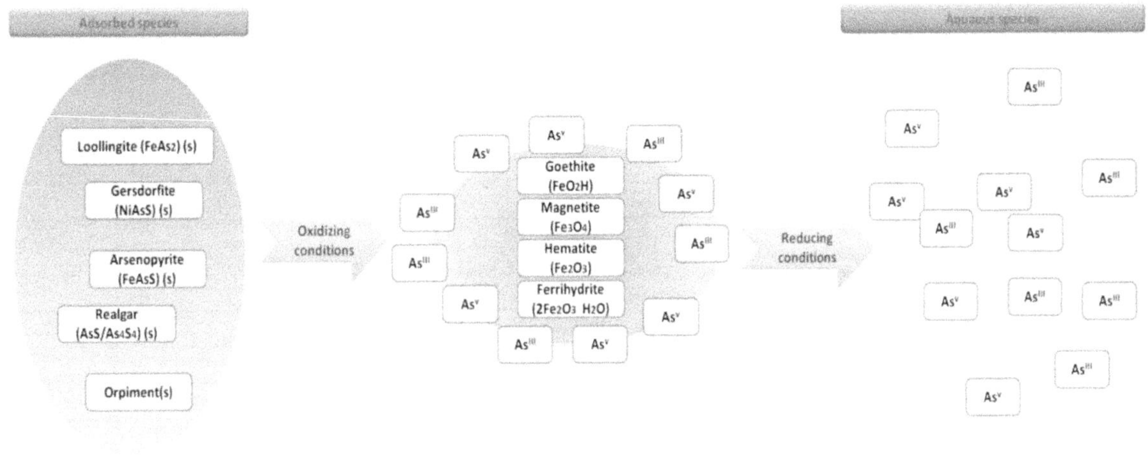


Chromium Cr is present in minerals and rocks. The common ore of Cr is chromite ($FeCr_2O_4$) which is associated with ultramafic rocks. It may enter groundwater from acid mine drainage. This metal may occur in groundwater as $Cr^{3+}$ or $Cr^{6+}$, as trace elements. The WHO has prescribed 50 μg $L^{-1}$ as the limit value of Cr for drinking water. Compounds that contain $Cr^{6+}$ are more soluble and mobile compared to $Cr^{3+}$, the latter makes only a small percentage of the total concentration of Cr in groundwater. They have a deleterious effect on liver, kidney and respiratory organs. Furthermore, due to greater solubility and mobility, $Cr^{6+}$ can persist in groundwater for longer duration and transported to longer distances compared with $Cr^{3+}$.

## 2.2. Anthropogenic Sources

Groundwater contamination could be from different anthropogenic sources. Table 2 contains a selective review of contaminants from point and nonpoint sources. Major sources are also listed. The list is composed of predominantly organic solvents and compounds associated with petroleum products and manufacturing and inorganic contaminants, such as heavy metals from industrial and mining activities. A list of environmentally important radioactive isotopes produced from various industries and at different stages from the nuclear fuel cycle, nuclear weapons test, research and medical waste is also given in Table 2.

Table 3 contains a categorization of sources of contamination developed by the USA Environmental Protection Agency (USEPA) (1976) and Office of Technology Assessment (1984) as referenced by Canter *et al.* (1988) (Canter *et al.*, 1988; "USEPA (1976). RCRA Subtitle I and Sections 7003, 9003(h), 9005 and 9006.,"). The categories are based on a functional classification of various activities.

The main sources of groundwater pollution in the environment are summarized in Figure 2, including analysis of key sources, pathways and fate of contaminants in the subsurface that can lead to groundwater pollution. Figure 2 highlights those sources of pollution and pathways which are thought to have the greatest impact on groundwater resources.

*Table 2. Sources of the major contaminants to groundwater*

| Source Type | Selected Major Contaminants | Persistence |
|---|---|---|
| **Point Sources** | | |
| Underground storage tanks | Fuels; petroleum hydrocarbons | M |
| | benzene, methyl-t-butyl ether | M |
| | Chlorinated solvents | H |
| | percholoroethylene; trichloroethylene | H |
| | Breakdown products<br>dichloroethylene, vinyl chloride | H |
| Landfills/hazardous waste sites | Fuels, solvents | M-H |
| | Heavy metals (Chromium, Cadmium, Lead, Arsenic, Copper, Mercury, etc) | H |
| | polychlorinated biphenyls, phenols, polycyclic aromatic hydrocarbons | H |
| | HCN | H |
| | Pesticides | L-M-H |
| | $CH_4$ | H |
| Septic systems | $NH_3$-N | L |
| | $NO_2^-$ and $NO_3^-$ | L-H |
| Surface wastewater impoundments | Inorganic and organic constituents | M-H |
| Liquid waste storage ponds (lagoons, leaching ponds, recharge basins) | Heavy metals, solvents, inorganic compounds | M-H |
| Radioactive wastes | Radioisotopes and radionuclides<br>Mining waste, uranium, phosphate, coal $^{222}Rn$, $^{226}Ra$, $^{230,232}Th$<br>Nuclear power plant $^{59,63}Ni$, $^{60}Co$, $^{90}Sr$, $^{93,99}Zr$, 99Tc, $^{107}Pd$, $^{129}I$, $^{134,137}Cs$<br>Nuclear weapons tests $^{3}H$, $^{60}Co$, $^{90}Sr$, $^{137}Cs$, $^{239,240}Pu$<br>Weapons manufacturing, research and medical waste $^{134,137}Cs$, $^{60}Co$, $^{151}Sm$, $^{152,154}Eu$, $^{237}Np$, $^{239,240,242}Pu$, $^{241,243}Am$ | H |
| Deep well injection | Variety of inorganic and organic compounds | L-M-H |
| Mining activities | Heavy metals<br>Chlorides<br>Sulfides<br>Acid mine drainage ($2 < pH < 4$, rich in $H_2SO_4$)<br>Organic solvents | M-H |
| Accidental spillage | Hydrocarbons | M |
| **Nonpoint Sources** | | |
| Agricultural activities | Insecticides (Chlordane; Diazinon, etc) | L-M-H |
| | Herbicides (Atrazine; 2-4D)<br>Heavy metals | L-M-H |
| | Fertilizers [ $NO_3^-$, $NH_3$]<br>Heavy metals | H |
| | Sewage sludge<br>Heavy metals, inorganic and organic compounds<br>Oxygen-consuming wastes | M-H |
| | Manure<br>Veterinary agents<br>Heavy metals | L-M-H |
| Natural soil/sediment constituents | Arsenic and Selenium | M |
| | Radon | H |
| Urban runoff infiltration | Heavy metals, inorganic and organic compounds<br>Petroleum products | L-M-H |
| Deicing activities | Chlorides, sodium, calcium | H |

*continued on following page*

*Table 2. Continued*

| Source Type | Selected Major Contaminants | Persistence |
|---|---|---|
| Industrial activities | Hydrocarbons<br>Chlorinated solvents<br>Heavy metals | M-H |
| | Gasworks<br>Aromatic hydrocarbons, phenols and $CN^-$ | H |
| | Metal plating and cleaning industries<br>Heavy metals, $CN^-$, organic solvents, spent plating wastes, acids and bases | L-M-H |
| | Metal smelting<br>Heavy metals, metal slag, smelting sludges, acids and oils | M-H |
| | Pesticide manufacturing sites<br>Pesticides, solvents, Hg, As and Cu | L-M-H |
| | Printing plants<br>Acids, bases, heavy metals, inorganics, ink, solvents, spent plating and sludges | L-M-H |
| | Petroleum refineries<br>A wide range of hydrocarbons | M |

L: Low; M: Medium; H: High.

*Table 3. Sources of groundwater contaminants and their potentials*

| Category | Sources |
|---|---|
| Category 1: Sources designed to discharge substances | 1 Sub-surface percolation from septic tanks/<br>cesspools |
| | 2 Injection wells<br>Hazardous waste<br>Nonhazardous waste (e.g., brine disposal)<br>Nonwaste (e.g., solution mining) |
| | 3 Land application<br>Wastewater (spray irrigation)<br>Wastewater by-products (biosolids)<br>Hazardous waste<br>Nonhazardous waste |
| Category 2: Sources designed to store, treat, and/<br>or dispose of substances; discharge through unplanned release | 1 Landfills<br>Industrial hazardous waste<br>Industrial nonhazardous waste<br>Municipal sanitary |
| | 2 Open dumps, including illegal dumping |
| | 3 Residential disposal |
| | 4 Surface impoundments<br>Hazardous waste<br>Nonhazardous waste |
| | 5 Materials stockpiles (nonwaste) |
| | 6 Graveyards |
| | 7 Animal burial |
| | 8 Above-ground storage tanks<br>Hazardous waste<br>Nonhazardous waste<br>Nonwaste |
| | 9 Underground storage tanks<br>Hazardous waste<br>Nonhazardous waste<br>Nonwaste |
| | 10 Containers<br>Hazardous waste<br>Nonhazardous waste<br>Nonwaste |
| | 11 Open burning and detonation sites |
| | 12 Radioactive disposal sites |
| | 13 Waste tailing |

*continued on following page*

*Table 3. Continued*

| Category | Sources |
|---|---|
| Category 3: Sources designed to retain substances during transport or transmission | 1 Pipelines<br>Hazardous waste<br>Nonhazardous waste<br>Nonwaste |
| | 2 Materials transport and transfer operations<br>Hazardous waste<br>Nonhazardous waste<br>Nonwaste |
| Category 4: Sources discharging substances as<br>a result of other planned activities | 1 Irrigation practices |
| | 2 Pesticide applications |
| | 3 Fertilizer applications |
| | 4 Animal feeding applications |
| | 5 De-icing salt applications |
| | 6 Urban run-off |
| | 7 Percolation of atmospheric pollutants |
| | 8 Mining and mine drainage<br>Surface mine related<br>Underground mine related |
| Category 5: Sources providing conduit or inducing discharge through altered flow patterns | 1 Production wells<br>Oil and gas wells<br>Geothermal and heat recovery wells<br>Water supply wells |
| | 2 Other wells<br>Monitoring wells<br>Exploration wells |
| | 3 Construction excavation |
| Category 6: Naturally occurring sources whose discharge is created and/or exacerbated by human activity | 1 Groundwater-surface-water interactions |
| | 2 Natural leaching |
| | 3 Salt water intrusion |

*Figure 2. Potential sources and pathways for groundwater contamination*

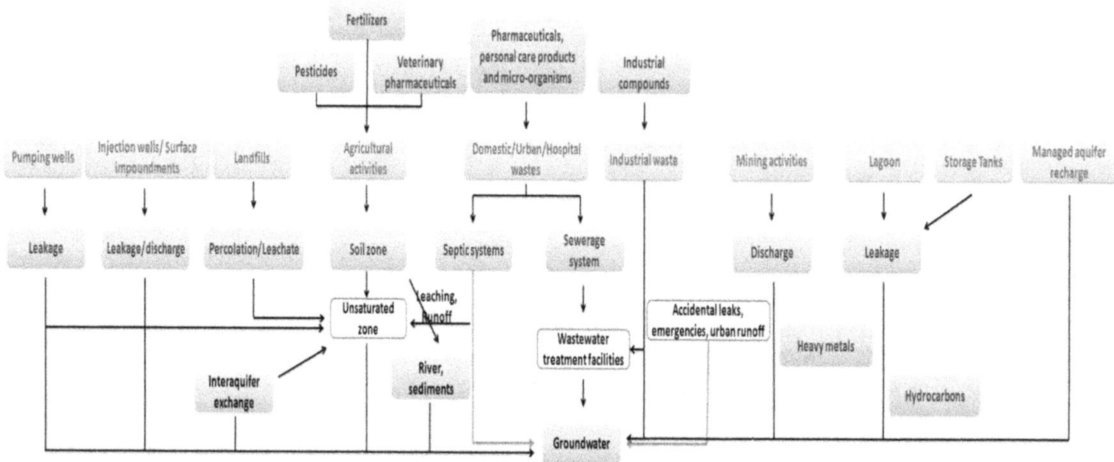

Direct pathways for pharmaceuticals, urban and industrial contaminants to reach groundwater including leaking sewers, discharge of wastewater treatment effluent, landfill leachate, leaking storage tanks and other discharges to the ground that bypass the soil zone, such as septic tanks are given in Figure 2.

### 2.2.1. Leaking Storage Tanks/Spills

Groundwater contamination from the storage and transport of commercial materials results from leaking storage tanks and spills. Tanks are widely used to store fuels and chemicals. Many of these materials have leaked over the years. Underground tanks are the most dangerous, because they can leak slowly for a long time without being discovered. The most common tank sources are gas tanks at filling stations as well as fuel and solvent storage tanks in industrial facilities. Most tanks installed before the 1970s were steel tanks that tended to corrode. Leakage is particularly frequent from bare steel tanks that are not protected against corrosion. Even if a leakage is small, it can pose a significant threat to groundwater quality. Many of these tanks eventually sprang leaks when corrosion went too far.

Underground transmission pipelines are another cause of groundwater pollution. Pipelines are subject to corrosion and structural fail tires with subsequent leaks that introduce a variety of contaminants into groundwater. What mainly leaks out of these are non-aqueous phase liquids (NAPLs). Leaked storage tanks and pipelines that lie within the saturated zone can contaminate groundwater directly.

Gasoline and petroleum products contain hydrocarbon components such as benzene, toluene and xylene that are highly soluble and mobile in groundwater and can be hazardous to humans. One gallon of gasoline is enough to render one million gallons of groundwater un-usable based on USEPA drinking water standards (Noonan and Curtij, 1990). In addition, immiscible compounds trapped in the pore spaces of the unsaturated soil continue to feed groundwater with contaminants as water moves into and through the subsurface or as the groundwater table fluctuates.

Spills and discharges on the ground of chemical products can migrate downward and contaminate groundwater. Spills and discharges vary from casual activities at industrial sites, such as leaks from pipes and valves, to accidents involving aboveground storage tanks, railroad cars and trucks. The discharged chemicals are usually entrained by Storm water runoff and transported to the subsurface where they reach the groundwater and degrade its quality.

### 2.2.2. Septic Systems

Septic systems remain important sources of groundwater pollution, particularly where groundwater tables are shallow and aquifers have high transmissivity. It is very difficult to effectively manage contamination from septic systems due to the huge number of these point sources and widespread use in rural settings.

Septic systems for subsurface disposal of human wastewater are used in rural areas not served by sewers and sewage treatment systems. Septic systems can serve a cluster of homes and/or offices. A typical septic system contains series of drainpipes in a home's plumbing system. These all connect to one pipe that runs outside to a buried septic tank, where solids settle and are trapped. The tank needs to be pumped out periodically to remove accumulated solids. From the tank, wastewater flows to a leaching field, usually a network of porous distribution pipes set in a porous material in the unsaturated zone.

In these devices, wastewater contains dissolved organic compounds that undergo redox reactions. Redox reactions in the tank are usually due to bacterial organisms in anaerobic conditions, including fermentation, methane generation and $SO_4^{-2}$ reduction. The water leaving the tank has high concentrations of organic compounds, $CO_2$ and $NH_4^+$.

In the leaching field, aerobic respiration takes place and nitrification is the key process. The concentrations of organic compounds and $NH_4^+$ decrease, $CO_2$ is evolved and $NO_3^-$ concentration increases (Figure 3). In fact, in most septic systems, $NO_3^-$ is a groundwater contaminant of concern. Some septic system designs include another anaerobic zone beyond the aerobic zone in the leaching field, where denitrification reduces $NO_3^-$ concentrations in the effluent.

While at close proximity to the sources concentrations may be high, the generally low volumes of leachate from this type of source means that total contaminants loading to groundwater are low on a regional scale and there is a high potential for dilution and attenuation.

It has recently come to light that wastewater effluent from household disposal systems can also contain a range of chemical compounds found in medications, food, personal care products and household products. The majority of them are considered as emerging contaminants of groundwater (Table 4).

*Figure 3. Nitrification of nitrogen compounds under aerobic conditions*

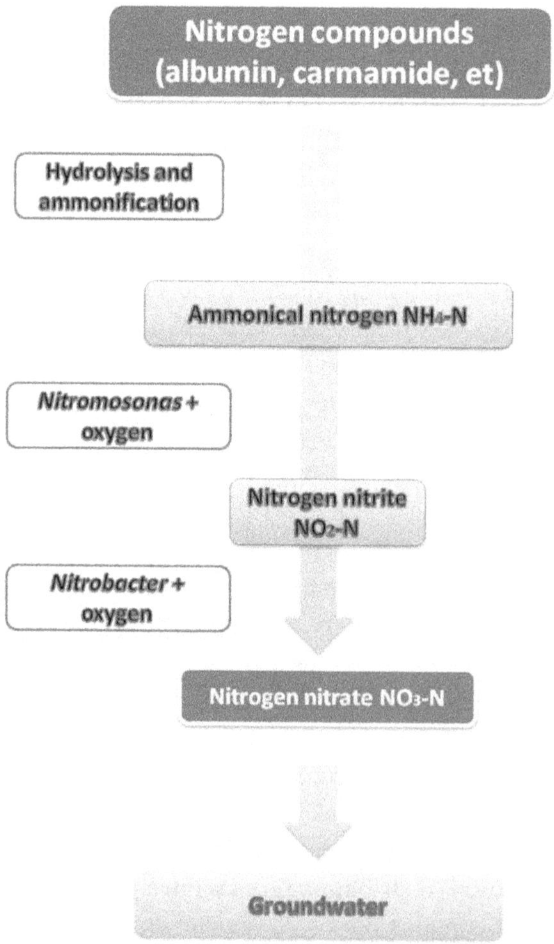

*Table 4. Maximum concentrations of emerging contaminants in groundwater from different regions and countries*

| Compound | Region/Country | Frequency (%) | Max concentration (ng L$^{-1}$) | Samples | Reference |
|---|---|---|---|---|---|
| Carbamazepine | USA | 1.5 | 420 | 1231 | (Barnes *et al.*, 2008) |
| | Europe | 42.1 | 390 | 164 | (Loos *et al.*, 2010) |
| | France | 42 | 167 | 218 | (Stuart and Lapworth, 2013) |
| | France | 13 | 30 | 45 | (Lapworth *et al.*, 2015) |
| | Germany | 12 | 900 | 105 | (Sacher *et al.*, 2001) |
| | Spain | 24.5 | 118 | 77 | (Teijon *et al.*, 2010) |
| | Switzerland | 19 | 45 | 47 | (Stuart and Lapworth, 2013) |
| | UK | 1.2 | 3600 | 2644 | (Stuart and Lapworth, 2013) |
| | UK | - | 3600 | 32 | (Stuart *et al.*, 2012) |
| | UK | 2.3 | 30 | 300 | (Lapworth *et al.*, 2015) |
| Codeine | Spain | 3.77 | 348.3 | 77 | (Teijon *et al.*, 2010) |
| Diclofenac | Spain | 3.7 | 477 | 77 | (Teijon *et al.*, 2010) |
| Ibuprofen | USA | 2.1 | 3110 | 47 | (Barnes *et al.*, 2008) |
| | Europe | 6.7 | 395 | 164 | (Loos *et al.*, 2010) |
| | France | 0.5 | 7 | 209 | (Stuart and Lapworth, 2013) |
| | Germany | 0 | <LQ | 105 | (Sacher *et al.*, 2001) |
| | Spain | 1.9 | 185 | 77 | (Teijon *et al.*, 2010) |
| | Spain | - | 145 | - | (Corada-Fernandez *et al.*, 2017) |
| | Switzerland | 0 | <LQ | 47 | (Stuart and Lapworth, 2013) |
| | UK | 0.3 | 290 | 2644 | (Stuart and Lapworth, 2013) |
| Naproxen | Spain | 3.8 | 263 | 77 | (Teijon *et al.*, 2010) |
| | Spain | - | 23 | - | (Corada-Fernandez *et al.*, 2017) |
| Sulfamethoxazole | USA | 23 | 1110 | 47 | (Barnes *et al.*, 2008) |
| | USA | 0.41 | 170 | 1231 | (Fram and Belitz, 2011) |
| | Europe | 24.2 | 38 | 164 | (Loos *et al.*, 2010) |
| | France | 18 | 18 | 147 | (Stuart and Lapworth, 2013) |
| | Germany | 1810 | 410 | 105 | (Sacher *et al.*, 2001) |
| | Spain | 15.1 | 117 | 77 | (Teijon *et al.*, 2010) |
| | Spain | - | 1 | - | (Corada-Fernandez *et al.*, 2017) |
| | Switzerland | 18 | 48 | 100 | (Stuart and Lapworth, 2013) |
| Caffeine | USA | 12.8 | 130 | 47 | (Barnes *et al.*, 2008) |
| | USA | 0.24 | 290 | 1231 | (Fram and Belitz, 2011) |
| | Europe | 82.9 | 189 | 164 | (Loos *et al.*, 2010) |
| | France | 24 | 150 | 45 | (Lapworth *et al.*, 2015) |
| | Spain | 100 | 44.9 | 4 | (Estévez *et al.*, 2012) |
| | Spain | 77.4 | 505 | 77 | (Teijon *et al.*, 2010) |
| | Spain | - | 27 | - | (Corada-Fernandez *et al.*, 2017) |
| | UK | - | 4500 | 722 | (Stuart *et al.*, 2012) |
| | UK | 6.3 | 1700 | 300 | (Lapworth *et al.*, 2015) |
| | UK | 27 | 4500 | 2644 | (Stuart and Lapworth, 2013) |
| | Zambia | - | 0.17 | 20 | (Sorensen *et al.*, 2015) |

*continued on following page*

*Table 4. Continued*

| Compound | Region/Country | Frequency (%) | Max concentration (ng L$^{-1}$) | Samples | Reference |
|---|---|---|---|---|---|
| Nicotine | Spain | 100 | 115.5 | 4 | (Estévez *et al.*, 2012) |
| | Spain | 71.7 | 144 | 77 | (Teijon *et al.*, 2010) |
| | UK | - | 8070 | 107 | (Stuart *et al.*, 2012) |
| Benzo(a)pyrene | Spain | - | 0.4 | 4 | (Estévez *et al.*, 2012) |
| | Spain | 17 | 2.26 | 77 | (Teijon *et al.*, 2010) |
| Benzo(b)fluoranthene | Spain | - | 0.4 | 4 | (Estévez *et al.*, 2012) |
| | Spain | 19 | 12.59 | 77 | (Teijon *et al.*, 2010) |
| Bisphenol A | USA | 29.8 | 2550 | 47 | (Barnes *et al.*, 2008) |
| | Austria | 59 | 930 | 110 | (Hohenblum *et al.*, 2004) |
| | Europe | 39.6 | 2299 | 164 | (Loos *et al.*, 2010) |
| | France | 8.9 | 1360 | 45 | (Lapworth *et al.*, 2015) |
| | UK | 8 | 9300 | 2644 | (Stuart and Lapworth, 2013) |
| | UK | - | 9300 | 209 | (Stuart *et al.*, 2012) |
| | UK | - | 1200 | - | (White *et al.*, 2016) |
| | UK | 2.7 | 100000 | 300 | (Lapworth *et al.*, 2015) |
| | Zambia | - | 1.1 | 20 | (Sorensen *et al.*, 2015) |
| Hexachlorobenzene | Spain | 75 | 13.4 | 4 | (Estévez *et al.*, 2012) |
| | Spain | 69 | 4.7 | 77 | (Teijon *et al.*, 2010) |
| Atrazine | Italy | - | 2700 | - | (Meffe and de Bustamante, 2014) |
| | France | 47 | 130 | 45 | (Lapworth *et al.*, 2015) |
| | Spain | - | 1 | 4 | (Estévez *et al.*, 2012) |
| | UK | 7.3 | 200 | 300 | (Lapworth *et al.*, 2015) |
| | UK | - | 180 | - | (White *et al.*, 2016) |
| | Zambia | - | 0.13 | 20 | (Sorensen *et al.*, 2015) |
| Chloropyrifos ethyl | Spain | 100 | 294 | 4 | (Estévez *et al.*, 2012) |
| | Spain | 85 | 20.78 | 77 | (Teijon *et al.*, 2010) |
| Diuron | France | 6.7 | 80 | 45 | (Lapworth *et al.*, 2015) |
| | Italy | - | 100 | - | (Meffe and de Bustamante, 2014) |
| | Spain | 50 | 8.1 | 4 | (Estévez *et al.*, 2012) |
| | Spain | 26.2 | 155 | 77 | (Teijon *et al.*, 2010) |
| | UK | 0.3 | 360 | 300 | (Lapworth *et al.*, 2015) |
| Dimethyl phthalate | UK | 2.7 | 350 | 300 | (Lapworth *et al.*, 2015) |
| | Zambia | - | 1 | 20 | (Sorensen *et al.*, 2015) |
| Diethyl phthalate | UK | 3 | 22000 | 300 | (Lapworth *et al.*, 2015) |
| Di-n-butyl phthalae | UK | 0.7 | 1500 | 300 | (Lapworth *et al.*, 2015) |
| Methylparaben | UK | - | 5000 | 44 | (Stuart *et al.*, 2012) |
| Propylparaben | UK | - | 5500 | 68 | (Stuart *et al.*, 2012) |

Chemicals in medications and food pass through humans and persist in the system effluent. For example, caffeine has been used as an indicator of septic system contamination. Hormones are particularly troubling chemicals that persist long enough to migrate through the disposal system to groundwaters and surface waters, where they can harm aquatic life.

A detailed study carried out in Canada showed septic systems samples contained elevated concentrations of several pharmaceutical compounds. As a result, when they leak they may be an important source of emergent organic contaminants of groundwaters. Of the compounds analysed pharmaceuticals, ibuprofen, gemfibrozil and naproxen were observed (Carrara *et al.*, 2008). Several case studies in the USA have reported a range of contaminants in groundwater impacted by septic systems. Common pharmaceuticals and personal care products reported include ibuprofen, paracetamol, salicylic acid, triclosan, carbamazepine, sulfamethoxazole, methylene blue active substances, caffeine and its degradation product paraxanthine and two fluorescent whitening agents. Other compounds include nicotine and its metabolite cotinine. Other substances like endocrine disturbing substances include bisphenol A, nonylphenol, nonylphenol mono- and diethoxycarboxylates, steroid hormones 17β-estradiol, estrone and their glucuronide and sulfate conjugates. These endocrine disturbing substances were detected in concentrations between 10-$10^3$ ng $L^{-1}$ (Godfrey *et al.*, 2007; Swartz *et al.*, 2006). In Cape Cod, Massachusetts, where large proportion of homes have septic systems, several ponds in the sand and gravel aquifer contained pharmaceuticals and estrogenic hormones (androstenedione, estrone and progesterone) as well as pharmaceuticals (carbamazepine, pentoxifylline, sulfamethoxazole and trimethoprim) at concentrations that may induce physiological responses in fish (Standley *et al.*, 2008).

In Spain, the urban groundwater of the central Beso's river basin (the La Llagosta aquifer) has become contaminated due to the infiltration of wastewater from septic tanks and sewage networks as well as industrial activities located in urban areas. The contaminated groundwater below the urban areas showed high contents of $NO_3^-$ (90-100 mg $L^{-1}$) and an increase in the concentrations of $Ca^{2+}$ and $Mg^{2+}$, which coincided with a decrease in pH. According to authors, the infiltration of domestic wastewater in urban areas resulted in the appearance of areas of high salinity and total hardness with a high $NO_3^-$ content. Furthermore, at some points, contaminants such as Ni, Pb, Zn and organic micropollutants (perchloroethene and trichloroethene) were detected. The evolution of the main cations ($K^+$, $Ca^{2+}$ and $Mg^{2+}$) and $NO_3^-$ showed a possible nitrification of the $NH_4^+$ and the oxidization of the dissolved organic matter, which gives rise to increased concentrations of $NO_3^-$, $CO_2$ and $H^+$ and other dissolved constituents (Navarro and Carbonell, 2007).

In another study, groundwater samples were collected from in New England (NE) and New York (NY). The NE network assessed the effect of a single large septic system that received discharge from an extended health care facility for the elderly. The NY network assessed the effect of many small septic systems used seasonally on a densely populated portion of Fire Island. The data collected from these two networks indicated that demographic factors affected micropollutant concentrations in these systems. Total concentrations for personal care/domestic use compounds, pharmaceutical compounds and plasticizer compounds generally ranged from 1 to over 20 μg $L^{-1}$ in the NE network samples. According to authors, high tris(2-butoxyethyl phosphate) plasticizer concentrations in wells beneath and downgradient of the leach beds (>20 μg $L^{-1}$) may reflect the presence of this compound in cleaning agents at the health-care facility. The highest micropollutant concentrations for the NY network were present in the shoreline wells and reflect groundwater that is most affected by septic system discharges. One of the shoreline wells had personal care/domestic use, pharmaceutical and plasticizer concentrations ranging from 0.4 to 5.7 μg $L^{-1}$. Estradiol equivalency quotient concentrations were also highest in a shore line well sample

(3.1 ng $L^{-1}$). Most micropollutant concentrations increased with increasing specific conductance and total nitrogen concentrations for shoreline well samples. Pharmaceutical and plasticizer concentrations for groundwater in the NY network were less than those observed in the NE samples. The difference in micropollutant concentrations in these networks could reflect demographic differences between the sites according to authors (Phillips *et al.*, 2015).

Estévez *et al.* (2012) studied contaminants of emerging concern and priority substances (2008/105/EC) in an aquifer in the Gran Canaria, Canary Islands, Spain. Several pollutants were detected: 42 pharmaceuticals, 20 pesticides, 12 polyaromatic hydrocarbons, 2 volatile organic compounds and 2 flame retardants. The most frequent compounds were caffeine, nicotine, chlorpyrifos ethyl, fluorene, phenanthrene and pyrene. Concentrations were always below 50 ng $L^{-1}$, although some pharmaceuticals and one pesticide, cholrpyrifos ethyl, were occasionally detected at higher concentrations. This priority substance for surface water exceeded the maximum threshold (0.1 μg $L^{-1}$) for pesticide concentration in groundwater (2006/ 118/EC). The average concentrations per chemical groups remained constant for all the sampling points, regardless of the water origin: pharmaceuticals (10 ng $L^{-1}$) > pesticides (5 ng $L^{-1}$) > polycyclic aromatic hydrocarbons (1 ng $L^{-1}$) > volatile organic compounds (1 ng $L^{-1}$). Other contaminants were present at all the groundwater sampling points (phenylephrine, nifuroxazide and miconazole), which indicate alternative contamination sources besides golf course irrigation, such as agricultural practices, septic tanks and sewerage breaks. Some chemicals other than pesticides that exceeded the 2006/118/EC limit for pesticides in groundwater were: benzalkonium chloride, nicotine and nifuroxacide in groundwater (Estévez *et al.*, 2012).

### 2.2.3. Pumping Activities

Among the naturally occurring sources of groundwater contamination, whose discharge is created and/or exacerbated by human activity, the best known is seawater intrusion in coastal aquifers due to pumping activities.

Under undisturbed natural conditions, a state of dynamic equilibrium is maintained whereby flow of fresh groundwater to the sea keeps the landward encroachment of seawater into the aquifer within certain limits. However, as pumpage from the aquifer develops, causing reduction of outflow to the sea and subsequent lowering of the groundwater table (or the piezometrtc surface in a confined aquifer) near the coast, the dynamic balance between the freshwater and seawater is disturbed, permitting seawater to advance inland and contaminate usable parts of the aquifer.

Pumping may precipitate the migration of more mineralized water from surrounding strata to the well. As long as the level of pumpage from the aquifer does not exceed its replenishment, a new equilibrium may be reached. Saltwater intrusion is usually caused when the hydrodynamic balance between the fresh water and the saline water is disturbed, such as when fresh groundwater is overpumped in coastal aquifers.

It is worth noting that, although the process of seawater intrusion is slow, it is practically irreversible. Field observations have shown that, once an area has been invaded by seawater, all pumping wells in that area will sooner or later be contaminated and abandoned.

In coastal areas, pumping has caused seawater to invade fresh-water aquifers. In parts of coastal west Florida, wild-flowing, abandoned artesian wells have salted large areas of fresh aquifers.

Saltwater intrusion can also occur when the natural barrier that separate fresh and saline water are destroyed, such as in the construction of coastal drainage canals that enable tidal water to advance inland and percolate into a freshwater aquifer.

According to some authors, contamination of groundwater by As in Bangladesh seems to be related to pumping activities that may affect As concentrations, but not by the oxidation of sulfides, or by slow reduction of iron hydroxides or sorbed arsenate by detrital organic carbon, as has been previously reported (Kresic, 2007).

A pollutant released from a source of contamination, within a groundwater recharge area, may reach a certain pumping well after many years. The path of such a pollutant and the evolution of its concentration are determined by a multitude of factors, such as the depth of the unsaturated soil, the flow pattern within the saturated soil and, therefore, the distance to be traveled by the pollutant within this zone toward the pumping well, the pattern of pumpage (rates and locations) as well as the properties of the two zones that affect the rate of transport and transformation of the pollutant.

## 2.2.4. Agriculture Activities

Agricultural Activities are one of the most important sources that cause contamination of groundwater due to greater use of inorganic fertilizers and pesticides in order to obtain higher crop yields.

### *2.2.4.1. Fertilizers and Pesticides*

Modern farming includes several practices that can lead to groundwater contamination: fertilizer and pesticide application, irrigation and animal waste storage. Pesticides and herbicides are usually sprayed on fields in an aqueous solution. Some of these compounds biodegrade rapidly, but some are persistent and contaminate groundwater over broad areas.

- **Fertilizers**

Fertilizers are the primary cause of groundwater beneath agricultural lands. Intensive, uncontrolled use of fertilizers is reported to be a major cause of groundwater contamination in developed countries. Both inorganic (chemically manufactured) and organic (from animal or human waste) fertilizers applied lo agricultural lands provide nutrients such as nitrogen, phosphorous and potassium that fertilize the land and stimulate plant growth. A portion of these nutrients usually leaches through the soil and reaches the groundwater table.

In several European countries with intensive farming activities such as France, Hungary, Russia, the UK, Germany, the Netherlands, a relationship has been found between the amount of nitrogen fertilizer applied to farmland and, $NO_3^-$ and phosphate concentration in groundwater, indicating that under current fertilization practices nitrogen input exceeds its uptake.

$NO_3^-$ concentration of groundwater due to the use of synthetic and organic fertilizers poses serious problems in several countries. Several studies conducted in Korea, India, Portugal, Philippines, China and Spain demonstrated groundwater pollution by $NO_3^-$, $SO_4^{-2}$, potassium and phosphorus due to excessive application of fertilizers (Agrawal *et al.*, 1999; Andrade and Stigter, 2009; Bouman *et al.*, 2002; Datta *et al.*, 1997; Kim *et al.*, 2015; Liu *et al.*, 2005; Sánchez Pérez *et al.*, 2003; Zhai *et al.*, 2017).

Only a portion of nitrogen is adsorbed by soil or used by plants and the rest is dissolved in water to form nitrates in a process called *nitrification.* Nitrates are mobile in groundwater and have summed on a regular basis. The contamination of groundwater by phosphorous and potassium is less than nitrogen. Nitrogen, in the form of $NO_3^-$, is more soluble in groundwater as compared with phosphorous while $K^+$

has low mobility due to cation exchange. Phosphate and potassium fertilizers are readily adsorbed on soil particles and seldom constitute a pollution problem. If the $NO_3^-$ concentration exceeds 50 mg $L^{-1}$, it becomes undesirable for drinking purposes.

- **Pesticides**

In addition to fertilizers, pesticides are also being extensively used. The annual growth rate world-over reached 12% in the 1960s but later decreased to around 3-4% (Chilton *et al.*, 1994).

Insecticides, herbicides and fungicides used for destroying unwanted animal pests, plants and fungal growth can also cause groundwater contamination. When applied to land, these chemicals degrade in the environment by a variety of mechanisms. Their parent compounds and their byproducts persist long enough to adversely impact the soil and groundwater.

The contamination of groundwater by pesticides is dependent on soil properties and application factors. The risk of pesticide contamination is higher in surface water than groundwater due to the adsorption of pesticides by clay mineral grains and organic matter degradation by bacteria. Some pesticides with higher solubility in water have significant mobility in some type of aquifer materials, such as clean sands and gravels. Preferential flow paths, e.g. macropores and fissures, in unsaturated soils lead also to fast movement of pesticides to water-table.

LeGrand (1970) suggested that chances of groundwater contamination are reduced at those sites that have (1) deep water-table, (2) high clay content, (3) water supply wells located on the upgradient side from the source of contamination and (4) large distances between wells and contamination sources (LeGrand, 1970).

In India, Shukla *et al.* (2006) determined the contamination levels of organochlorine pesticides in the groundwater of Hyderabad City. Water samples were collected from 28 domestic well supplies of the city. All the samples analyzed were found to be contaminated with four pesticides i.e. DDT, β-Endosulfan, α-Endosulfan and Lindane. DDT was found to range between 0.15 and 0.19 μg $L^{-1}$, β-Endosulfan ranges between 0.21 and 0.87 μg $L^{-1}$, α-Endosulfan ranges between 1.34 and 2.14 μg $L^{-1}$ and Lindane ranges between 0.68 and 1.38 μg $L^{-1}$. These concentrations of pesticides in the water samples were found to be above their respective Acceptable Daily Intake (ADI) values for Humans. The authors suggested that this contamination is due to possible transfer of organochlorine pesticides from agricultural activities carried out near Hyderabad (Shukla *et al.*, 2006).

The pollution of agricultural land due to herbicides was assessed by Carabias *et al.* (2000) in the Guarena and Almar river basins, situated in the provinces of Zamora and Salamanca, Spain. Surface and groundwaters, taken from different locations in the basins, were analyzed over a 6-month period. The presence of six out of the fifteen herbicides monitored-chlortoluron, atrazine, terbutryn, alachlor, diflufenican and fluazifop-butyl- was detected in several samples at levels ranging from the detection limit to 1.2 mg $L^{-1}$ (Carabias Martinez *et al.*, 2000).

Andrade *et al.* (2009) studied an area in the Mondego River alluvial body in central Portugal, where agriculture is the main land use, with predominantly maize, rice and some vegetable crops supported by river water irrigation. The results showed that shallow groundwater is easily contaminated with herbicides. Molinate and 3,4 dichloroaniline (metabolite of propanil) are the most detected herbicide compounds in these water samples; propanil is rarely detected due to its low persistence in soil. Total concentrations of the analyzed herbicide compounds above the regulatory limit of 0.5 mg $L^{-1}$ were observed in 32% of the analyzed water samples, with a maximum value of 16.09 mg $L^{-1}$ (Andrade and Stigter, 2009).

#### 2.2.4.2. Sludge Use

With regard to municipal liquid waste, soil application of sewage effluent and sludge is perhaps the largest contributor to groundwater contamination in faming activities. Sludge has been applied to land for many years to recharge groundwater and provide nutrients that fertilize the land and stimulate plant growth. However, land application of sewage effluent can introduce bacteria, viruses and organic and inorganic chemicals into groundwater.

#### 2.2.4.3. Irrigation

When crops are irrigated, most of the applied water is transpired or evaporates. Evaporation and transpiration of irrigated water leave the remaining water with higher solute concentrations. Therefore, irrigation can lead to the accumulation of salts and other contaminants in soils and in local surface waters.

In arid and semi-arid areas intensive irrigation leads to rise in water-table, thereby increasing the soil and groundwater salinity due to this evapotranspiration losses. A case study of soil salinization due to rise in water-table from San Joaquin Valley in California, USA is described by Kehew (2001) (Kehew, 2001).

Chen *et al.* (2010) performed 394 water samples from potential contaminated irrigation wells during a period of five years (from 2002 to 2007) in Huantai County in North China. They demonstrated that the average $NO_3^-$ concentration in 2002 and 2007 were 8.08 and 14.68 mg $L^{-1}$, respectively, which indicated there was a significant increasing from 2002 to 2007 related to land use change and irrigation water resource (Chen *et al.*, 2010).

#### 2.2.4.4. Animal Wastes

Concentrated animal feed lots for cattle, pigs and other animals generate large amounts of animal wastes that can contaminate groundwater. Common contaminants in these settings include $NO_3^-$, veterinary pharmaceuticals (antibiotics) and steroid hormones. The use of veterinary antibiotics in concentrated animal feeding operations is a growing concern as an important source of environmental contamination. This is of particular relevance in the USA and Europe where intensive rearing of livestock is becoming more common.

Veterinary antibiotics were investigated in various environmental compartments including waste lagoons, groundwater below lagoons, as well as shallow groundwater from areas where animal waste had been applied to fields (Stuart *et al.*, 2012).

A significant proportion of the antibiotic given passes through the livestock and is excreted and stored in waste lagoons. It poses a potential threat to groundwater resources through leakage from the lagoons or when the waste is spread on the land. In dairy waste lagoons, the endogenous estrogens E2 and estrone as well as the androgens testosterone and androstenedione have been detected at high concentrations (Kolodziej *et al.*, 2004).

Application of manure and bio-solids from sewage sludge processing has the added benefit of enhancing soil nutrient levels. However, the incomplete removal of contaminants during wastewater treatment may result in residual concentrations of toxic materials. Two types of halogenated hydrocarbons (perflurochemicals and polychlorinated alkanes) are considered important potential groundwater contaminants from applications to soil due to their relatively high concentrations in wastewater and their relatively high solubility (Clarke and Smith, 2011).

It has been reported the occurrence of a range of veterinary antimicrobials at low concentrations in groundwater that are attributed to manure application (Hu *et al.*, 2010; Sarmah *et al.*, 2006). Furthermore, Buerge *et al.* (2011) reported the occurrence of saccharin in groundwater in Switzerland, after application of pig manure (Buerge *et al.*, 2011).

Manure derived contaminants can reach groundwater via indirect routes such as surface water-groundwater exchange from runoff rather than downward migration due to attenuation mechanisms in the soil and unsaturated zone. Surface waters contain higher concentrations and a larger range of contaminants than groundwaters, and are therefore an important source and pathway of groundwater pollution.

### 2.2.5. Mining Activities

Groundwater can be contaminated by the drainage from mines operations. Drainage of both active and abandoned surface and underground mines can produce a variety of groundwater pollution problems. Mining activities is responsible for groundwater pollution mainly due to infiltration of leachate from mine waste and tailings. Surface mines and mine tailings produce highly mineralized runoff frequently referred to acid mine drainage. This runoff can percolate into the ground and degrade the quality of groundwater.

Water that percolates through the mine workings and tailings often has unusual concentrations of inorganic materials. Sulfide and coal mines, where most Zn, Cu and Pb come from very acidic leachate because of iron and sulfide oxidation reactions. Entry of oxygen which results in the oxidation of sulphide ores resulting in strongly acidic ($H_2SO_4$) waters. Acidity depends on the amount of reactive pyrite present in the coal. The acidic leachate, in turn, can mobilize various metals from interacting surfaces. Therefore, water seepage can leach toxic metals from exposed ores and raw materials and introduce them to groundwater. Compared to colliery waters, drainage waters from metal mines have lower concentration of $SO_4^{-2}$ but higher amount of heavy metals (Hg, Pb, Cd, Cu, etc) due to lower pH.

Water from polymetallic (Zn-Pb-Cu) mines for example Rangpo mine in Sikkim, India, has low pH (around 5.0) and higher amount of dissolved solids, Pb, Zn and Cu (3,616 mg $L^{-1}$, 277 mg $L^{-1}$, 35 mg $L^{-1}$ and 1.80 mg $L^{-1}$respectively). Acid mine drainage is also a significant problem in Eastern coal mining and some western base metal mines in USA. Discharge of mercury from gold mining activities has contaminated streams and basement rock aquifers in northern and central Precambrian shield in Brazil and Guiana creating serious health risks for settlements. High As concentration is reported from acid mine drainage in Obusi gold mines in Ghana and peat deposits in The Netherlands. In the Netherlands, oxidation of pyrite in peat deposits is regarded to be the source of As in water (Singhal and Gupta, 2010).

Lei *et al.* (2010) studied the acidification and the heavy metals distribution and evolution of groundwater in the black swan nickel sulfide mine (Western Australia). The groundwater samples were collected from the drilling holes situated in the vicinity of tailings storage facility (TSF) and in the background of the mine (away from TSF). The results disclosed that the TSF groundwater is remarkably acidified (pH mean$\approx$5, pH min=3) and the average contents of heavy metals (Co, Cu, Zn, Cd, Al and Mn) are of 1-2 orders of magnitude higher in TSF groundwater than in background groundwater. According to the authors, this is may be due to the percolation of tailings wastewater from mill process, which leads the tailings to oxidize and the deep groundwater to acidify and contaminate with heavy metals. They mentioned that the wastewater and tailing slurry in tailing storage facility can provide water for oxidation metal sulfides and forming acid mine drainage, therefore even if a metal sulfide mine is located in arid territory, there could be a potentiality of producing acid mine drainage and releasing heavy metals, contaminating the deep groundwater (Lei *et al.*, 2010).

Herbert (2006) investigated groundwater contamination from a mine rock dump in Dalarna, Sweden from the onset of snowmelt runoff (April) until October. The results demonstrated that considerable variation in solute concentration (Al, Cu, Fe, $SO_4^{-2}$ and Zn) and acidity occurs in groundwater; the greatest change in solute concentrations occurs during the melting of the snow cover, when sulfide oxidation products are flushed from the rock dump. During this period, groundwater flow is concentrated. Groundwater acidity varied by a factor of four closest to the rock dump during the sampling period, but these variations were attenuated with distance from the rock dump. Over a distance of 145 m, groundwater pH increased from 2.5 to 4.0. This study demonstrated the considerable variations in solute concentration and acidity in groundwater as a result of the episodic releases of oxidation products from mine waste deposits, which are apparently greatest following snowmelt runoff, when sulfide oxidation products are flushed from the rock dump. Major nutrients (i.e. $NO_3^-$, $NH_4^+$) are also flushed from soils. The period of snowmelt runoff was associated with a high groundwater table and surface runoff, such that acidity and heavy metals were rapidly mobilized through the upper permeable soil horizons and on the ground surface (Herbert, 2006).

Another hazardous situations are the acid mine drainage emanating from Rand gold mine in South Africa, which contains uranium, one of the most toxic element to aquatic life (Winde, 2006). Brown *et al.* (1998) demonstrated that Soils of the Ranger Uranium Mines in Northern Australia are capable of retaining uranium, present in water irrigated on the area. Conversely, conservative contaminants such as $SO_4^{-2}$, calcium and magnesium are not retained by the soil and readily reach the groundwater. Elevated levels of these contaminants migrate towards Magela Creek, most rapidly in regions of high hydraulic conductivity, where they are diluted by the surficial flow of the creek that occurs during the monsoonal rains of the wet season of tropical northern Australia (Brown *et al.*, 1998).

Finally, Cr contamination in groundwater was reported in India due to mining activities of chromite in Sukinda area in Orissa (Tiwary *et al.*, 2005) .

## 2.2.6. Municipal Waste and Wastewater Treatment

Underground disposal practices of municipal liquid waste can cause groundwater contamination. Hospital wastewater forms an important source for a range of specific contaminants, including x-ray contrast related substances and some therapeutic drugs. Diffuse leakage from reticulated sewerage systems may also pose a significant risk for groundwater pollution.

The range of molecules potentially released into the environment depends largely on the origin of the water treated. A recent example which has illustrated the potential for molecules to by-pass sophisticated treatment processes has been the widespread occurrence of metaldehyde, the active ingredient of slug pellets, in treated drinking water sources (Lapworth *et al.*, 2012).

Contamination is most extensive within shallow wells sited in areas of low cost housing, due to inadequate sanitation, household waste disposal and significantly poor well protection and construction. Among disposal practices of domestic liquid waste, septic tanks and cesspools contribute the most wastewater to the ground and are the most frequently reported sources of groundwater contamination. Septic tanks and cesspools contribute filtered sewage effluent directly to the ground, which can introduce high concentrations of $NO_3^-$, organic chemicals and possibly bacteria and viruses into groundwater. $NO_3^-$ is the major contaminant resulting from municipal wastes and leaky sewers and is responsible of groundwater contamination as demonstrated in several studies (Grimmeisen *et al.*, 2017; Selvakumar *et al.*, 2017).

One of the major municipal sources of groundwater pollution is runoff from roadway containing deicing agents. In many urban areas, large quantities of salts and deicing agents are applied to roads during winter season. These agents facilitate the melting of ice. However, they can percolate with the water and cause groundwater contamination of shallow aquifers. In addition, the high solubility of these salts in water and the relatively high mobility of the resulting contaminants such as sodium and chloride ions in groundwater can cause an expansion of the contamination.

Stockpiles of materials and waste tailings can also be a source of groundwater contamination. Precipitation falling on uncovered or unlined Stockpiles or waste tailings causes leachate generation and seepage into the ground. The leachate can transport heavy metals, salts and other inorganic and organic constituents as pollutants to groundwater.

Several studies were conducted to evaluate groundwater pollution by municipal wastes. Osenbruck *et al.* (2007) used isotopic ($\delta^{18}O$, $\delta\ ^{2}H$ and $\delta\ ^{34}S\text{-}SO_4$) and chemical tracers (boron) to assess the sources and transport processes of the micropollutants carbamazepine, galaxolide and bisphenol A in groundwater underlying the city of Halle (Saale), Germany. Data showed that their ubiquitous presence in urban groundwater resulted from a combination of local river water infiltration, sewer exfiltration and urban stormwater recharge. The authors found that the presence of the micropollutants carbamazepine, galaxolide and bisphenol A in urban areas severely affected urban groundwater and high concentrations of these micropollutants may enter the groundwater if hydraulic conditions allow the infiltration of river water. Additionally, sewer exfiltration contributed significantly to the ubiquitous pollution of groundwater with these micropollutants. Sorption processes of carbamazepine and galaxolide during transport in groundwater were much stronger in context with sewer exfiltration as compared to river infiltration, most probably due to higher contents of organic carbon and higher residence times in the colmation layer around sewer leakages (Osenbruck *et al.*, 2007).

White *et al.* (2016) used a combination of micro-organic contaminants and inorganic hydrochemistry to trace recharge pathways and quantify the variability of groundwater quality in the city of Doncaster, UK. A total of 23 micro-organic contaminants were detected during this study. Four of the compounds detected are EUWater Framework Directive priority substances: atrazine, simazine, naphthalene and bis(2-ethylhexyl)phthalate. Long-term changes in inorganic hydrochemistry show possible changes in contaminant input or the dissolution of minerals. $NO_3^-$ was detected above 50 mg $L^{-1}$ but on the whole $NO_3^-$ concentrations have declined in the intervening years either due to a reduction of $NO_3^-$ application at the surface or a migration of peak $NO_3^-$ concentrations laterally or to greater depth (White *et al.*, 2016).

Sorensen *et al.* (2014) characterized a broad range of emerging organic contaminants (n > 1,000) in groundwater sources in Kabwe, Zambia. Groundwater sources are distributed across the city to encompass peri-urban, lower cost housing, higher cost housing and industrial land uses. A total of 27 compounds were identified including the omnipresent N,N-Diethyl-m-toluamide (DEET). This insect repellent was ubiquitous within groundwater at concentrations up to 1.8 μg $L^{-1}$. Other compounds were detected in less than 15% of the sources and included the bactericide triclosan (up to 0.03 μg $L^{-1}$), chlorination by-products trihalomethanes (up to 50 μg $L^{-1}$) and the surfactant 2,4,7,9-tetramethyl-5-decyne-4,7-diol (up to 0.6 μg $L^{-1}$). Emerging contaminants were most prevalent in shallow wells sited in low cost housing areas. This was attributed to localized vulnerability associated with inadequate well protection, sanitation and household waste disposal. The average five-fold increase in median DEET concentration following the onset of the wet season highlights that more mobile contaminants can transit rapidly from the surface to the bedrock aquifer. According to authors, the deep groundwater resources that provide the majority of the water supply to the city may be more vulnerable than previously considered (Sorensen *et al.*, 2015).

### 2.2.7. Landfills

Landfill sites have been shown to be important sources for toxic contaminants in groundwater for many years and have a continued legacy regarding historical landfill practices regarding landfill construction, waste management and in some cases poor choices of landfill location. In many developing countries there is presently limited effective regulation regarding groundwater protection from landfill sources. This, combined with the appearance of emergent pollutants and large increase in use of pharmaceuticals in recent years as well as the increased chemical domestic agent suggests that this may be an important source of groundwater contamination for many years to come.

Landfills may contain municipal solid wastes, construction debris or industrial wastes like incinerator ash and paper mill sludge. Infiltrating water can move down through the refuse, picking up dissolved compounds on its way. Leachate is the water that percolates out of the base of the refuse. The leachate will continue to migrate downward under the influence of gravity until it reaches the saturated zone, contaminating the shallow groundwater aquifer. Landfills excavated to the saturated zone may also permit direct contact of contaminants with groundwater. When a polluted surface water body recharges into uncontaminated groundwater, it will contaminate the groundwater (Figure 2). Buried waste is also subject to leaching by percolating rain water and surface water or by groundwater contact with the fill.

Older dumps have generated many leachate plumes. In more permeable settings, these plumes have reached kilometers in length and may last for many decades.

The generated leachate can contain high levels of BOD, COD, chloride, alkalinity, trace elements and even toxic constituents that can degrade the quality of groundwater. It usually has high dissolved solids content and it may also pick up large number of dissolved organic contaminants. The total dissolved solids may be as much as 30,000 mg $L^{-1}$. Therefore, it can cause severe damage to groundwater quality. Leachate from landfills may contain large amounts of short and long-chain fatty acids and can also contain caffeine, nicotine, phenols, sterols, pesticides, polycyclic aromatic hydrocarbons, chlorinated solvents and phthalates. Many studies reported a wide range of pharmaceuticals in groundwater down gradient of landfills receiving domestic and industrial waste (Lapworth *et al.*, 2012). In addition, a variety of gases such as $CH_4$ and $CO_2$, are also produced in landfills due to the biochemical decomposition of organic matter. $CH_4$ can be very dangerous as it is combustible and can cause severe damage.

The biochemical decomposition of ammonia and hydrogen sulfide that migrate through the unsaturated zone into adjacent terrains can cause potential hazards. For example, The generated $NO_3^-$ in groundwater from a landfill located in fractured quartzite in Delhi is reported to be more than 110 mg $L^{-1}$, which necessitates abatement solutions to protect groundwater quality (Dey *et al.*, 2003).

Two studies showed the persistence of some organic compounds in groundwater down gradient of landfills and detected a range of industrial compounds (detergents, antioxidants, fire retardants and plasticizers) as well as pharmaceuticals and personal care products (antibiotics, anti-inflammatories and barbituates), caffeine and the nicotine metabolite cotinine. Some of them have been detected in significant concentrations between 10-$10^4$ ng $L^{-1}$ (Barnes *et al.*, 2004; Buszka *et al.*, 2009).

Bruner *et al.* (1998) studied groundwater samples taken near a closed municipal landfill at Norman, *Oklahoma*. Groundwater samples were highly toxic in the area near the landfill, indicating a plume of toxicants. The groundwater between the landfill and the Canadian River was contaminated, with toxicity diminishing as distance from the landfill increases. The small stream which runs adjacent to the landfill also revealed high toxicity, probably as a result of interaction between the stream and the shallow alluvial aquifer (Bruner *et al.*, 1998).

Lopes *et al.* (2012) studied the concentrations of some heavy metals in the surrounding area of a landfill-Londrina, Brazil. The results showed that Ni, Zn, Cd and Cu in the groundwater are below the limits established by the potable water quality standards in Brazil, except for Pb whose concentration in groundwater were higher if compared to Brazilian legislation (Lopes *et al.*, 2012).

Jensen *et al.* (1999) studied the distribution of heavy metals in landfill-leachate polluted groundwater in Denmark. The groundwater samples were contaminated with heavy metals at concentrations within the range of metal concentrations found in landfill leachates (Cd: 100 µg $L^{-1}$, Ni: µg $L^{-1}$, Zn: 1,000 µg $L^{-1}$, Cu: 1,000 µg $L^{-1}$ and Pb: 1,000 µg $L^{-1}$) and divided into colloidal (screen-filtration and cross-flow ultrafiltration), organic (anion-exchange) and dissolved inorganic species of the heavy metals. According to authors, the distribution of the heavy metals between the different size-fractions showed that a substantial, but highly varying part of the heavy metals was associated with the colloidal fractions (Cd: 38±45%, Ni: 27±56%, Zn: 24±45%, Cu: 86±95% and Pb: 96±99%). Speciation by the use of an anion-exchange technique showed that the heavy metals complexed strongly with the organic matter in leachate-polluted groundwater, especially with respect to Cd, Cu and Pb. More than 60% of Cd, Cu and Pb were found to be organic species. The results suggested that, except for Zn, heavy metals in leachate-polluted groundwater are strongly associated with small-size colloidal matter, which was to a large extent organic matter (Jensen *et al.*, 1999).

The sign of groundwater contamination was noticed in many surrounding wells in Ruseifa municipal landfill resulting in the high number of fecal coliform bacteria and total coliform bacteria and the increase in inorganic parameters such as chloride. The pollution of groundwater wells in the landfill area was attributed to the leachate bodies which flow through the upper part of Wadi Es Sir or Amman-Wadi Es Sir Aquifer (Al-Tarazi *et al.*, 2008).

### 2.2.8. Industrial Activities

One of the major pollutants in freshwaters is industrial effluents. It is, arguably, the dominant problem in overdeveloped nations. Industrial effluents may contain a variety of hazardous components, such as radioactive materials, heavy metals, hydrocarbons and toxic organic compounds. Thermal pollution, due to discharge of cooling waters, may also have a significant effect on the biology of water bodies. A summary of the major types of groundwater pollutants and their industrial sources is given in Table 2.

The wastewater discharged from industries carries with it a variety of dissolved and suspended impurities, the composition of which depends on the type of industries and the processes used. For example, high concentrations of Cr, Ni, Cd, Pb and Zn in groundwater are reported from several places due to industries in North India (Kakar, 1990).

Contamination of groundwater by organic compounds due to inadvertent release of industrial effluents is of great concern and poses safety hazards to drinking water. Petroleum hydrocarbons are the most common organic contaminants but they do not pose significant risk to groundwater because of their lower aqueous solubility. However, halogenated hydrocarbons e.g. carbon tetrachloride and ethylene dibromide are of greater concern due to their persistence and toxicity. They are used in different ways for making plastics, as solvents and many types of manufacturing processes.

Most of the organic compounds are immiscible in water with limited dissolution between aqueous and organic phases. Such liquids are known as non-aqueous phase liquids "NAPLs". If the density of NAPL is less than that of water, the liquid is classified as a light non-aqueous phase liquid (LNAPL) like acetone and benzene etc., which represent mainly petroleum products. If the density of NAPL is greater than that of water, it is classified as dense nonaqueous phase liquid (DNAPL) like chlorinated hydrocarbons, phenol and chloroform etc. The infiltrating water will dissolve the soluble components of LNAPL and transport them to the water-table forming a plume while the volatile components will be transported to the upper parts of the aquifer by molecular diffusion causing secondary contamination in the unsaturated zone by volatilization or by dissolution in groundwater.

Groundwater samples from an industrialized area near Milan were analyzed to identify the main pollutants and to quantify two classes of chemicals: polychloro-1,3-butadienes (PCBD) and some aromatic amines. The water contained several halogenated aromatic and aliphatic compounds and heavy contamination due to PCBD, probably arising from contaminated land where a disused chemical plant is located (Fattore *et al.*, 1998).

In India, groundwater has become contaminated with As due to the discharge of industrial effluent after production of the insecticide Paris-Green (Copper acetoarsenite $Cu(CH_3COO)_2$, $3Cu(AsO_2)_2$) by a local factory at the P. N. Mitra Lane, Behala. For several years this factory had been producing 20 tons of Paris-Green per year and had been dumping its effluent in that area. More than seven thousand people were using this As contaminated tube-well water for drinking and house-hold purposes. Many people of the area were hospitalized and symptoms of As toxicity were visible amongst a large number of the population. Analytical study revealed that soil around the area of effluent dumping point, which is at the middle of the locality, contains a very high concentration of As and Cu. According to the authors, the effluent treatment for As removal was not adequate and As percolated to the underground aquifers. Consequently, As concentration in the groundwater was very high. Both arsenite and arsenate were present in groundwater (Chatterjee *et al.*, 1993).

Kumar *et al.* (2017) studied the spatio-chemical contamination sources and health risk assessment arising from the consumption of groundwater contaminated with trace and toxic elements in the Chhaprola Industrial Area, Gautam Buddha Nagar, Uttar Pradesh, India. In this study 33 tubewell water samples were analyzed. Concentration of some trace and toxic elements such as Al, As, B, Cd, Cr, Mn, Pb and U exceeded their corresponding WHO (2011) guidelines and BIS (2012) standards. Multivariate statistics (PCA and CA) indicated that natural and anthropogenic activities like industrial effluent and agricultural runoff are responsible for the degrading of groundwater quality in the research area. The hazard quotient (HQ) value exceeded the safe limit of 1 which for As, B, Al, Cr, Mn, Cd, Pb and U at few locations while hazard index (HI) > 5 was observed in about 30% of the samples, which indicated potential health risk from these tubewells for the local population if the groundwater is consumed (Kumar *et al.*, 2017).

Kumar (2014) studied the natural and anthropogenic influences on groundwater quality in Vaniyambadiarea of Vellore district, India. Groundwater samples were collected and analyzed for physico-chemical parameters and the results showed a dominance in the following order of $Na^+ > Mg^{2+} > Ca^{2+} > K^+$ and $HCO_3^- > Cl^- > SO_4^{-2} > NO_3^-$. In contrast to this anion dominance were changed to $Cl^- > HCO_3^- > SO_4^{-2} > NO_3^-$ in samples collected near the tannery industries. Generally, the water type was $Na^+$ $Cl^-$ to $Ca^{2+}$ $Mg^{2+}$ $HCO_3^-$ type with an intermediate $Ca^{2+}$ $Mg^{2+}$ $Cl^-$, suggesting the mixing of fresh groundwater with tannery effluent and cation exchange. According to Kumar natural and anthropogenic inputs are equally influencing the groundwater quality. Further investigations provided that silicate weathering is the dominant geogenic source of groundwater solute content, whereas tannery effluent is the anthropogenic source. This study demonstrated that groundwater in the Vaniyambadi area is under serious threat from both natural and anthropogenic contamination (Kumar, 2014).

Grassi *et al.* (2014) identified sources of B and As contamination in surface water and groundwater downstream of the Larderello geothermal-industrial area (Tuscany-Central Italy). The study area, which comprises a Northern sector of the Larderello geothermal field, has in time been contaminated by both surface geothermal manifestations and anthropogenic activity. The latter refers to the disposal of spent geothermal fluids and borogypsum sludge, by-product of colemanite treatment with sulphuric acid, which until the late '70s were discharged in the Larderello area into the Possera Creek, a Southern tributary of the Cecina River. As-rich sediments were found at shallow depths in the area of the Cecina-Possera confluence and in the upper part of the aquifer skeleton. These sediments contributed to increase up to 76 μg $L^{-1}$ As content of groundwater of the Cecina-Possera confluence area which, draining water from the Possera Creek, represents the aquifer root zone. This zone determines the B and As contents of groundwater which flows more or less parallel to the Cecina River, undergoes progressive dilution during its westward flow and locally supplies the same river. According to authors, most of the study stream water and groundwater in the study area cannot be exploited because mean B and As contents (respectively in the range 1.2-15.6 mg $L^{-1}$ and 1.1-75.9 μg $L^{-1}$) that are often well above the permissible limits for drinking water (1 mg $L^{-1}$ for B and 10 μg $L^{-1}$ for As). The As content of both runoff waters and groundwaters is greatly affected by their interaction with As-rich sediments likely containing remnants of borogypsum waste and/or involved in the adsorption of As from past discharges of acidic waste (Grassi *et al.*, 2014).

Cr contamination in soil and water is mainly due to its use in numerous industrial processes like Cr plating and alloying, hide tanning and chemical manufacturing. A Cr plume from chrome plating wastes is reported to extend nearly 1,300 m down gradient in the glacial deposits of Long Island, New York (Kehew, 2001).

In Spain, the Besos River Basin has supported rapidly growing chemical, metallurgical and textile industries that have damaged the quality of the water supply. High contents of Fe, Cr and Ni characterize the chemical composition of the analyzed wastes. These were also the heavy metals most frequently found at toxic levels in the groundwaters (Navarro and Font, 1993).

Ma *et al.* (2013) investigated volatile organic compound contamination in groundwater at an industrial site in Beijing. All the studied 13 volatile organic compounds were tested in the four wells in the industrial contamination sites, with their concentrations in the range of 12-73,660 μg $L^{-1}$. In addition, while benzene and toluene were heavily concentrated up to 74,000 μg $L^{-1}$ and 6,000 μg $L^{-1}$, respectively, 1,2,3-trichlorobenzene and bromobenzene had relatively low contamination levels (below 25 μg $L^{-1}$) (Ma *et al.*, 2013).

Zhang *et al.* (2012) evaluated petroleum contamination of groundwater and identified its sources in the Hunpu, a wastewater-irrigated area located in the southwest of Shenyang, China. The analysis revealed the presence of biogenic and degraded petrogenic hydrocarbons. Petroleum pollution and degradation levels were significantly higher in October 2009 (total aliphatic hydrocarbons (TAH): 909.3-10,343.1 μg $L^{-1}$) than those in May of the same year (TAH: 357.0 to 6,802.1 μg $L^{-1}$). Aside from biogenic hydrocarbons input, the contamination of degraded petroleum products was discovered in the water in Hunpu wastewater irrigated region through concentration and geochemical analyses. The hydrocarbon concentration in water was higher near oil wells and the Xihe River Reach (Zhang *et al.*, 2012).

### 2.2.9. Injection Wells/Surface Impoundments

With regard to industrial liquid waste, surface impoundments and injection wells are probably the largest contributors to groundwater contamination. The use of surface impoundments and injection wells has become an attractive wastewater effluent disposal option for many facilities.

Some liquid wastes are disposed of in injection wells. This approach is based on forcing fluids out of the well screen into the surrounding formation. The wells should be designed to inject into a formation that is isolated from any useful aquifers or surface water ecosystems. The types of wastes most commonly injected are brines and other waters recovered from oil fields, fluids from solution mining and treated wastewaters. Contaminants may enter the aquifer system from the surface by leakage through injection wells because of waste disposal through the deteriorated or improperly constructed wells. Leakage of contaminants through the bottom of a surface impoundment or migration of fluids from an injection well into a hydrologically connected usable aquifer can cause groundwater contamination. The extent and severity of groundwater contamination from these sources are further complicated by the fact that, in addition to being hazardous, many of the organic and inorganic chemicals in industrial wastewater effluent and sludge are persistent.

Oil and gas mining operations can cause groundwater contamination. These operations generate a substantial amount of wastewater, often referred to as brine. The brine is usually disposed of in surface impoundments or injected in deep wells. Therefore, it can reach groundwater and its constituents, such as ammonia, boron, calcium, dissolved solids, sodium, $SO_4^{-2}$ and trace metals, can subsequently degrade the quality of groundwater.

### 2.2.10. Managed Aquifer Recharge

Managed aquifer recharge refers to the use of surface water sources, including treated wastewater, to artificially recharge an aquifer. It is a particularly useful management tool in semi-arid regions where water resources are scarce. It is used to replenish aquifers, use them as natural temporary storage systems and in some cases manage river flow. However, artificial recharge can in some cases short circuit natural attenuation mechanisms in the soil and subsurface leading to potential long term contamination of groundwater resources.

Managed aquifer recharge continues to be a very important potential source of groundwater contamination, particularly when groundwater residence times are short, and poses a threat to adjacent groundwater bodies as well as surface water resources. This source is clearly an important input of contaminants into the environment globally, especially in regions where wastewater treatment is poorly regulated and rudimentary or non-existent.

### 2.2.11. Interaquifer Exchange

In stratified geological formation, contaminated groundwater can mix with uncontaminated groundwater through a process known as inter-aquifer exchange, in which one water-bearing unit communicates hydraulically with another. In interaquifer exchange, two aquifers are hydraulically connected. Contamination occurs when contaminants are transferred from a contaminated aquifer to a clean aquifer. Interaquifer exchange is common when a deep well penetrates more than one aquifer to provide increased yield or when an improperly cased or abandoned well serves as a direct connection between two aquifers of different potential heads and different water quality. The hydraulic connection (well or fractures) can allow contaminants from aquifers with the greatest hydraulic head to move to aquifers of less hydraulic head.

Once contaminants enter a surface water course they may stay in the surface water body or by lateral or vertical hydraulic exchange, through the hyporheic zone, can be transferred to a groundwater body. Groundwater interacts with surface water ranging from small streams, lakes and wetlands in headwater areas to major river valleys and seacoasts. Because of the interchange of water between these two components of the hydrologic system, development or contamination of one commonly affects the other. Where the regional groundwater table is below the surface water level, for example in more arid regions, the surface water may recharge directly into the unsaturated zone beneath the river course. Through such mechanism surface waters can be considered a source and pathway of groundwater pollution. This process is particularly important in aquifers below and adjacent to water courses.

Surface water generally contains less dissolved solids than groundwater, although at certain times where groundwater runoff is the major source of stream flow, the quality of both surface water and groundwater is similar. During periods of surface runoff, streams may contain large quantities of suspended materials and, under some circumstances, a large amount of dissolved solids. Most commonly, however, during high rates of flow streams have a low dissolved-mineral concentration. Human-induced influences on surface-water quality reflect not only waste discharge directly into a stream, but also include contaminated surface runoff. Another major influence on surface-water quality is related to the discharge of groundwater into a stream. If the adjacent groundwater is contaminated, stream quality tends to deteriorate. Fortunately in the latter case because of dilution, the effect in the stream generally will not be as severe as it is in the ground.

## 2.2.12. Air Deposition

Links exist between air pollution and water pollution that can cause changes in water quality. There are many sources of air pollution, but combustion of fossil fuels is one the largest once leading to high concentration of airborne pollutants. Combustion of fuels releases huge quantities of nitrogen, sulfur and carbon oxides in the air every year. The elevated carbon dioxide air levels lead to a higher carbon dioxide concentration in water resulting in lower pH. In fact, the average pH of the ocean decreased from 8.12 in the late 1980s to 8.09 in 2008. A further decrease in the ocean pH will increase the solubility of calcium carbonate minerals such as aragonite and calcite that comprise the shells of many marine organisms (Boyd, 2015). It could have also drastic effects on marine biodiversity. Sulfuric and nitric acids formed from NOx and SOx reach the earth's surface in precipitation called acid rain. These chemicals result in greater acidity of fresh water. However, recent research claims that acid rain is causing alkalization of many stream waters by the acceleration of weathering of minerals (limestone, calcium silicate and feldspars) in watersheds as a result of acid deposition (Kaushal *et al.*, 2013). Further investigation into the cause of stream alkalization is needed.

The high-level radioactive waste containing various radionuclides may enter groundwater flow systems by accidental escape in the air from atomic/nuclear reactors and radioactive waste management systems. One such example is from the Chernobyl disaster in Ukarine, on 26 April 1986, which caused contamination of groundwater from radioactive nuclides i.e. $^{137}Cs$ and $^{90}Sr$. The analytical data show that $^{137}Cs$ and $^{90}Sr$ concentrations in groundwater samples taken during 1992-1997 from wells in Kyiv and Chernobyl regions reached 100 Bq $dm^{-3}$ indicating preferential flow of contaminated water into the subsurface up to depths of 100 m and more (Shestopalov *et al.*, 2006).

## 2.2.13. Construction and Land Uses

Anywhere the land surface is disrupted provides a potential site for erosion and sediment creation. Sediments and suspended solids are classified as pollutants. Sediments consists of mostly inorganic materials washed into a stream as a result of construction, demolition, logging and mining operations. They often lead to very high rates of soil loss. However, the major source of sediment in water bodies is soil particles eroded from the land by rainfall and runoff. In fact, erosion always occurs when the soil is unprotected from direct rainfall or from water running over its surface. When there are no plants to hold the soil, the sun dries it and the wind blows it away, or water washes it away.

Erosion of soil at construction sites not only cause water quality problems offsite, but may be regarded as the loss of a valuable natural resources. Because construction involves land clearing and grading, this activity is a major source of sediment contamination of streams.

Areas that are under construction release the most significant amounts of sediment because there is little vegetation to stop erosion and efficient drainage systems are often set up to sweep water, sediment and other pollutants directly into the municipal drainage system and waterways. Areas under construction typically have 35-45 tons per acre per year of erosion compared to 1-10 tons per acre per year of cropland (Johnson and Juengst, 1997). Land-use activities such as recreational off-road vehicle use can destabilize soil formations, thus increasing erosion. Sediment alters the morphology of water bodies and often has undesirable consequences on ecosystems. For example, sediments interfere with fish spawning because they can cover gravel beds and block light penetration, making food harder to find. Organic sediments can create depletion of water oxygen and anaerobic conditions that may affect the aquatic life cycle and may lead to health challenges.

Furthermore, dissolved nutrients, salts and pesticides used in crop production find their way into surface and ground waters where they can affect biota and prevent further uses of the contaminated water. Sediments can also clog stream channels, exacerbating flood risks, while also transporting organic materials and adsorbed agricultural chemicals to streams. Finally, the sediments increase the turbidity of streams and lakes and smother benthic habitats.

## 2.2.14. Miscellaneous Sources of Groundwater Contamination

In addition to the sources of contamination mentioned above, there are other causes of deterioration of groundwater quality: deforestation and radioactive waste disposal. Deforestation may cause rise in water-table and subsequent increase in the salinity of groundwater. Radioactive waste is generated in all the stages of the nuclear fuel cycle. This includes mining, milling and refining of uranium ore, fuel fabrication and fuel consumption in reactors, waste solidification and burial of solidified waste.

Large volume of high-level radioactive waste is produced from nuclear reactors for power production, weapons manufacturing and research. The reactor waste is currently temporarily stored in iron, concrete or steel containers, which are put underground, preferably in the unsaturated zone above the water-table. A list of environmentally important radioactive isotopes produced from various industries and at different stages from the nuclear fuel cycle, nuclear weapons test, research and medical waste is given in Table 2. They can make groundwater unfit for drinking purposes.

## SUMMARY

In many regions in the world, groundwater represents an important source of fresh water. It is now established that several contaminants enter groundwater from a number of sources and pathways. These sources are both natural and anthropogenic. Contamination of groundwater resources by a variety of anthropogenic pollutants from both point and nonpoint sources represents a key global environmental problem. The most frequently identified contaminant sources are industrial manufacturing, agricultural activities, municipal landfills and wastes. Frequently detected contaminants included nitrates, volatile organic compounds, arsenic and fluorides. Other contaminant species include solvents, fuel hydrocarbons, heavy metals, pesticides, disinfectants, detergents and radionuclides.

In this chapter, the main sources and pathways for contaminants in groundwater are reviewed. It identifies challenges that need to be met to minimize risk to drinking water and ecosystems. Particular attention is paid to the occurrence of known and potential endocrine disrupting substances in groundwater.


## REFERENCES


Agrawal, G. D., Lunkad, S. K., & Malkhed, T. (1999). Diffuse agricultural nitrate pollution of groundwaters in India. *Water Science and Technology*, *39*(3), 67–75. doi:10.1016/S0273-1223(99)00033-5

Al-Tarazi, E., Abu Rajab, J., Al-Naqa, A., & El-Waheidi, M. (2008). Detecting leachate plumes and groundwater pollution at Ruseifa municipal landfill utilizing VLF-EM method. *Journal of Applied Geophysics*, *65*(3), 121–131. doi:10.1016/j.jappgeo.2008.06.005

Andrade, A. I. A. S. S., & Stigter, T. Y. (2009). Multi-method assessment of nitrate and pesticide contamination in shallow alluvial groundwater as a function of hydrogeological setting and land use. *Agricultural Water Management*, *96*(12), 1751–1765. doi:10.1016/j.agwat.2009.07.014

Barnes, K. K., Christenson, S. C., Kolpin, D. W., Focazio, M. J., Furlong, E. T., Zaugg, S. D., ... Barber, L. B. (2004). Pharmaceuticals and Other Organic Waste Water Contaminants Within a Leachate Plume Downgradient of a Municipal Landfill. *Ground Water Monitoring and Remediation*, *24*(2), 119–126. doi:10.1111/j.1745-6592.2004.tb00720.x

Barnes, K. K., Kolpin, D. W., Furlong, E. T., Zaugg, S. D., Meyer, M. T., & Barber, L. B. (2008). A national reconnaissance of pharmaceuticals and other organic wastewater contaminants in the United States--I) groundwater. *The Science of the Total Environment*, *402*(2-3), 192–200. doi:10.1016/j.scitotenv.2008.04.028 PMID:18556047

Bouman, B. A. M., Castañeda, A. R., & Bhuiyan, S. I. (2002). Nitrate and pesticide contamination of groundwater under rice-based cropping systems: Past and current evidence from the Philippines. *Agriculture, Ecosystems & Environment*, *92*(2), 185–199. doi:10.1016/S0167-8809(01)00297-3

Boyd, C. E. (2015). *Water Pollution. In Water Quality: An Introduction* (pp. 313–338). Cham: Springer International Publishing.

Brown, P. L., Guerin, M., Hankin, S. I., & Lowson, R. T. (1998). Uranium and other contaminant migration in groundwater at a tropical Australian Uranium Mine. *Journal of Contaminant Hydrology*, *35*(1), 295–303. doi:10.1016/S0169-7722(98)00129-6

Bruner, M. A., Rao, M., Dumont, J. N., Hull, M., Jones, T., & Bantle, J. A. (1998). Ground and Surface Water Developmental Toxicity at a Municipal Landfill: Description and Weather-Related Variation. *Ecotoxicology and Environmental Safety*, *39*(3), 215–226. doi:10.1006/eesa.1998.9999 PMID:9570913

Buerge, I. J., Keller, M., Buser, H. R., Muller, M. D., & Poiger, T. (2011). Saccharin and other artificial sweeteners in soils: Estimated inputs from agriculture and households, degradation, and leaching to groundwater. *Environmental Science & Technology*, *45*(2), 615–621. doi:10.1021/es1031272 PMID:21142066

Buszka, P. M., Yeskis, D. J., Kolpin, D. W., Furlong, E. T., Zaugg, S. D., & Meyer, M. T. (2009). Waste-Indicator and Pharmaceutical Compounds in Landfill-Leachate-Affected Ground Water near Elkhart, Indiana, 2000–2002. *Bulletin of Environmental Contamination and Toxicology*, *82*(6), 653–659. doi:10.100700128-009-9702-z PMID:19290448

Canter, L. W., Knox, R. C., & Fairchild, D. M. (1988). Groundwater quality protection. M. Lewis Publishers Inc.

Carabias Martinez, R., Rodriguez Gonzalo, E., Fernandez Laespada, M., & Sanchez San Roman, F. (2000). Evaluation of surface- and ground-water pollution due to herbicides in agricultural areas of Zamora and Salamanca (Spain). *Journal of Chromatography. A*, *869*(1-2), 471–480. doi:10.1016/S0021-9673(99)01188-7 PMID:10720260

Carrara, C., Ptacek, C. J., Robertson, W. D., Blowes, D. W., Moncur, M. C., Sverko, E., & Backus, S. (2008). Fate of pharmaceutical and trace organic compounds in three septic system plumes, Ontario, Canada. *Environmental Science & Technology*, *42*(8), 2805–2811. doi:10.1021/es070344q PMID:18497127

Chandrasekharam, D. (2005). Arsenic pollution in groundwater of West Bengal, India: Where we stand? *Remediation and Management*, 25-29.

Chatterjee, A., Das, D., & Chakraborti, D. (1993). A study of ground water contamination by arsenic in the residential area of behala, calcutta due to industrial pollution. *Environmental Pollution*, *80*(1), 57–65. doi:10.1016/0269-7491(93)90010-L PMID:15091873

Chen, S., Wu, W., Hu, K., & Li, W. (2010). The effects of land use change and irrigation water resource on nitrate contamination in shallow groundwater at county scale. *Ecological Complexity*, *7*(2), 131–138. doi:10.1016/j.ecocom.2010.03.003

Chilton, P. J., Lawrence, A. R., & Stuart, M. E. (1994). *The impact of tropical agriculture on groundwater quality in groundwater quality*. London: Chapman & Hall.

Clarke, B. O., & Smith, S. R. (2011). Review of 'emerging' organic contaminants in biosolids and assessment of international research priorities for the agricultural use of biosolids. *Environment International*, *37*(1), 226–247. doi:10.1016/j.envint.2010.06.004 PMID:20797791

Corada-Fernandez, C., Candela, L., Torres-Fuentes, N., Pintado-Herrera, M. G., Paniw, M., & Gonzalez-Mazo, E. (2017). Effects of extreme rainfall events on the distribution of selected emerging contaminants in surface and groundwater: The Guadalete River basin (SW, Spain). *The Science of the Total Environment, 605-606*, 770–783. doi:10.1016/j.scitotenv.2017.06.049 PMID:28679121

Datta, P. S., Deb, D. L., & Tyagi, S. K. (1997). Assessment of groundwater contamination from fertilizers in the Delhi area based on 180, N03− and K+ composition. *Journal of Contaminant Hydrology, 27*(3), 249–262. doi:10.1016/S0169-7722(96)00099-X

Dey, A., Bhowmick, A., & Chakraborty, D. (2003). Contamination through landfills in NCT, Delhi. In *Recent Trends in Hydrogeochemistry* (pp. 153–159). New Delhi: Capital Publ. Co.

Estévez, E., Cabrera, M. C., Molina-Díaz, A., Robles-Molina, J., & Palacios-Díaz, M. P. (2012). Screening of emerging contaminants and priority substances (2008/105/EC) in reclaimed water for irrigation and groundwater in a volcanic aquifer (Gran Canaria, Canary Islands, Spain). *The Science of the Total Environment, 433*(Supplement C), 538–546. doi:10.1016/j.scitotenv.2012.06.031 PMID:22858460

Fattore, E., Muller, L., Davoli, E., Castelli, D., & Benfenati, E. (1998). Industrial pollutants in ground waters from northern Milan. *Chemosphere, 36*(9), 2007–2017. doi:10.1016/S0045-6535(97)10085-6 PMID:9532728

Fram, M. S., & Belitz, K. (2011). Occurrence and concentrations of pharmaceutical compounds in groundwater used for public drinking-water supply in California. *The Science of the Total Environment, 409*(18), 3409–3417. doi:10.1016/j.scitotenv.2011.05.053 PMID:21684580

Godfrey, E., Woessner, W. W., & Benotti, M. J. (2007). Pharmaceuticals in on-site sewage effluent and ground water, Western Montana. *Ground Water, 45*(3), 263–271. doi:10.1111/j.1745-6584.2006.00288.x PMID:17470115

Grassi, S., Amadori, M., Pennisi, M., & Cortecci, G. (2014). Identifying sources of B and As contamination in surface water and groundwater downstream of the Larderello geothermal – industrial area (Tuscany–Central Italy). *Journal of Hydrology (Amsterdam), 509*(Supplement C), 66–82. doi:10.1016/j.jhydrol.2013.11.003

Grimmeisen, F., Lehmann, M. F., Liesch, T., Goeppert, N., Klinger, J., Zopfi, J., & Goldscheider, N. (2017). Isotopic constraints on water source mixing, network leakage and contamination in an urban groundwater system. *The Science of the Total Environment, 583*(Supplement C), 202–213. doi:10.1016/j.scitotenv.2017.01.054 PMID:28104328

Herbert, R. B. Jr. (2006). Seasonal variations in the composition of mine drainage-contaminated groundwater in Dalarna, Sweden. *Journal of Geochemical Exploration, 90*(3), 197–214. doi:10.1016/j.gexplo.2005.12.002

Hohenblum, P., Gans, O., Moche, W., Scharf, S., & Lorbeer, G. (2004). Monitoring of selected estrogenic hormones and industrial chemicals in groundwaters and surface waters in Austria. *The Science of the Total Environment, 333*(1-3), 185–193. doi:10.1016/j.scitotenv.2004.05.009 PMID:15364528

Hu, X., Zhou, Q., & Luo, Y. (2010). Occurrence and source analysis of typical veterinary antibiotics in manure, soil, vegetables and groundwater from organic vegetable bases, northern China. *Environmental Pollution, 158*(9), 2992-2998. doi: 10.1016/j.envpol.2010.05.023

Jensen, D. L., Ledin, A., & Christensen, T. H. (1999). Speciation of heavy metals in landfill-leachate polluted groundwater. *Water Research*, *33*(11), 2642–2650. doi:10.1016/S0043-1354(98)00486-2

Johnson, C. D., & Juengst, D. (1997). *Polluted urban runoff: A source of concern. No. I-02-97-5M-20-S.* Madison, WI: University of Wisconsin Extension.

Kakar, Y. (1990). Groundwater pollution due to industrial effluents. *Central Ground Water Board, Govt. of Indian Archives*, *2*, 1–12.

Kaushal, S. S., Likens, G. E., Utz, R. M., Pace, M. L., Grese, M., & Yepsen, M. (2013). Increased river alkalization in the eastern U.S. *Environmental Science & Technology*, *47*(18), 10302–10311. PMID:23883395

Kehew, A. (2001). *Applied Chemical Hydrogeology*. Prentice Hall.

Kim, H., Kaown, D., Mayer, B., Lee, J. Y., Hyun, Y., & Lee, K. K. (2015). Identifying the sources of nitrate contamination of groundwater in an agricultural area (Haean basin, Korea) using isotope and microbial community analyses. *The Science of the Total Environment*, *533*, 566–575. doi:10.1016/j.scitotenv.2015.06.080 PMID:26204420

Kolodziej, E. P., Harter, T., & Sedlak, D. L. (2004). Dairy wastewater, aquaculture, and spawning fish as sources of steroid hormones in the aquatic environment. *Environmental Science & Technology*, *38*(23), 6377–6384. doi:10.1021/es049585d PMID:15597895

Kresic, N. (2007). *Hydrogeology and Groundwater Modeling* (2nd ed.). Boca Raton, FL: CRC Press.

Kumar, M., Ramanathan, A. L., Tripathi, R., Farswan, S., Kumar, D., & Bhattacharya, P. (2017). A study of trace element contamination using multivariate statistical techniques and health risk assessment in groundwater of Chhaprola Industrial Area, Gautam Buddha Nagar, Uttar Pradesh, India. *Chemosphere*, *166*(Supplement C), 135–145. doi:10.1016/j.chemosphere.2016.09.086 PMID:27693874

Kumar, P. J. (2014). Evolution of groundwater chemistry in and around Vaniyambadi Industrial Area: Differentiating the natural and anthropogenic sources of contamination. *Chemie der Erde - Geochemistry, 74*(4), 641-651. doi: 10.1016/j.chemer.2014.02.002

Lapworth, D. J., Baran, N., Stuart, M. E., Manamsa, K., and Talbot, J. (2015). Persistent and emerging micro-organic contaminants in Chalk groundwater of England and France. *Environmental Pollution, 203*, 214-225. doi: 10.1016/j.envpol.2015.02.030

Lapworth, D. J., Baran, N., Stuart, M. E., and Ward, R. S. (2012). Emerging organic contaminants in groundwater: A review of sources, fate and occurrence. *Environmental Pollution 163*, 287-303. doi: 10.1016/j.envpol.2011.12.034

LeGrand, H. (1970). *Movement of agricultural pollutants with groundwater. In Agricultural practices and water quality* (pp. 303–313). Ames, IA: Iowa State University Press.

Lei, L., Song, C., Xie, X., Li, Y., & Wang, F. (2010). Acid mine drainage and heavy metal contamination in groundwater of metal sulfide mine at arid territory (BS mine, Western Australia). *Transactions of Nonferrous Metals Society of China*, *20*(8), 1488–1493. doi:10.1016/S1003-6326(09)60326-5

Liu, G. D., Wu, W. L., & Zhang, J. (2005). Regional differentiation of non-point source pollution of agriculture-derived nitrate nitrogen in groundwater in northern China. *Agriculture, Ecosystems & Environment*, *107*(2), 211–220. doi:10.1016/j.agee.2004.11.010

Loos, R., Locoro, G., Comero, S., Contini, S., Schwesig, D., Werres, F., ... Gawlik, B. M. (2010). Pan-European survey on the occurrence of selected polar organic persistent pollutants in ground water. *Water Research*, *44*(14), 4115–4126. doi:10.1016/j.watres.2010.05.032 PMID:20554303

Lopes, D. D., Silva, S. M. C. P., Fernandes, F., Teixeira, R. S., Celligoi, A., & Dall'Antônia, L. H. (2012). Geophysical technique and groundwater monitoring to detect leachate contamination in the surrounding area of a landfill – Londrina (PR – Brazil). *Journal of Environmental Management*, *113*(Supplement C), 481–487. doi:10.1016/j.jenvman.2012.05.028 PMID:22749799

Ma, X., Tan, Z., Pang, L., & Liu, J. (2013). Determination of Vocs in groundwater at an industrial contamination site using a homemade low-density polyethylene passive diffusion sampler. *Journal of Environmental Sciences (China)*, *25*(11), 2338–2343. doi:10.1016/S1001-0742(12)60322-2 PMID:24552064

Meffe, R., & de Bustamante, I. (2014). Emerging organic contaminants in surface water and groundwater: A first overview of the situation in Italy. *The Science of the Total Environment*, *481*, 280–295. doi:10.1016/j.scitotenv.2014.02.053 PMID:24602913

Nath, B., Chakraborty, S., & Jana, J. (2006). Arsenic in the groundwater of the Bengal delta plain: hydrogeochemical studies from the districts of Nadia and South 24 Parganas, West Bengal, India. *Conference on Groundwater for Sustainable Dev*.

Navarro, A., & Carbonell, M. (2007). Evaluation of groundwater contamination beneath an urban environment: The Besos river basin (Barcelona, Spain). *Journal of Environmental Management*, *85*(2), 259–269. doi:10.1016/j.jenvman.2006.08.021 PMID:17223248

Navarro, A., & Font, X. (1993). Discriminating different sources of groundwater contamination caused by industrial wastes in the Besós River Basin, Barcelona, Spain. *Applied Geochemistry*, *8*(Supplement 2), 277–279. doi:10.1016/S0883-2927(09)80050-4

Naymik, T. G. (1993). Contaminant Hydrogeology: C. W. Fetter. Macmillan, 1993, xvii + 458p., $61.25 (ISBN 0-02-337135-8). *Geochimica et Cosmochimica Acta*, *57*(15), 3820. doi:10.1016/0016-7037(93)90162-P

Noonan, D. C., & Curtij, J. T. (1990). Gronnwater remediation and petroleum: A guide for underground storage tanks. M. L. P. Chelsea.

Osenbruck, K., Glaser, H. R., Knoller, K., Weise, S. M., Moder, M., Wennrich, R., ... Strauch, G. (2007). Sources and transport of selected organic micropollutants in urban groundwater underlying the city of Halle (Saale), Germany. *Water Research*, *41*(15), 3259–3270. doi:10.1016/j.watres.2007.05.014 PMID:17575997

Phillips, P. J., Schubert, C., Argue, D., Fisher, I., Furlong, E. T., Foreman, W., ... Chalmers, A. (2015). Concentrations of hormones, pharmaceuticals and other micropollutants in groundwater affected by septic systems in New England and New York. *The Science of the Total Environment*, *512-513*, 43–54. doi:10.1016/j.scitotenv.2014.12.067 PMID:25613769

Sacher, F., Lange, F. T., Brauch, H. J., & Blankenhorn, I. (2001). Pharmaceuticals in groundwaters analytical methods and results of a monitoring program in Baden-Wurttemberg, Germany. *Journal of Chromatography. A*, *938*(1-2), 199–210. doi:10.1016/S0021-9673(01)01266-3 PMID:11771839

Sánchez Pérez, J. M., Antiguedad, I., Arrate, I., Garcialinares, C., & Morell, I. (2003). The influence of nitrate leaching through unsaturated soil on groundwater pollution in an agricultural area of the Basque country: A case study. *The Science of the Total Environment*, *317*(1), 173–187. doi:10.1016/S0048-9697(03)00262-6 PMID:14630420

Sarmah, A. K., Meyer, M. T., & Boxall, A. B. (2006). A global perspective on the use, sales, exposure pathways, occurrence, fate and effects of veterinary antibiotics (VAs) in the environment. *Chemosphere*, *65*(5), 725–759. doi:10.1016/j.chemosphere.2006.03.026 PMID:16677683

Selvakumar, S., Chandrasekar, N., & Kumar, G. (2017). Hydrogeochemical characteristics and groundwater contamination in the rapid urban development areas of Coimbatore, India. *Water Resources and Industry*, *17*(Supplement C), 26–33. doi:10.1016/j.wri.2017.02.002

Shestopalov, V. M., Rudenko, Y. F., Bohuslavsky, A. S., & Bublias, V. N. (2006). Chernobyl-born radionuclides: groundwater protectability withrespect to preferential flow zones. In H. Vereecken, A. Binley, G. Cassiani, A. Revil, & K. Titov (Eds.), *Applied Hydrogeophysics* (pp. 341–376). Dordrecht: Springer Netherlands. doi:10.1007/978-1-4020-4912-5_12

Shukla, G., Kumar, A., Bhanti, M., Joseph, P. E., & Taneja, A. (2006). Organochlorine pesticide contamination of ground water in the city of Hyderabad. *Environment International*, *32*(2), 244–247. doi:10.1016/j.envint.2005.08.027 PMID:16183122

Singhal, B. B. S., & Gupta, R. P. (2010). Groundwater Contamination. In B. B. S. Singhal & R. P. Gupta (Eds.), *Applied Hydrogeology of Fractured Rocks* (2nd ed.; pp. 221–236). Dordrecht: Springer Netherlands. doi:10.1007/978-90-481-8799-7_12

Sorensen, J. P. R., Lapworth, D. J., Nkhuwa, D. C. W., Stuart, M. E., Gooddy, D. C., Bell, R. A., ... Pedley, S. (2015). Emerging contaminants in urban groundwater sources in Africa. *Water Research*, *72*(Supplement C), 51–63. doi:10.1016/j.watres.2014.08.002 PMID:25172215

Standley, L. J., Rudel, R. A., Swartz, C. H., Attfield, K. R., Christian, J., Erickson, M., & Brody, J. G. (2008). Wastewater-contaminated groundwater as a source of endogenous hormones and pharmaceuticals to surface water ecosystems. *Environmental Toxicology and Chemistry*, *27*(12), 2457–2468. doi:10.1897/07-604.1 PMID:18616377

Stuart, M., & Lapworth, D. (2013). Emerging Organic Contaminants in Groundwater. In S. C. Mukhopadhyay & A. Mason (Eds.), *Smart Sensors for Real-Time Water Quality Monitoring* (pp. 259–284). Berlin: Springer Berlin Heidelberg. doi:10.1007/978-3-642-37006-9_12

Stuart, M., Lapworth, D., Crane, E., & Hart, A. (2012). Review of risk from potential emerging contaminants in UK groundwater. *The Science of the Total Environment*, *416*, 1–21. doi:10.1016/j.scitotenv.2011.11.072 PMID:22209399

Swartz, C. H., Reddy, S., Benotti, M. J., Yin, H., Barber, L. B., Brownawell, B. J., & Rudel, R. A. (2006). Steroid estrogens, nonylphenol ethoxylate metabolites, and other wastewater contaminants in groundwater affected by a residential septic system on Cape Cod, MA. *Environmental Science & Technology*, *40*(16), 4894–4902. doi:10.1021/es052595+ PMID:16955883

Teijon, G., Candela, L., Tamoh, K., Molina-Diaz, A., & Fernandez-Alba, A. R. (2010). Occurrence of emerging contaminants, priority substances (2008/105/CE) and heavy metals in treated wastewater and groundwater at Depurbaix facility (Barcelona, Spain). *The Science of the Total Environment*, *408*(17), 3584–3595. doi:10.1016/j.scitotenv.2010.04.041 PMID:20593552

Tiwary, R. K., Dhakate, R., Ananda Rao, V., & Singh, V. S. (2005). Assessment and prediction of contaminant migration in ground water from chromite waste dump. *Environmental Geology*, *48*(4), 420–429. doi:10.100700254-005-1233-2

USEPA. (1976). RCRA Subtitle I and Sections 7003, 9003(h), 9005 and 9006.

White, D., Lapworth, D. J., Stuart, M. E., & Williams, P. J. (2016). Hydrochemical profiles in urban groundwater systems: New insights into contaminant sources and pathways in the subsurface from legacy and emerging contaminants. *The Science of the Total Environment*, *562*(Supplement C), 962–973. doi:10.1016/j.scitotenv.2016.04.054 PMID:27155350

Winde, F. (2006). Long-term Impacts of Gold and Uranium Mining on Water Quality in Dolomitic Regions — examples from the Wonderfonteinspruit catchment in South Africa. In B. J. Merkel & A. Hasche-Berger (Eds.), *Uranium in the Environment: Mining Impact and Consequences* (pp. 807–816). Berlin: Springer Berlin Heidelberg. doi:10.1007/3-540-28367-6_83

Zhai, Y., Zhao, X., Teng, Y., Li, X., Zhang, J., Wu, J., & Zuo, R. (2017). Groundwater nitrate pollution and human health risk assessment by using HHRA model in an agricultural area, NE China. *Ecotoxicology and Environmental Safety*, *137*(Supplement C), 130–142. doi:10.1016/j.ecoenv.2016.11.010 PMID:27918944

Zhang, J., Dai, J., Chen, H., Du, X., Wang, W., & Wang, R. (2012). Petroleum contamination in groundwater/air and its effects on farmland soil in the outskirt of an industrial city in China. *Journal of Geochemical Exploration*, *118*(Supplement C), 19–29. doi:10.1016/j.gexplo.2012.04.002

# Chapter 12
# Occurrence and Fate of Selected Heavy Metals in a Conventional Municipal Wastewater Treatment Plant in Kisumu City, Kenya:
## A Case Study

**Victor Odhiambo Shikuku**
*Kaimosi Friends University College, Kenya*

**George O. Achieng'**
*Maseno University, Kenya*

## ABSTRACT

*The objective of this work was to investigate the occurrence and fate of five heavy metals in water, sludge, and sediments from a conventional municipal wastewater treatment facility in Kisumu City, Kenya. The effluent quality was compared with the effluent quality parameters stipulated by the National Environmental Management Authority (NEMA) to assess the efficiency of the plant and potential effect of the discharged effluent on the recipient river. The levels of the heavy metals recorded in the sludge samples were significantly higher than those in the corresponding water samples. The order of the metal percentage removal efficiency (%R) from the treatment plant was Mg>Cu>Mn>Fe>Zn. It is concluded that the plant is a point source for Zn loading into the recipient waters which poses potential risk to end users downstream. The heavy metal-laden sludge was within permissible limits for utilization in agricultural lands.*

DOI: 10.4018/978-1-5225-5754-8.ch012

## 1. INTRODUCTION

Heavy metals have been defined differently by different authors. However, heavy metals are generally considered as elements having atomic masses ranging between 63.5 and 200.6, specific gravity above 5.0, are toxic or carcinogenic and cannot be metabolized by living organisms (Srivastava and Majumder, 2008). Unlike organic pollutants, heavy metals are persistent environmental contaminants due to their non-biodegradable nature and tend to be biomagnified in living organisms. Of specific interest in wastewater treatment processes include Zn, Hg, Cd, Cu, Pb, Cr and Ni.

Accumulation of heavy metals in different environmental compartments, especially water resources, and the associated negative impacts have been directly or indirectly associated with industrial processes, especially in developing countries. Most of the aforementioned heavy metals are characteristic of most effluents from industrial sources and constitute finished products that end in waterwater at the end of their use or through improper disposal (Olujimi *et al.,* 2012).

In urban settings, the major source waters to wastewater treatment plants (WWTPs) include household effluents, hospital effluents, business effluents (e.g. car washes), and traffic-related emissions (e.g. motor exhaust, brake linings, gasoline/oil leakage etc) which find way into the sewage *via* storm water.

The toxic effects of heavy metals ions on living organisms are well documented and are therefore briefly mentioned here.

### 1.1. Toxicity of Selected Heavy Metals

Some heavy metals are essential for metabolic processes at low concentrations and become toxic above certain tolerable limits. Other heavy metals are wholly toxic with no metabolic benefits reported. For instance, Zn is essential for human health. It is a regulator of various biochemical processes and other physiological functions of living tissue. However, excess zinc causes apparent health dysfunctions, such as anemia, skin irritations, vomiting and stomach cramps among others (Oyaro et al., 2007). Noteworthy, Zn is the most profuse of heavy metals found in extremely elevated levels in municipal wastewater (Carletti *et al.*, 2008). Cu is also reported to play essential role in animal metabolism. But as zinc, above certain limits copper causes serious toxicological effects, such as vomiting, cramps even death (Paulino et al., 2006).

Lead intake can potentially damage the central nervous system (CNS), kidney, liver, reproductive system, and brain functions. The symptoms associated to lead poisoning include anemia, insomnia, headache, dizziness, hallucination and renal damages among others (Naseem and Tahir, 2001). Mercury on the other hand is also a neurotoxin that can cause damage to the CNS, as Pb. High exposure to Hg causes kidney dysfunction, impairment of pulmonary glands and dyspnoea (Namasivayam and Kadirvelu, 1999).

### 1.2. Heavy Metals Removal From Water

Heavy metals are increasingly becoming environmental priority contaminants due to the aforementioned problems. Consequently, more stringent regulations and policies on allowable limits of heavy metals in wastewater from discharge point sources and in treated portable and wastewater are being adopted. Such policies demand more efficient technologies for removal of these toxic heavy metals from the wastewater to for environmental protection. Conventional wastewater treatment processes consist of a combination of physical, chemical, and biological processes and operations to eliminate solids, organic matter and nutrients from wastewater. Common terms employed to describe different scales of treatment, in order of increasing treatment level, are preliminary, primary, secondary, and tertiary and/or advanced wastewater treatment. The next section discusses the convectional processes for the removal of heavy metals from wastewater. Their advantages and limitations in application are also evaluated.

## 1.3. Conventional Heavy Metal Wastewater Treatment Methods

### 1.3.1. Chemical Precipitation

Chemical precipitation is the most widely applied technique in industry due to its relative simplicity and inexpensiveness and yields commendable effectiveness (Ku and Jung, 2001). Here, selective chemicals react with heavy metal ions forming insoluble precipitates, such as hydroxide and sulfide precipitation, separable from the wastewater by sedimentation.

### 1.3.2. Adsorption

Adsorption is universally recognized as an efficient and cost effective method for heavy metal sequestration in wastewater treatment regimes. Adsorption as a technique possesses the inherent advantages of flexibility in design besides high removal efficiency of various pollutants and possibility of adsorbent recovery, regeneration by desorption process and reuse for numerous cycles. Activated carbon (AC) is the most widely used adsorbent in the removal of heavy metal from wastewater. The large micropore and mesopore volumes and high surface area of activated carbons imbue them with high adsorption capabilities. Application of activated carbons derived from various sources for heavy metals removal from water has been widely published and reviewed (Fu and Wang, 2011). However, due to the recent depletion of source materials for commercial coal-based AC, the capital costs for AC has been on an exponential rise. Consequently, there has been increased research in search of alternative, abundant and inexpensive carbonaceous materials as precursors for development of activated carbons. The use of such materials has previously been critically reviewed (Dias et al. 2007).

### 1.3.3. Coagulation and Flocculation

Coagulation is the destabilization of colloids by neutralizing the forces separating them. Coagulants commonly used in the conventional WWTPs for removal of particulates and impurities by charge neutralization include aluminium, ferric chloride and ferrous sulfate which have been empirically demonstrated to possess excellent heavy metal elimination potential (El Samrani et al. 2008). Nevertheless, coagulation suffers the limitation of separating hydrophobic colloids and suspended particles and effectiveness is significantly affected by pH. This challenge is reported to be effectively overcome by use of amphoteric polyelectrolyte as a coagulant (Chang and Wang, 2007).

Flocculation is the binding of particles into large clumps by use of polymers. The flocculated suspended particles form larger particles which are then separated by filtration. Polyferric sulfate (PFS) and polyacrylamide (PAM) are among the commonly used flocculants in most conventional WWTPs. Unfortunately, the inability of these coagulation-flocculation to effectively remove heavy metals of wastewater is well documented (Chang et al. 2009b; Chang and Wang, 2007). Essentially, the cumulative effect of WWTP processes is designed and expected to limit the discharge of heavy metals into the recipient water systems.

Furthermore, besides the release of treated water as an output, a large mass of sludge is also generated; generally, between 1-4% of the total dry weight (Olujimi *et al.,* 2012). The sludge is also saturated with the heavy metals derived from the influent to the WWTP and therefore effective management of heavy metal-laden sludge is of great environmental concern. Some of the available sludge disposal options include: disposal as landfills, incineration, wet air oxidation, and agricultural application among others. The available sludge reuse options and treatment strategies are summarized in section 1.5.

## 1.4. Fate of Heavy Metals in Wastewater Treatment Plants

In order to determine the appropriate treatment options required to meet effluent discharge quality regulations a detailed knowledge of the levels of priority substances entering wastewater treatment plants (WWTP) is essential in order to prescribe possible sequestration measures related to process configuration. Furthermore, large wastewater treatment plants generally serve large settlement and industrial areas; as a result, wastewaters of varied compositions, enter these plants. The influent quality depends on the type of wastewater entering the WWTP. Consequently, relatively high and varied amounts of different heavy metal loads undergo the wastewater treatment and are removed at different extent. Figure 1 shows the different types of wastewater.

A comprehensive knowledge of the occurrence and behavior of heavy metals and nutrients through the activated sludge process is of fundamental significance in order to both quantify and predict their fate. A study by Karvelas *et al.* (2003) on the relative distribution of individual heavy metals in the treated effluent and the sludge streams showed that 47-63% of Cd, Cr, Pb, Fe, Ni and Zn remained in the treated effluent whereas Mn and Cu were largely accumulated (>70%) in the sludge. In a separate study, Carletti *et al.* (2008) documented that metals were concentrated in waste activated sludge and

*Figure 1. Types of wastewater*

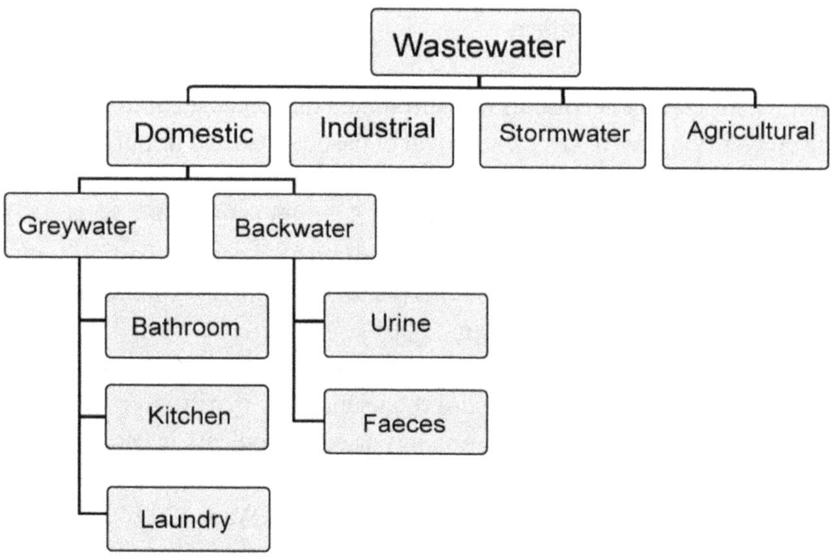

accumulated after sludge stabilization due to volatile solids degradation. The efficiency of removal of the examined heavy metals were in the order As < Cd = Cr =Zn < Pb < Hg < Ni = Al < Cu < Fe in Italian wastewater treatment plants. Accumulation of toxic metals, such as mercury, cadmium and nickel, in the sludge present a challenge for agricultural application of the sludge. The results showed that aluminum and iron are always bound to the particulate matter, while other heavy metals show a half and half distribution between the solid and liquid phases. Hg, Cu and Pb tend to exist in a soluble form in industrial wastewaters.

It seems therefore that, in municipal wastewater, heavy metals are associated to the suspended particulate while in industrial wastewaters they are associated to the soluble fraction. This observed trend becomes important when designing the pre-treatment stages in plants handling municipal or mixed wastewaters. The correlation between the metals, if present, can also be used to a predictive model for occurrence of the metals. As such, the presence of one metal can be used as a control parameter for the presence of others thus reducing the time and costs for periodic analysis. The correlation is however not straight forward and may not exist at all in other cases. The reported removal efficiency of metals was within acceptable limits. Similar results have been published by other authors (Kurniawan *et al.* 2006). A six months (January-June) study on the efficiency of a pre-urban WWTP in Thohoyandou in South Africa showed no removal efficiency of Fe because Fe concentration was higher in the effluent than in the influent. The effluent concentrations of Fe varied between 0.49–1.33 mg/L, values exceeding the stipulated limit of 0.3 mg/L for Fe in wastewater effluent. Similar observation was reported for Cr. The study concluded the WWTP released poor quality effluent injurious to both environment and human health (Edokpayi *et al.,* 2015). These results corroborate previous findings by Ogola and co-workers (2009) which demonstrated that WWTPs in Limpopo Province of South Africa seldom treated the wastewater to acceptable standards, a report further confirmed by Pindihama et al. (2011). The reports suggest that insufficient investment in wastewater treatment infrastructure, shortage of skilled personnel and poor planning or corruption are possible reasons for the poor performances of WWTPs in that region (Edokpayi *et al.,* 2015).

## 1.5. Sludge Treatment and Management

The need for solid waste management as part of environmental management cannot be overemphasized. The principles of environmental management include reduction, maximum reuse and adequate treatment of solid wastes. Like other agricultural wastes, sewage sludge should be considered as high-value biosolids for agricultural application due to its nutrient loading and high organic matter content (Bauerfeld et al., 2008). However, sludge is also a sink for toxic substances and pathogens since a significant fraction of disease-causing microorganisms are not destroyed by wastewater treatment processes but are enmeshed in primary and secondary sludge (Bitton, 2005). Generally, sludges emerging from municipal wastewater treatment plants in emerging economies or developing countries tend to have a low heavy metal concentrations but higher pathogen levels (Jimenez et al., 2004). Therefore, it is fundamental that the pathogen content in sludge is significantly reduced and sanitized before disposal or recycling especially if there is likelihood of human contact. That includes before reuse of sludge as fertilizer in agricultural production. Figure 2 summarizes the proposed technologies for stabilization and sludge treatment for various applications.

*Figure 2. Sludge treatment technologies and reuse and disposal options*

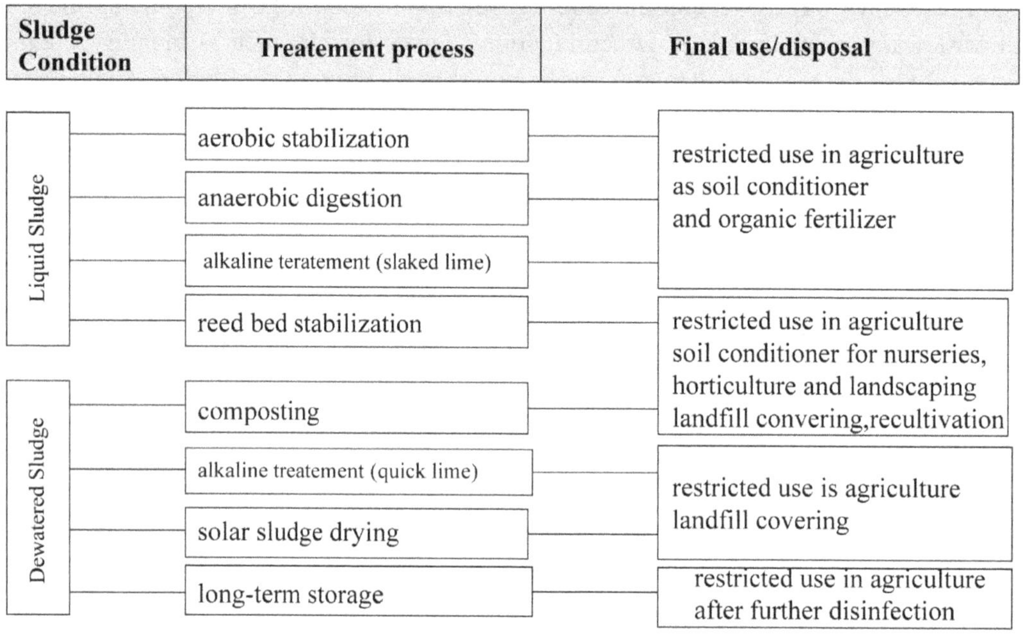


The aim of this study was to evaluate of the occurrence and fate of selected of heavy metals in a conventional municipal WWTP in Kisumu City-Kenya and investigate the overall efficiency of the WWTP at removing heavy metals of interest from wastewater and suitability of the heavy-metal laden sludge for agricultural use.

## 2. MATERIALS AND METHODS

### 2.1. Description of WWTP and Sample Collection

The plant under study is located in Kisumu County-Kenya, characterized by domestic, multiple small and large scale industries discharging effluent to the streams that supply source of water (raw water) for the treatment plant. Water samples were collected by grab sampling, at each of the six sampling sites, depicted in Figure 3, within seven days, a sufficient time necessary for the influent water to traverse all the treatment steps from site 1 to 5. The sampling sites are as follows; site 1-source water, site 2-sedimentation pond, site 3-facultative pond, site 4-maturation pond, site 5-final effluent and site 6-recipient river water. Sludge and underlying sediment samples were simultaneous collected from each pond in Zip-Lock bags and labeled accordingly. To eliminate deterioration of samples, sampling procedures and preservation protocols were done according to *Standard Methods for the Examination of Water and Wastewater* (Eaton *et al.,* 1995) details of which will not be reiterated here.

*Figure 3. Schematic diagram of the WWTP treatment stages and sampling points*

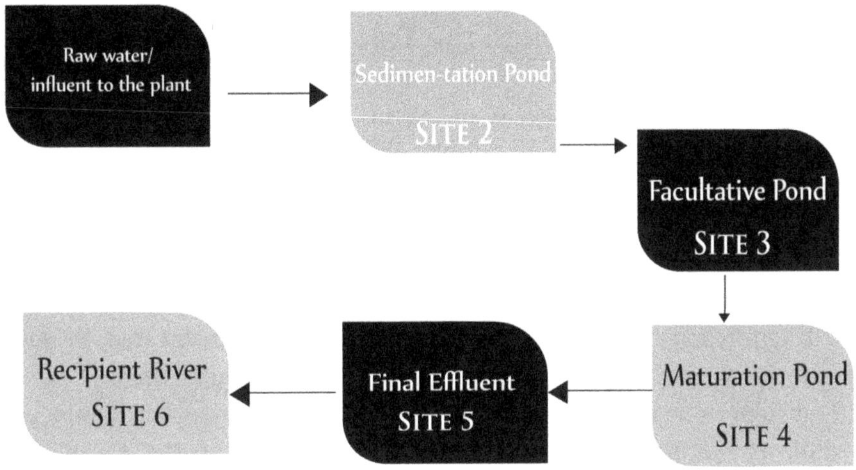


## 2.2. Sample Preparation

The samples were digested with concentrated nitric acid ($HNO_3$) for quantitative analysis of total heavy metal content using Atomic Absorption Spectrometric (AAS) analysis, following the standard methods for the examination of water and wastewater (APHA, 1998). Sludge samples were initially air dried and the dried sludge was mineralized then sieved through 2 mm sieve for uniform and control of particle size. For digestion, about 5 g of the dried sludge samples were treated with aqua ragia solution (concentrated $HNO_3$ and HCl) at 343 K for 60 min. Noteworthy, there were no field blanks sampled to reflect possible contamination or introduction of target analytes into the environmental samples by sampling procedures. Nevertheless, none of the target ions were detected in the laboratory blank samples.

Percent removal efficiency (%R) by each of the WWTP's operational step was calculated for each metal ion by the equation:

$$\%R = \left\{ \left( 1 - \frac{C}{C_0} \right) \times 100 \right\} \tag{1}$$

where C is the mean concentration in effluent from the treatment process, and $C_0$ is the mean concentration in effluent from the preceding treatment process. To obtain the heavy metal content in the sludge and sediments, the AAS values were converted using the expression:

$$SC = \left( \frac{AAS_{reading} - B_{reading}}{M} \right) \times DF \tag{2}$$

Where; SC is the sludge/sediment concentration (μg/g), AAS is atomic absorption spectrometer values, B is the blank values, M is the mass of the sludge/soil (g) and DF is the dilution factor.

## 3. RESULTS AND DISCUSSION

### 3.1. Heavy Metals Removal Through the Treatment Processes

The mean concentrations of the studied heavy metals in the wastewater, sediment and sludge samples are presented in Table 1. As it can be seen, all the selected metal ions (Cu, Pb, Zn, Fe, Mg and Mn) were detected in the source water to the plant with exception for Pb which was below the instrumental detection limit (0.001 mg/L). Generally, all the determined heavy metals were present in relatively higher concentrations in corresponding settled-solid samples than in the water samples.

The results indicate that the average concentrations of the heavy metals in the water samples decreased and varied significantly ($p<0.05$) along each subsequent treatment step. Noteworthy, the levels of the heavy metals in the sediment samples were significantly higher than in their corresponding water samples. This implies that the heavy metals were mostly sequestered from the influents by sorption onto the sediments and dissolution in the liquid fraction of the sludge. Furthermore, the relative abundance of heavy metal levels in the water ($Pb<Cu<Fe<Zn<Mn<Mg$) did not follow the same exact order in the corresponding solid samples ($Pb<Cu<Mn<Zn<Mg<Fe$) at the various steps of treatment. The sequence of fractionation observed in the present work differs from some findings reported in literature (Karvelas *et al.*, 2003; Penradee *et al.*, 2008). This is attributable to the variance in the point sources of the heavy metals in the respective influents and operational factors such as amount, source, type and age of the applied sludge. Magnesium and Iron were the highest recorded metal ions in water and sediment, respectively. As observed, Fe was partitioned more in the solid phase and therefore was most abundant in the sludge than all the metals analyzed. The same was observed for Zn partitioning in solid phase relative to Mn despite their reversed concentrations in the liquid phase. These variances in distribution suggest that the partitioning dynamics are metal-specific and that the metal ion speciation varied with pH changes along the treatment stages from site 1 to 5. Generally, the concentrations of the heavy metals in the effluent were significantly lower than in the influent, depictive of the overall effectiveness of wastewater treatment facility. This finding concurs with a previous study by Penradee et al. (2008) for similar treatment systems. Furthermore, all the heavy metals levels reported in the treated effluent discharged to the recipient river were arguably far much lower than the limits stipulated by the local environmental authority (NEMA) except for Mg and Mn with no limits cited. Since we could find no documentation from the local authorities on permissible limits of heavy metals in sludge for use in agriculture, the suitability of the studied sludge samples for agricultural use were benchmarked against the limits established by other countries listed in Table 2.

From Table 2, it could be seen that the heavy metal characteristics of the sludge from the WWTP was within the stipulated standards set by most European countries except for Cu that exceeded the set limits set by Spain. The sludge is therefore concluded to be appropriate for application in agricultural fields. Related studies indicate that heavy metal contents in sludge vary significantly with season. Unfortunately, as far as our knowledge is concerned, there are no previous reports on the heavy metal contents of sludge from the WWTP for monitoring the evolution of the metal contents in the sludge due to seasonal variations in heavy metal input to the plant. Due to this paucity of information, this work therefore represents the first comparative report of empirical data on suitability of the sludge from the municipal WWTP in Kisumu City, based on heavy metal content, for agricultural application.

*Table 1. Mean levels of heavy metals in water (mg/L) and solid (μg/g) in different sites*

| Site No. | Site Description | Metal Ions | | | | | | | | | | | |
|---|---|---|---|---|---|---|---|---|---|---|---|---|---|
| | | Cu | | Pb | | Zn | | Fe | | Mg | | Mn | |
| | | water | solid | water | solid | water | solid | water | solid | water | solid | water | solid |
| 1 | Influent into the plant | 0.13±0.04 | - | <0.001 | - | 0.11±0.03 | - | 0.41±0.04 | - | 30.25±2.22 | - | 0.99±0.08 | - |
| 2 | Pond 1: Sedimentation | 0.10±0.01 | 77.99±2.86 | <0.001 | 25.53±0.99 | 0.64±0.03 | 676.30±132.94 | 0.66±0.02 | 49538.27±3231.78 | 30.18±2.45 | 6457.24±427.65 | 0.86±0.04 | 348.96±6.75 |
| 3 | Pond 2: Facultative | 0.08±0.01 | 15.36±4.84 | <0.001 | 14.40±1.68 | 0.30±0.02 | 60.95±19.51 | 0.52±0.01 | 15330.56±2594.60 | 20.79±1.14 | 3633.05±284.04 | 0.87±0.01 | 220.26±2.65 |
| 4 | Pond 3: Maturation | 0.07±0.01 | 8.04±0.30 | <0.001 | 5.08±0.16 | 0.28±0.02 | 9.46±1.38 | 0.41±0.02 | 7601.46±804.51 | 5.94±0.20 | 3015.29±628.80 | 0.73±0.04 | 139.47±4.96 |
| 5 | Effluent from plant | 0.11±0.04 | - | <0.001 | - | 0.22±0.03 | - | 0.38±0.03 | - | 22.90±1.22 | - | 0.81±0.10 | - |
| 6 | Recipient River | 0.10±0.03 | - | <0.001 | - | 0.20±0.01 | - | 0.51±0.03 | - | 4.83±0.46 | - | 0.80±0.01 | - |
| NEMA Limits | | 1.0 | | 0.1 | | 0.5 | | 10 | | * | | * | |

* ·No limit is quoted for this parameter

*Table 2. Maximum permissible levels of heavy metals in the sludge for agricultural usage (μg/g)*

| Country | Pb | Cu | Zn | Reference |
|---|---|---|---|---|
| EU | 750-1200 | 1000-1750 | 2500-4000 | 19 |
| USA | 840 | 4300 | 7500 | 20 |
| Italy | 500 | 600 | 2500 | 21 |
| Germany | 900 | 800 | 2500 | 22 |
| France | 800 | 1000 | 3000 | 23 |
| Spain | 300 | 50 | 1100 | 24 |

## 3.2. Comparison of Heavy Metals Concentration in Influent and Effluent

The overall removal efficiency (%R) of the plant was obtained by a comparison of the heavy metal content in the source water (site 1) and effluent samples (site 5) assuming steady-state conditions and insignificant effect of evapotranspiration on water storage relative to inflow and outflow. Since previous findings indicate that the heavy metals input in a wastewater plant varies even if audited hourly or monthly, the present work reports the assessment within a week (Chipasa, 2003). Table 3 depicts the percent positive effectiveness (%R) of the plant in the step-wise and the overall reduction of heavy metal (Pb, Cu, Fe, Mg, Mn and Zn) loads in the wastewater. It is realized that the overall variation of Mg and Cu concentrations in the released effluent were more pronounced than Mn and Fe. The variation of Mg concentration was the most pronounced whilst Fe was the least. Furthermore, Zn had a negative mean value attributable to desorption of the metal from the sediments back into the water possibly due to saturation of the sludge due to age and concomitant competition between the metal ions present and other adsorbates for the limited adsorption sites. Consequently, the plant becomes a point source for Zinc loading into the recipient river. Similar phenomenon was reported for nitrates removal by a wastewater treatment facility (Mohebi-Fard *et al.*, 2015). In view of the potential health hazards of Zn exposure above tolerable limits, the observed loading of Zn from the WWTP to the recipient river is of great concern and presents a great risk to the downstream users of river.

The heavy metal removal efficiency from the treatment plant was in the order: Zn (-127.77%) < Fe (3.66%) < Mn (16.64%) < Cu (24.26%) < Mg (46.97%). This order seemed to be somewhat directly related to the initial concentration of the metal ion in the source water (influent) except Fe. The cause for the discrepancy in Fe removal remains unclear and relatively difficult to account for. This indicates that

*Table 3. Percent positive effectiveness (%R) of the plant in step-wise reduction of heavy metal loads in the water*

| | | Metal Ions | | | | | |
|---|---|---|---|---|---|---|---|
| Site No. | Site Description | Cu | Pb | Zn | Fe | Mg | Mn |
| 1 | Pond 1: Sedimentation 1 | 38.46 | - | -481.82 | -60.98 | 0.23 | 13.13 |
| 2 | Pond 2: Facultative 1 | 20.00 | - | 53.13 | 21.21 | 31.11 | -1.16 |
| 3 | Pond 3: Maturation 1 | 12.50 | - | 6.67 | 21.15 | 71.43 | 16.09 |
| Mean Positive Effectiveness (%R) | | 24.26 | - | -127.77 | 3.66 | 46.97 | 16.64 |

the metal-specific mechanisms of heavy metals removal through the plant require further investigation. This direct correlation between initial metal ion concentrations in influent with removal extent has been previously reported by other researchers (Chipasa, 2003). Furthermore, since the removal efficiencies also depend on other operation, chemical, physical and biological factors such as pH, which affects the metal ions solubility, dissolved organic matter, metal ion species, initial metal ion concentration, the wastewater composition among other factors which are time-dependent and exhibit high temporal fluctuations, the future performance of the plant in heavy metal removal remains unpredictable and the present findings may not be reproducible. Periodic monitoring, though not cost-effective, can be done. Better still, the use of semi-permeable membrane devices (SPMDs) and Artificial Mussels (AMs) to provide time-integrated estimate of heavy metal levels is strongly recommended as a practical and cost-effective tool for monitoring heavy metals removal efficiency from the sewage treatment plant.

## 4. CONCLUSION

The levels of the heavy metals recorded in the sediment samples were significantly higher than in the corresponding water samples, implying that adsorption was the key mode of metal elimination from the wastewater. Lead (Pb) was the least abundant metal in both water and sediment whereas Mg and Fe were the highest in water and sediment, respectively. All the detected heavy metals had mean concentrations within the allowable limits regulated by National Environmental Management Authority (NEMA) with the exception of Zn. The plant was proven to be a point source for Zn loading into the recipient river and regular replacement of the sludge to enhance uptake of Zn and periodic assessment of removal efficiency is hereby recommended. The selected heavy metals contents in the sludge met the general regulations applied world over and the sludge could be recommended for agricultural use subject to further studies. There still exists a paucity of information on effectiveness of the WWTP under study to remove other crucial heavy metals such as Hg, As, Cd and Cr. Furthermore, development of policies on allowable limits of heavy metal content in sludge for agricultural usage by the local authority is recommended.

## REFERENCES

Abdel-Shafy, H., Hegemann, W., & Guldner, C. (1996). Fate of heavy metals *via* chemical–biological upgrading of sewage sludge treatment plants. *Environmental Management and Health*, *713*(3), 28–36. doi:10.1108/09566169610117895

Achieng, G. O., Ongeri, D. M. K., & Omwoyo, W. N. (2013). Impact of anthropogenic activities on copper and lead levels in Kisumu city soils. *Sky Journal of Soil Science and Environmental Management*, *2*(5), 47–52.

APHA-AWWA-WPCF. (1998). Standard methods for the Examination of Water and Wasewater. American Public Health Association.

Barakat, M. A. (2011). New trends in removing heavy metals from industrial wastewater. *Arabian Journal of Chemistry*, *4*(4), 361–377. doi:10.1016/j.arabjc.2010.07.019

Bazrafshan, E., Moein, H., Kord Mostafapour, F. & Nakhaie, S. (2013). Application of electrocoagulation process for dairy wastewater treatment. *Journal of Chemisry,*1-8.

Bitton, G. (2005). *Wastewater Microbiology* (3rd ed.). Hanaken, NJ: Wiley-Liss. doi:10.1002/0471717967

Carletti, G., Fatone, F., Bolzonella, D., & Cecchi, F. (2008). Occurrence and fate of heavy metals in large wastewater treatment plants treating municipal and industrial wastewaters. *Water Science and Technology, 57*(9), 1329–1336. doi:10.2166/wst.2008.230 PMID:18495995

Chang, Q., & Wang, G. (2007). Study on the macromolecular coagulant PEX which traps heavy metals. *Chemical Engineering Science, 62*(17), 4636–4643. doi:10.1016/j.ces.2007.05.002

Chang, Q., Zhang, M., & Wang, J. X. (2009). Removal of $Cu^{2+}$ and turbidity from wastewater by mercaptoacetyl chitosan. *Journal of Hazardous Materials, 169*(1-3), 621–625. doi:10.1016/j.jhazmat.2009.03.144 PMID:19414213

Chipasa, K. B. (2003). Accumulation and fate of selected heavy metals in a biological wastewater treatment system. *Waste Management (New York, N.Y.), 23*(2), 135–143. doi:10.1016/S0956-053X(02)00065-X PMID:12623088

Council Directive 86/278/EEC on the protection of the environment, and in particular of the soil, when sewage sludge is used in agriculture, *Official Journal*, No. L181, 04/07/1986, p. 6.

Dias, J. M., Alvim-Ferraz, M. C. M., Almeida, M. F., Rivera-Utrilla, J., & Sanchez-Polob, M. (2007). Waste materials for activated carbon preparation and its use in aqueous phase treatment: A review. *Journal of Environmental Management, 85*(4), 833–846. doi:10.1016/j.jenvman.2007.07.031 PMID:17884280

Eaton, A. D., Clesceri, L. S., & Greenberg, A. E. (1995). *Standard Methods for the Examination of Water and Wastewater*. Washington, DC: American Public Health Association.

Edokpayi, J. N., Odiyo, J. O., Msagati, T. A. M., & Popoola, E. O. (2015). Removal efficiency of faecal indicator organisms, nutrients and heavy metals from a Peri-Urban wastewater treatment plant in thohoyandou, Limpopo Province, South Africa. *International Journal of Environmental Research and Public Health, 12*(7), 7300–7320. doi:10.3390/ijerph120707300 PMID:26132481

El Samrani, A. G., Lartiges, B. S., & Villiéras, F. (2008). Chemical coagulation of combined sewer overflow: Heavy metal removal and treatment optimization. *Water Resources, 42*, 951–960. PMID:17961629

Environmental Management & Coordination (Water Quality) Regulations. (2006). *Legal Notice 120.* Third Schedule, Standards for Discharge of Effluent into the Environment.

EPA. (1994). *Land application of sewage sludge. A guide for land applies on the requirement of the Federal standards for the use or disposal of sewage sludge, 40 CFR 503*. EPA.

Gulyas, G., Fazekas, B., Varga, R., & Karpati, A. (2010). Biological purification of chemically pre-treated dairy industry. *Current World Environment, 5*(2), 373–378.

Jimenez, B., Barrios, J. A., Mendez, J. M., & Diaz, J. (2004). Sustainable sludge management in developing countries. *Water Science and Technology, 49*(10), 251–258. PMID:15259962

Karvelas, M., Katsoyiannis, A., & Samara, C. (2003). Occurrence and fate of heavy metals in the wastewater treatment process. *Chemosphere*, *53*(10), 1201–1210. doi:10.1016/S0045-6535(03)00591-5 PMID:14550351

Kimosop, S. J., Getenga, Z. M., Orata, F. O., Okello, V. A., & Cheruiyot, J. K. (2016). Residue levels and discharge loads of antibiotics in wastewater treatment plants and hospital lagoons within Lake Victoria Basin, Kenya. *Journal of Environmental Monitoring and Assessment*, *188*(9), 532. doi:10.100710661-016-5534-6 PMID:27553946

Ku, Y., & Jung, I. L. (2001). Photocatalytic reduction of Cr(VI) in aqueous solutions by UV irradiation with the presence of titanium dioxide. *Water Resources*, *35*, 135–142. PMID:11257867

Kurniawan, T. A., Chan, G. Y. S., Lo, W. H., & Babel, S. (2006). Physicoechemical treatment techniques for wastewater laden with heavy metals. *Chemical Engineering Journal*, *118*(1-2), 83–98. doi:10.1016/j.cej.2006.01.015

Lazzari, L., Sperni, L., Bertin, P., & Pavoni, B. (2000). Correlation between inorganic (heavy metals) and organic (PCBs and PAHs) micropollutant concentrations during sewage sludge composting processes. *Chemosphere*, *41*(3), 427–435. doi:10.1016/S0045-6535(99)00289-1 PMID:11057605

Mohebi-Fard, E., Reyahi-Khoram, M. & Sobhan-Ardakani, S. (2015). Performance evaluation of the wastewater treatment plant of Pelareh Dairy Industry, Iran. *Journal of Advances in Environmental Health Research,* 3(4), 250-257.

Mostafa, A. A. (2013). Treatment of cheese processing wastewater by physicochemical and biological methods. *International Journal of Microbiology*, *4*(3), 321–332.

Musungu, C. P., Jondiko, I. O., Lalah, J., Ongeri, D., Chepkui, R., & Kiema, F. (2013). The extent of nutrient removal by wastewater treatment plants along the Nyalenda Wigwa Stream and the River Kisat (Kenya). *Ecohydrology & Hydrobiology*, *13*(4), 236–240. doi:10.1016/j.ecohyd.2013.10.002

Namasivayam, C., & Kadirvelu, K. (1999). Uptake of mercury (II) from wastewater by activated carbon from unwanted agricultural solid by-product: Coirpith. *Carbon*, *37*(1), 79–84. doi:10.1016/S0008-6223(98)00189-4

Naseem, R., & Tahir, S. S. (2001). Removal of Pb(II) from aqueous solution by using bentonite as an adsorbent. *Water Resources*, *35*, 3982–3986. PMID:12230182

Ogola, J. S., Chimuka, L., & Maina, D. (2009). Occurrence and fate of trace metals in and around treatment and disposal facilities in Limpopo Province, South Africa (a case of two areas). *Proceedings of the IASTED International Conference on Environmental Management and Engineering*.

Olanrewaju, O. O., Olalekan, S. F., Odendaal, J., Adegbenro, P. D., & Ogheneochuko, O. (2012). Preliminary Investigation into Occurrence and Removal of Arsenic, Cadmium, Mercury and Zinc in Wastewater Treatment Plants in Cape Town and Stellenbosch. *Polish Journal of Environmental Studies*, *21*, 1755–1765.

Oliveira, A., Bocio, A., Trevilato, T., Takayanagui, A., Domingo, J., & Segura-munoz, S. (2007). Heavy metals in untreated/treated urban effluent and sludge from a biological wastewater treatment plant. *Environmental Science and Pollution Research International*, *14*(7), 483–489. doi:10.1065/espr2006.10.355 PMID:18062480

Oyaro, N., Juddy, O., Murago, E. N. M., & Gitonga, E. (2007). The contents of Pb, Cu, Zn and Cd in meat in Nairobi, Kenya. *International Journal* of *Food. Agriculture & Environment*, *5*, 119–121.

Pathak, U., Das, P., Banerjee, P., & Datta, S. (2016). Treatment of wastewater from a dairy industry using rice husk as adsorbent: Treatment efficiency, isotherm, thermodynamics, and kinetics modelling. *Journal of Thermodynamics*, *2016*, 1–7. doi:10.1155/2016/3746316

Paulino, A. T., Minasse, F. A. S., Guilherme, M. R., Reis, A. V., Muniz, E. C., & Nozaki, J. (2006). Novel adsorbent based on silkworm chrysalides for removal of heavy metals from wastewaters. *Journal of Colloid and Interface Science*, *301*(2), 479–487. doi:10.1016/j.jcis.2006.05.032 PMID:16780853

Penradee, C., Kyoung-Woong, K., & Suthipong, S. (2008). Metal Contents and Its Variation in Wastewater and Sewage Sludge: A Case Study of Bangkok Central Wastewater Treatment Plants. *Proceedings of the International Symposia on Geoscience Resources and Environments of Asian Terranes*, 489-492.

Pindihama, G. K., Gumbo, J. R., & Oberholster, P. J. (2011). Evaluation of a low-cost technology to manage algal toxins in rural water supplies. *African Journal of Biotechnology*, *10*(86), 19883–19889. doi:10.5897/AJBX11.024

Planquart, P., Bonin, G., Prone, A., & Massiani, C. (1999). Distribution, movement, and plant availability of trace metals in soils amended with sewage sludge composts: Application to low loadings. *The Science of the Total Environment*, *241*(1-3), 161–179. doi:10.1016/S0048-9697(99)00338-1

Ravva, S. V., & Sarreal, C. Z. (2014). Survival of Salmonella enterica in aerated and non-aerated wastewaters from dairy lagoons. *International Journal of Environmental Research and Public Health*, *11*(11), 11249–11260. doi:10.3390/ijerph111111249 PMID:25358096

Srivastava, N. K., & Majumder, C. B. (2008). Novel biofiltration methods for the treatment of heavy metals from industrial wastewater. *Journal of Hazardous Materials*, *151*(1), 1–8. doi:10.1016/j.jhazmat.2007.09.101 PMID:17997034

Taha, A. W., Dakroury, A. M., Sayed, G. E., & El-Salam, S. A. (2007). Assessment removal of heavy metals ions from wastewater by cement kiln dust (CKD). *Eleventh International Water Technology Conference*, 879-893.

Verma, R., & Suthar, S. (2015). Impact of density loads on performance of duckweed bioreactor: A potential system for synchronized wastewater treatment and energy biomass production. *American Institute of Chemical Engineers*, *34*, 1596–1604.

Voutsa, D., Zaxariadis, G., Gantidis, N., Samara, C., & Kouimtzis, T. (1996). Evaluation of municipal and industrial wastewater sludges for agricultural purposes. *Fresenius Environmental Bulletin*, *5*, 1–6.

# Chapter 13
# Environmental Recycling System (ERS):
## An Emerging Approach to Solid Waste Management

**Shashi Kant Dubey**
*Hindustan College of Science and Technology, India*

**Athar Hussain**
*Chaudhary Brahm Prakash Government Engineering College, India*

**Somesh Ajnavi**
*Hindustan College of Science and Technology, India*

## ABSTRACT

*A lot of solid waste is generated in every country of the world. Among various disposal methods used for management of solid waste, open dumping is mostly used in many countries, especially developing countries. Due to open dumping, the environment becomes polluted. It also creates aesthetic problems at the site. Industrial and agricultural wastes are simple to handle being of specific characteristics. This chapter describes in detail the composition of municipal solid waste, various problems arising due to solid waste, approaches used for solid waste management, and environmental recycling system (ERS). ERS, its components, and methods for segregation of waste have also been discussed in detail in this chapter.*

## INTRODUCTION

Unplanned growth may result into lack of space, generation of pollution and solid waste disposal problems. Municipal solid waste is considered as a complex waste which contains different physical and chemical constituents. The type and composition of solid waste is dependent upon type of the consumer products, culture, climate, regulation of country, life quality etc. While selecting any method for management of waste, the information about the constituents of waste and its composition becomes necessary. It also

DOI: 10.4018/978-1-5225-5754-8.ch013

determines the frequency of collection of waste and method used for disposal of waste. Generally solid waste in any country includes organic matter, plastic, waste of wood, paper, glass waste, metallic waste etc. Global average for generation of solid waste is around 0.6 kg/capita/day. However waste generation rate as high as 2.7 to 4.0 kg/capita/day has been reported in case of developed countries.

It has been observed that with urbanization of any country composition of waste also changes. It has increased amount of waste including packaging waste, paper waste, plastic etc. It has been reported that in Delhi, capital of India, about 6000 tons of waste is generated currently which has been estimated to increase up to 15000 tons per day by 2021. It has also been found that organic waste is about 60% of total waste while remaining 40% is inorganic waste. There are various categories of solid waste-

1. Organic like food, vegetables, meat etc.
2. Combustible waste having high calorific value
3. Non combustible or inert waste
4. Dust or sweeping waste
5. Hazardous waste
6. Waste arising of construction

## BACKGROUND

Municipal solid waste is considered to be most difficult to handle being mixed type of waste (Wang and Nie 2001). According to Medina (1997) an increase in income of the people may induce increased consumption of resources. Types of waste and quantity increases with increase in consumption by the populations. According to an estimate 70% of the world's populations will inhabit the cities in near future. Burning of municipal solid waste releases $PM_{2.5}$ and $PM_{10}$ which have adverse impacts on human health. Up to 50% of the solid waste which is produced in transition economies is found lying on the roads or plots.

As reported by World Resources Institute, local administration in most of the developing countries has to allocate more than 30% of expenses on collection and disposal of solid waste. In most of the countries sanitary landfills have not been constructed. 3R approach is practiced for reduction, reusing and recycling of the waste produced during the operations. This approach is responsible for decrease of amount of solid waste which needs to be disposed in the sanitary landfills (MoE Japan, 2006).

*Table 1. composition of municipal solid waste*

| *S. No.* | *Component (type of waste)* | *Source* |
|---|---|---|
| *1.* | Food wastes | Residential, commercial and Institutional activities |
| *2.* | Rubbish | Residential, commercial and Institutional activities |
| *3.* | Ashes and residues | Residential, commercial and Institutional activities |
| *4.* | Demolition and construction wastes | Construction and demolition activities |
| *5.* | Special wastes | Residential, commercial, Institutional, municipal services |

Many cities of developing countries are using new modern waste management systems. Waste materials are source of income for urban poor people. It has been reported that up to 2% of the urban population in Asia depends on waste picking (Medina, 2000). Medina (2004) reported that scavenging is a good example of sustainable development. Scavenging aims at conservation of resources, reduction of pollution and so protection of the environment. Collaboration is required for household awareness of recycling, improvement of waste handling and disposal operations including characterization and segregation of waste (Buenrostro and Bocco, 2003). This chapter aims at exploring environmental recycling system as approach for management of municipal solid waste.

## MAIN FOCUS OF THE CHAPTER

### Problems Due to Solid Waste

In developing countries collection of waste is a major problem. Waste separation is often not practiced. Shortage in number of waste collection vehicles increases the problem. The waste can be seen within the locality where the population resides. In open areas solid waste is burned for smooth accommodation of waste so it also creates air pollution. It becomes breeding ground for mosquitoes and spread of life threatening diseases. Bad smell arising at the site creates problem to residents. It also creates aesthetic problem to the visitors. These sites have effect on residents as well as tourist who visit there. In summers fire can also break itself due to release of methane by decomposition of the solid waste.

In rainy season problems are increased many folds as the insects are generated from the waste and spread many water borne diseases and health problems. During rainy season run off water generated from dumping site moves to sources of ground water and surface water which are nearby the point of disposal. Leachate is also generated at site of dumping of waste which creates pollution of soil and underground water. Organic compounds including ammonia, chloride salts, potassium and other toxic compounds are found in leachate which pollutes ground water. Many residents near such dumping sites use tube wells or submersible pumps to extract ground water and use it for daily activities.

1. **Spread of Diseases:** As open decomposition of solid waste is a site for generation of flies the same may be the cause for spread of diseases. It also attracts rats and other rodents which are also responsible for diseases like plague. Solid waste work as breeding ground for mosquitoes of ades responsible for spread of dengue and yellow fever.
2. **Contamination of Environment:** Poor handling of municipal solid waste creates the problem of contamination of ground and surface water. Leachate may be generated by decomposition of waste due to which water gets toxicity. Air pollution and fire hazard are also common from mishandling of solid waste at disposal sites.

### Waste Management Approach

A system used for management of municipal waste contains five components namely collection of municipal solid waste, transfer or transportation, separation, recovery of recyclables and disposal of residuals. Management of solid waste is aimed at prevention of diseases in population, prevention of environmental pollution and financial benefits. The management practice needs to be sustainable for environment.

Municipal waste management systems of any area should be operated by local administration for best management practices. Despite of amount of solid waste produced being less transition economies have improper solid waste management practices. The practices used for handling of solid waste in developing countries are old and unscientific. Burning of municipal solid waste releases PM $_{2.5}$ and PM $_{10}$ which have adverse impacts on human health. A solid waste management system has following components-

1. **Source or Generation Point:** Solid waste is generated at the point where any material has no identified use further.
2. **Storage of Solid Waste:** Storage is the facility to keep the waste upto the time of its collection. A storage facility may include- small or large containers at domestic level depending upon frequency of collection and containers at community level called communal bins. Determination of size of storage bins depends upon number of users generating the waste and frequency of collection of waste.
3. **Collection of the Waste:** It includes transfer of the waste from storage area to disposal site. This facility is designed keeping in mind about the size of storage so that unhygienic conditions should not be created at the site of waste collection.
4. **Transportation of Solid Waste:** Methods for transfer of solid waste includes vehicles which are either manual powered, animal powered or motorized.
5. **Disposal of Solid Waste:** Disposal of solid waste intends for minimization of hazards related to mismanagement of solid waste. It includes various methods like open land dumping, composting, burning or incineration and recycling.
6. **Land Filling:** Mostly the solid waste is disposed by the means of land filling. Waste is filled in a trench and covered by soil with thickness of 0.5 m to prevent disturbance by animals and flies generation. It requires very large area of land for making land filling site. However, if managed properly a land filling site may provide suitable solution to problem of solid waste.
7. **Incineration:** Incineration is suitable for the waste which has good calorific value. However it creates air pollution at the site. Incineration may be used at the site where land is not available for landfilling. The incineration can be carried out in reactors or pits which are later covered with layer of soil to avoid any disturbance at the site. However, smoke generation at the site is harmful for the health of residents.
8. **Composting:** Composting is favorable for the waste which has high organic matter content like vegetable waste. It can be suitably dumped into gardens or agricultural fields. It requires regular monitoring of composting process parameters like moisture level and chemical composition. Finally, microbial degradation converts the waste into fertilizer. However, this method requires experienced operating staff to manage the facility.

## SOLUTIONS AND RECOMMENDATIONS

### Environmental Recycling System: Solution to the Problem

Existing methods of solid waste management are major cause of degradation of environment. Also, old practices of dumping of solid waste in open areas are unsustainable due to lack of land and expenditure related to procurement of land. It has been reported that effectiveness of management programme for solid waste relies upon the public participation. Because of less public participation problem of uncollection of municipal solid waste is seen in developing countries (Pappu et al., 2007). Table 1 shows recyclable components of solid waste. Besides recyclable waste there is also non recyclable waste like biodegradable waste which can be handled best by the method of composting. Composting is a good method to produce manure so the nutrients can be added to the field where they are required.

There are many methods for disposal of municipal waste like incineration in incinerators specially designed for the purpose, filling inside the land plots and composting. All these methods are either costly or are harmful for the environment in one or another way. Municipal solid waste management programme is an institutional management programme along with being a technical programme. Recovery needs to be the approach for a better solid waste management practice.

3R approach is very suitable for addressing the problem of solid waste management. Solid waste management programme based on recovery aims at sustainable development of the area. Countries which have proper financial systems in place provide financial resources for recycling of waste. In transition economies however recycling is viewed as a source of income for certain sectors of society.

Many researchers have proposed integrated waste management for dealing with municipal solid waste in an sustainable manner which includes technical, financial and management aspects along with participation of public. In transient economies, population increases at a rapid rate than the waste disposal system itself. Public participation can bring change in the area of solid waste management in these types of countries. Citizens need to segregate the waste which can be recycled from total waste. Increasing population, increasing quantum of solid waste and increasing pollution is cause of major concern. Cost of land, labor and transport has also been increased. Due to this unavailability of landfill and open dumping sites problem increases.

On the other hand developed countries use programmes for recycling of waste for collection and segregation of waste. Transition economies can tie up with public for such activities. These people can visit the city and dump sites. The collaboration is shown in Figure 1.

In Figure 1, solid line shows a relationship between institutions contributing to sustainable recycling. In this figure a dashed line shows influence on sustainable recycling by a given factor upon institutional interaction. Public need to be aware about segregation of solid waste before collection and its benefits. They must be educated for recovery of the materials. Government may arrange the funds required for such operations. Infect burden of the disease, health of public and sanitary aspects of urban area in developing countries, requires proper policies regarding material recovery for achievement of millennium development goals.

*Table 2. Recyclable items in solid waste (for U.S.)*

| S.No. | Items | Recycled (% of original) |
|---|---|---|
| 1. | Aluminum | 67 |
| 2. | Paper | 49 |
| 3. | Plastic (including PET, HDPE, PVC, LDPE, Polypropylene, Polystyrene) | 50 |
| 4. | Glass | 35 |
| 5. | Tires | 15 |
| 6. | Glass | 36 |

*Figure 1. Collaboration for sustainable solid waste management (Source: Troschinetz et al., (2009)*

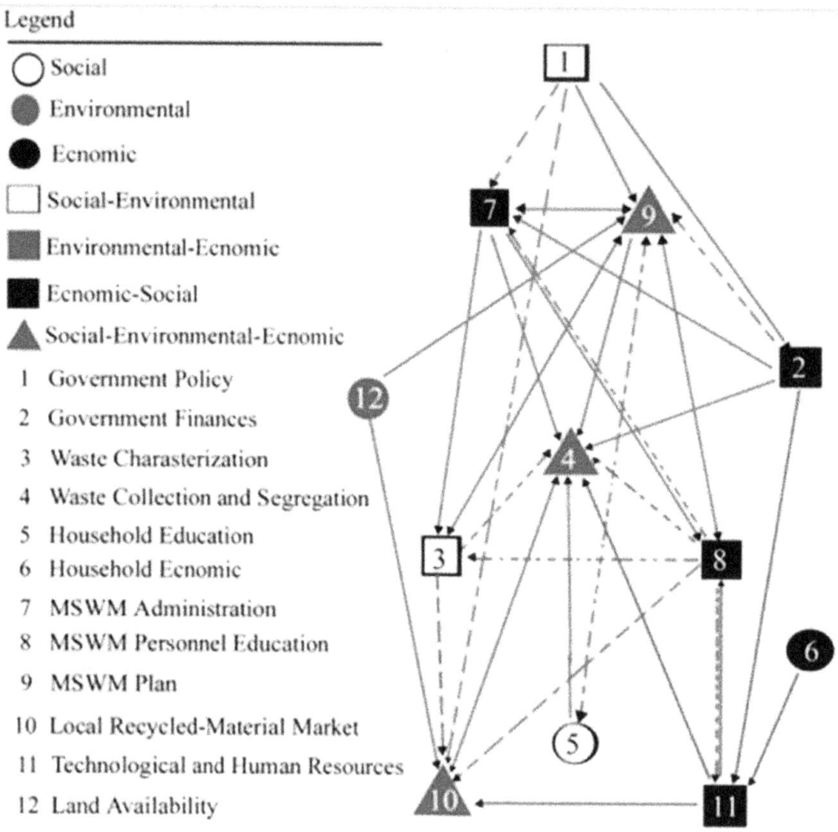


An integrated solid waste management approach is adopted to manage various aspects related to environmental impact of municipal solid waste. Recycling of the waste has many benefits like, reduction of environmental stresses, reduction of expenses over waste disposal along reduction of dependence on resources. Recycling also creates jobs and increases GDP of the country. Motorbikes with trailers can be used for collection of waste in developing countries. Which are cheap in cost. Moreover motorbikes can access all the types of roads in all the areas.

More transfer stations may be constructed for improvement of waste collection facility along with an increase in number of garbage collection trailers. Rules and regulations regarding recycling can be made more workable for encouragement of recycling.

## Components of ERS

### Waste Collection System

With increase in amount of waste produced, collection of waste becomes more difficult. Collection of waste consist of picking up of waste from various sites, transportation of these wastes to transfer stations. Rossel and Jorge (1999) reported that waste management planning strategies should be such that it minimize the waste generation. Also, cleaner technologies must be used to address the problem of pollution during disposal of waste. Collection of waste is disturbed if the vehicles are more expensive.

## Recycling of Resources

Upon fulfillment of landfills incineration and recycling are most practicable. Recycling is found to be associated with creation of jobs, reduction of pollution and environmental sustainability. Practices currently used for disposal of waste are environmentally sensitive. Recovery of materials conserves material as well as environment. Figure 2 shows collection and transportation of municipal solid waste.

## Waste Segregation/Separation

Segregation of waste can be done at the place it is produced or the place where it is transferred. The materials found in segregation reduce the load on the resources. It can also reduce the amount of solid waste to be land filled. Hazardous type of waste can also be segregated for prevention of toxic compounds.

*Figure 2. Collection and transportation of recyclable waste Source: Sanneh et. al., (2011)*

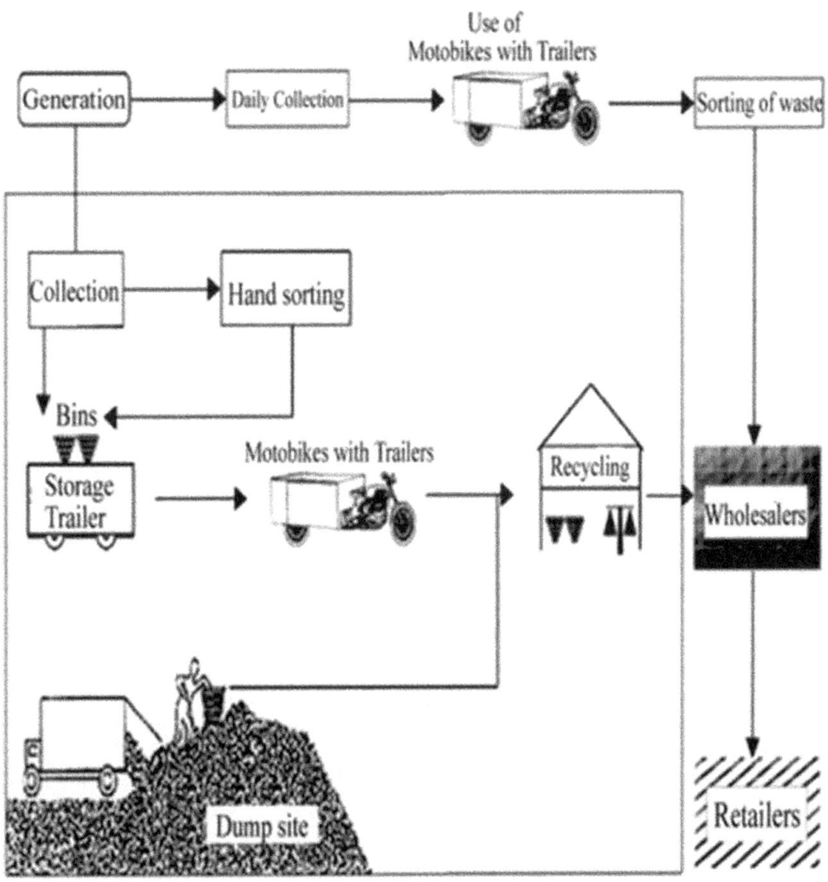

The most effective way of segregation at the source is manual sorting in households prior to collection. In many cities (e.g., Bangalore, Chennai, etc., in India), such systems are now routinely used. The municipality generally provides separate, easily identifiable containers into which the householder deposits segregated recyclable materials such as paper, glass, metals, etc. Usually, separate collections are carried out for the recyclable material.

In case the segregation is not done prior to collection, it could be sorted out through mechanical techniques (on Recovery sites) such as air separation, magnetic separation, etc., to recover the wastes.

### *Air Separation*

This technique has been in use for a number of years in industrial operations for segregating various components from dry mixture. Air separation is primarily used to separate lighter materials (usually organic) from heavier (usually inorganic) ones. The lighter material may include plastics, paper and paper products and other organic materials.

Generally, there is also a need to separate the light fraction of organic material from the conveying air streams, which is usually done in a cyclone separator. In this technique, the heavy fraction is removed from the air classifier (i.e., equipment used for air separation) to the recycling stage or to land disposal, as appropriate. The light fraction may be used, with or without further size reduction, as fuel for incinerators or as compost material. There are various types of air classifiers commonly used, some of which are listed below:

1. **Conventional Chute Type:** This is one of the simplest types of air classifiers. In this type, when the processed solid wastes are dropped into the vertical chute, the lighter material is carried by the airflow to the top while the heavier materials fall to the bottom of the chute.

The control of the percentage split between the light and heavy fraction is accomplished by varying the waste loading rate, airflow rate and the cross section of chute. A rotary air lock feed mechanism is required to introduce the shredded wastes into the classifier.

2. **Zigzag Air Classifier:** Shredded wastes are introduced at the top of the column at a controlled rate, and air is introduced at the bottom of the column. As the wastes drop into the air stream, the lighter fraction is fluidised and moves upward and out of column, while the heavy fraction falls to the bottom. Best separation can be achieved through proper design of the separation chamber, airflow rate and influent feed rate.
3. **Open Inlet Vibrator Type:** In this type of air classifier, the separation is accomplished by a combination of the following actions:
    a. **Vibration:** This helps to stratify the material fed to the separator into heavy and light components. Due to this agitation, the heavier particles tend to settle at the bottom as the shredded waste is conveyed down the length of the separator.
    b. **Inertial Force:** In this action, the air pulled in through the feed inlet imparts an initial acceleration to the lighter particle, while the wastes travel down the separator as they are being agitated.
    c. **Air Pressure:** This action refers to the injection of fluidising air in two or more high velocity and low mass flow curtains across the bed.

A final stripping of light particles is accomplished at the point where the heavy fraction discharges from the elutriators. It has been reported that the resulting separation is less sensitive to particle size than a conventional vertical air classifier, be it of straight or zigzag design. An advantage of this classifier is that an air lock feed mechanism is not required and wastes are fed by gravity directly into the separator inlet.

## Selection of Air Separation Equipment

The factors that are to be considered for selecting air separation equipment include the following:

- Characteristics of the material produced by shredding equipment including particle size, shape, moisture content and fibre content.
- Material specification for light fraction.
- Methods of transferring wastes from the shredders to the air separation units and feeding wastes into the air separator.
- Characteristics of separator design including solids-to-air ratio, fluidising velocities, unit capacity, total airflow and pressure drop.
- Operational characteristics including energy requirement, maintenance requirement, simplicity of operation, proved performance and reliability, noise output, and air and water pollution control requirements.
- Site considerations including space and height access, noise and environmental limitations.

### *Magnetic Separation*

The most common method of recovering ferrous scrap from shredded solid wastes involves the use of magnetic recovery systems. Ferrous materials are usually recovered either after shredding or before air classification. When wastes are mass-fired in incinerators, the magnetic separator is used to remove the ferrous material from the incinerator residue. Magnetic recovery systems have also been used at landfill disposal sites.

The specific locations, where ferrous materials are recovered will depend on the objectives to be achieved, such as reduction of wear and tear on processing and separation equipment, degree of product purity achieved and the required recovery efficiency. Equipment used for magnetic separation various types of equipment are in use for the magnetic separation of ferrous materials.

The most common types are the following:

1. **Suspended Magnet:** In this type of separator, a permanent magnet is used to attract the ferrous metal from the waste stream. When the attracted metal reaches the area, where there is no magnetism, it falls away freely. This ferrous metal is then collected in a container. This type of separation device is suitable for processing raw refuse, where separators can remove large pieces of ferrous metal easily from the waste stream.
2. **Magnetic Pulley:** This consists of a drum type device containing permanent magnets or electromagnets over which a conveyor or a similar transfer mechanism carries the waste stream. The conveyor belt conforms to the rounded shape of the magnetic drum and the magnetic force pulls the ferrous material away from the falling stream of solid waste.

### *Eddy Current Separation*

The edge current separator provides a mechanism for sorting a waste mixture of similar density grades. This technology relies on the induction of eddy currents in metal objects in response to an electromagnetic field. Eddy currents are created when conductive objects are located in or exposed to a spatially or temporally alternating magnetic field.

Eddy currents, their form independent of the circuits form, flow in closed loops within the conductor. According to Lenz's law, the induced electric current produces a magnetic field opposite of the field to which it is exposed.

A force is produced against the conductive object which trusts the object out of the magnetic field. Less conductive objects require less force. With increasing density, greater hurling force is necessary due to the mass inertia of individual objects.

Extensive use of this technology began in the United States, where aluminium used beverage cans were reprised.

### *Optical Separation*

This treatment version for the waste stream was especially developed for sorting mixed, crushed glasses into the different colors. For separation manufactures have developed electro-optical sorters which recognize the color of the glasses based on their opacity and with the aid of a blast of compressed air, deflect the particle from its flight path into appropriate catch bins.

While sorting of glass it is practicable to achieve a purity grade of 98%. To achieve this degree of separation it is necessary to sieve, to reduce the particle size and to individualize the glasses.

Singular glass particles are optically inspected in the following order: flint glass, green glass, brown glass and nontransparent.

## SOLUTION TO NON RECYCLABLE WASTE (WASTE TO ENERGY SYSTEMS)

### Incineration

Waste to energy means burning of waste under controlled conditions. In this method waste is sorted to recover the waste for recycling. The remaining waste is burned at high temperature (usually over 800 °C). Energy produced during the incineration process can be used to produce steam, which can be further used for electricity production. These types of units also provide water for industrial and domestic requirement. Ash from these plants can be used for construction projects. This type of system has been recorded to reduce green house gas emissions by up to 40%.

### Gasification and Pyrolysis

Gasification is heating of waste in presence of limited amount of oxygen however pyrolysis occurs in absence of oxygen. Before these processes non combustible materials are separated from the waste used for gasification and pyrolysis. Gasification produces producer gas which contain about 1% methane whereas in pyrolysis methane is produced in larger amount.

## OTHER WASTE MANAGEMENT PRACTICES ESPECIALLY FOR PLASTIC WASTE

Plastics cannot be recycled indefinitely; each recycling cycle reduces the strength and utility of the plastic. Energy recovery from plastics shall be considered when recycling is no longer possible. Plastic waste is accepted as fuel in cement kilns; residence times and temperatures are adequate to pre-empt the production of dioxins and furans.

### Use of Waste Plastic in Construction of Bituminous Roads (CPCB, 2013)

The use of plastic waste in bitumen roads has been found to have several advantages including decreasing the susceptibility of the road to infiltration. There are also no observed deleterious impacts on the strength or properties of the road.

### Process Description

- Waste plastic bags collected from roads, garbage trucks, dump sites and compost plants, rag-pickers, waste-buyers and households can be used in this process. Here plastic waste is sorted as per thickness and type of plastic. Polyethylene with a thickness greater than 60 micron is sent for recycling; Polyetylene below 60 microns is to be used in this process. Plastic waste should be cleaned by de-dusting or washing if required.
- Collected Plastic is cut and sized into fine pieces as far as possible.
- Lower micron plastic mixes easily with bitumen at higher temperatures (160°C - 170°C). Bitumen is heated up to its melting temperature, around 160°C-170°C. Sized plastic pieces are added slowly to the hot bitumen at a temperature around 160°C - 170°C.The mixture is stirred for about 20-30 minutes at a constant temperature of 160°C-170°C. Polymer-bitumen mixtures of different compositions are prepared and tested for their application in road construction.

### Conversion of Plastic Waste Into Liquid Fuel (CPCB, 2013)

A research-cum-demonstration plant at Nagpur, Maharashtra converts waste plastics into liquid fuel. The process adopted is based on random de-polymerization of waste plastics into liquid fuel in presence of a catalyst. The entire process is undertaken in closed reactor vessels followed by condensation, if required. Waste plastics while heating upto 2700°C to 3000°C convert into liquid-vapour state, which is collected in condensation chambers in the form of liquid fuel while the tarry liquid waste is topped-down from the heating reactor vessel. Organic gas is generated, which is vented due to lack of storage facility. However, the gas can be used in dual fuel diesel-generator set for generation of electricity. This technology is at an experimental stage and may be considered only after it is well established as technically, financially and environmentally viable option of plastic waste processing and/ or disposal.

### Plastic to Liquid Fuel (Case Study): (CPCB, 2013)

Initiative at New Moti Bagh, New Delhi Under a 'zero waste' project at the GPRA Complex of the Ministry of Urban Development at New Moti Bag, a small facility to convert plastic waste into liquid fuel has been installed. The facility handles 50 kg of plastic waste per batch. The plastic waste, collected from the staff quarters, is segregated and put through a pyrolytic conversion at about 400°C in the presence of a catalyst. The large polymeric molecules break down into smaller fractions leading to yield of liquid fuel (in the range of light diesel oil / furnace oil), gaseous fuel and char, all of which can be used for different purposes. The Light Diesel Oil (LDO) / furnace oil is of key interest for commercial purposes. However, its environmental implications in terms of emissions etc., needs to be further established in due course of its operation.

## CONCLUSION

Conventional methods of solid waste disposal are not able to manage solid waste in proper way. Integrated waste management approach can be used for prevention of problems due to solid waste (Sanneh et. al., 2011). Local authorities may have constraints of land, money and man power to come up with the problem. Authorities need financial resources, legal framework, and labor for better management of solid waste. Recycling of recyclables is the major strategy which can be implemented to give the desired output.

## REFERENCES

Buenrostro, O., & Bocco, G. (2003). Solid waste management in municipalities in Mexico: Goals and perspectives. *Resources, Conservation and Recycling*, *39*(3), 251–263. doi:10.1016/S0921-3449(03)00031-4

CPCB. (2013). *Overview of plastic waste management*. Retrieved from http://www.cpcb.nic.in/waste/plasticwaste/ Plastic_waste-1.pdf

Medina, M. (1997). The effect of income on municipal solid waste generation rates for countries of varying levels of economic development: A model. *Journal of Solid Waste Technology Management*, *24*(3), 149–155.

Medina, M. (2000). Scavenger co-operatives in Asia and Latin America. *Resources, Conservation and Recycling*, *31*(1), 51–69. doi:10.1016/S0921-3449(00)00071-9

Medina, M. (2004). *Scavenger cooperatives in Asia and Latin America*. Global Development Network. Retrieved from http://www.gdnet.org/pdf/medina.pdf

MoE Japan. (2006). *Annual report on the environment in Japan*. Retrieved from https://www.env.go.jp/en/wpaper/2006/fulltext.pdf

Pappu, A., Saxena, M., & Asolekar, S. R. (2007). Solid waste generation in India and their recycling potentials in building materials. *Building and Environment*, *42*(6), 2311–2320. doi:10.1016/j.buildenv.2006.04.015

Rossel, S. A., & Jorge, M. F. (1999). Cuban strategy for management and control of waste. In *Proceedings of Recovery, recycling, re-integration (R '99) congress*. EMPA.

Sanneh, E. S., Hu, H. A., Chang, Y. M., & Sanyang, E. (2011). Introduction of a recycling system for sustainable municipal solid waste management: A case study on the greater Banjul area of the Gambia. *Environment, Development and Sustainability*, *13*(6), 1065–1080. doi:10.100710668-011-9305-9

Troschinetz, A. M., & Mihelcic, J. R. (2009). Sustainable recycling of municipal solid waste in developing countries. *Waste Management (New York, N.Y.)*, *29*(2), 915–923. doi:10.1016/j.wasman.2008.04.016 PMID:18657963

Wang, H., & Nie, Y. (2001). Municipal solid waste characteristics and management in China. *Journal of the Air & Waste Management Association*, *51*(2), 250–263. doi:10.1080/10473289.2001.10464266 PMID:11256500

# Chapter 14
# Status of Electronic Waste Management in India: A Review

**Sanjay Kumar Koli**
*Chaudhary Brahm Prakash Government Engineering College, India*

**Athar Hussain**
*Chaudhary Brahm Prakash Government Engineering College, India*

## ABSTRACT

*Electronics waste is becoming a major global issue. Huge accumulation of e-waste and its recycling through primitive means for extraction of precious metals are a real concern in the developing countries due to the presence of hazardous materials in e-waste. The major portion of e-waste generated domestically as well as illegally imported is recycled in a crude manner leading to pollution of the environment. Current practices of e-waste management in India encounters many challenges like the difficulty in inventorization, ineffective regulations, pathetic and unsafe conditions of informal recycling, poor awareness of consumers, and reluctance on part of stakeholders to address the issues. As a result, toxic materials enter waste stream with no special precautions to avoid the known adverse impacts on the environment and human health. Resources are wasted when economically valuable materials are dumped. This chapter highlights the hazards caused due to improper handling of e-wastes and also describes some appropriate measures to be adopted for its management and safe disposal.*

## 1. INTRODUCTION

India is the fifth biggest producer of e-waste in the world; discarding 1.7 million tones (Mt) of electronic and electrical equipment in 2014. In India e-waste collection, transportation, segregation, dismantling, recycling and disposal is done manually by untrained labors in informal sector (Kumar and Karishma, 2016). Globally, WEEE is growing by about 40 million tons a year. In developed countries, e-waste constitutes 1-2% of the total solid waste generation. In US, it accounts for 1-3% of the total municipal waste generation. In European Union (EU), total amount of e-waste generation ranges from 5-7 million

DOI: 10.4018/978-1-5225-5754-8.ch014

tons per annum or about 14-15 kg per capita and is expected to grow at the rate of 3-5% per year. Developing countries are also in the challenging phase as they are already facing the continuum of hazardous e-waste mountains. In South Africa and China for example, the report predicts that by 2020, e-waste from old computers will jumped by 200 to 400 percent from 2007 levels and by 500% in India (Bhat et al.2012). About 80,000 people working for recycling sector, and scrap markets exists where piles of e-wastes are separated for recycling. They separate copper from wires after burning them. Plastic and PVC codes produce noxious smoke which is irritable to eyes and cause respiratory problems. In addition, acid treatment is given to isolate metals also corrosive acids releases from used batteries of cell phones and computers. However, currently they are building an e-waste recycling plant in Bangalore which was estimated as having the capacity to handle 60,000 tons of e-wastes annually. In India about 24% of e-waste was produced from Mumbai, and, Delhi, Bangalore and Chennai are 21.2%, 10.1% and 9.1% respectively (Sivaramanan,2013).

## 2. E-WASTE GLOBAL SCENARIO

As far as global e-waste management is concerned, Switzerland is the first country to implement the organized e-waste management system in the world. Extended Producer Responsibility (EPR) and Advance Recycling Fee (ARF) are the backbone of e-waste management system in Switzerland and other developed countries. Advanced countries like USA, UK, France & Germany generate 1.5 to 3 million tons of e-waste annually and are among the largest generators of e-waste. But these countries also have standardized e-waste management processes in place. Proper e-waste management, from efficient sourcing and collection right upto extraction and disposal of material, has ensured that this huge pile of junk turns into a lucrative business opportunity. Due to very stringent environmental standards, the cost of collection, pre-processing, recycling and disposal are pretty high. So for every organized recycler in the first world countries, there are quite a few who pose as recyclers and are mere brokers who ship these obsolete items to developing countries like India and China in the pretext of donation or second hand goods. With very ambiguous laws related to environmental protection, India, China and a few African countries have become dumping sites to the first world countries .There are many countries that have already started the “take back” system for electronic products and they also have dedicated laws on e-waste management. In USA, National Electronics Action Plan has been initiated by US Environment Protection Agency to address the various issue related to electronic waste. Two very important frameworks for protecting environment from e-waste have been put forward by European Union i.e., WEEE Directives and Restriction of use of Certain Hazardous Substances(RoHS),which are also implemented by other countries. According to EU directives (2003), it is mandatory for all 27 countries of European Union to recycle their e-waste (Jadhav,2013).

The global quantity of e-waste in 2014 is comprised of 1.0 Mt lamps, 3.0 Mt of Small IT, 6.3 Mt of screens and monitors, 7.0 Mt of temperature exchange equipment (cooling and freezing equipment), 11.8 Mt large equipment, and 12.8 Mt of small equipment. The amount of e-waste is expected to grow to 49.8 Mt in 2018, with an annual growth rate of 4 to 5 per cent(balde,2014).

Most of the e-waste was generated in Asia: 16 Mt in 2014. This was 3.7 kg for each inhabitant. The highest per inhabitant e-waste quantity (15.6 kg/inh.) was generated in Europe. The whole region (including Russia) generated 11.6 Mt. The lowest quantity of e-waste was generated in Oceania, and was 0.6 Mt. However, the per inhabitant amount was nearly as high as Europe’s (15.2 kg/inh.). The lowest

*Table 1. Global quantity of e-waste generated(balde,2014)*

| Year | E-waste generated (Mt) | Population (billion) |
|---|---|---|
| 2010 | 33.8 | 6.8 |
| 2011 | 35.8 | 6.9 |
| 2012 | 37.8 | 6.9 |
| 2013 | 39.8 | 7.0 |
| 2014 | 41.8 | 7.1 |
| 2015 | 43.8 | 7.2 |
| 2016 | 45.8 | 7.3 |
| 2017 | 47.8 | 7.4 |
| 2018 | 49.8 | 7.4 |

amount of e-waste per inhabitant was generated in Africa, where only 1.7 kg/inh. was generated in 2014. The whole continent generated 1.9 Mt of e-waste. The Americas generated 11.7 Mt of e-waste (7.9 Mt for North America, 1.1 Mt for Central America, and 2.7 Mt for South America), which represented 12.2 kg/inhabitants. The graph shows the increase of e-waste generation linearly as our population is increasing as described in Figure 1.

## 3. E-WASTE PRODUCTION AND MANAGEMENT IN INDIAN SCENARIO

The story of current Indian e-waste management is different from the worldwide. Practices e-waste is a serious issue because of the informal recycling activities. Therefore, quantification of e-waste in India is very difficult and there is no mechanism and policy to check the flow of e-waste in the system. Business sector is mainly responsible for the waste generation. In addition to this, about 1050 tones per year of computer waste comes from retailers and manufacturers. This is important to note that in spite of global agreements, e-waste from developed nations is imported to developing nations like India. Sixty-five cities in India generate more than 60% of total e-waste generated in India. Ten states generate 70% of the total e-waste generated in India. Maharashtra followed by Tamil Nadu, Andhra Pradesh, Uttar Pradesh, West Bengal, Delhi, Karnataka, Gujarat, Madhya Pradesh and Punjab in the list of e-waste generated states in India (Shagun et al.,2013).

The composition of e-waste is divided into various categories such as ferrous metal, non-ferrous metal, plastics, glass and others. The classification of different constituents of electronic waste as represented in literature is summarized below in Table 2.

A number of countries have come out with their own definition, interpretation and usage. But, the most widely accepted definition is as per European Union (EU) directive, followed in member countries of EU and other countries of Europe. Definition as per EU directive is electrical or electronic equipment which is waste including all components, subassemblies and consumables, which are part of the product at the time of discarding presents various definitions provided by the different organizations. The definitions pertaining to e-waste from different organizations is summarized in Table 3 (Mohan and Bhamawat,2008).

*Figure 1. Global quantity of e-waste generation*

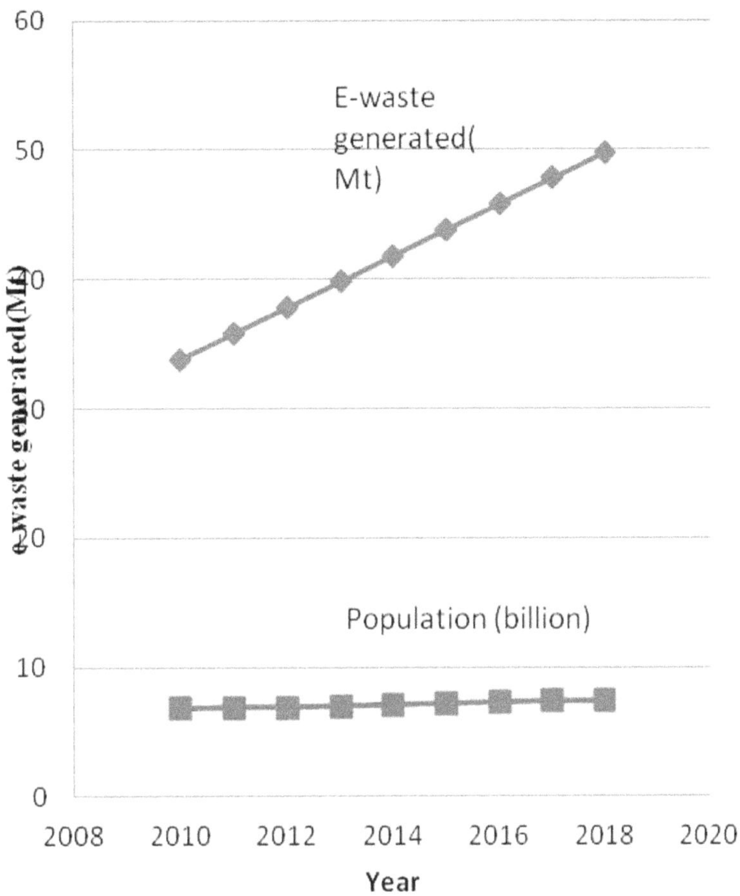


*Table 2. Composition of e-waste (source: http://ewasteguide.info/e-waste-composition)*

| Ferrous Metal | Iron and steel |
|---|---|
| Non- ferrous metal | Aluminium, copper, lead, cadmium, mercury, gold, silver, palladium, arsenic, indium, selenium |
| Plastic | Brominated and non-brominated plastic |
| Glass | Lead glass and normal glass |

In Indian contest, as given in the report of Rajya Sabha (2011), ferrous metal contributes (43%) in large household appliances. There are four categories which are large household appliances, small household appliances, ICT & consumer electronics and lamps. It is seen that percentage of ferrous metal is decreasing day by day as the percentage of plastics is increasing to make the products lighter and thinner. The problem while using plastics is this that they use the blended plastic (mixed plastic) which is not separable when we dismantle the laptops or monitors. The data of Indian e-waste scenario is prescribed report 2011 is tabulated in Table 4.

*Table 3. E-waste definitions prescribed by different organizations (Source: Mohan and Bhamawat,2008)*

| Organization | Definitions |
|---|---|
| Basel Action Network | E-waste encompasses a broad and growing range of electronic devices, ranging from large household devices such as refrigerators, air conditioners, cell phones, personal stereos, and consumer electronics to computers, which have been discarded by their users. |
| *OECD (USA) | Any appliance using an electric power supply that has reached its end of- life is known as electronic waste. |
| StEP | E-waste refers to the reverse supply chain, which collects products no longer desired by a given consumer and refurbishes for other consumers, recycles, or otherwise processes wastes. |
| Sinha | An electrically powered appliance that no longer satisfies the current owner for its original purpose. |
| Maharashtra Pollution Control Board | Electronic Waste (E-Waste) comprises of waste electronic goods, which are not fit for their originally intended use. These range from household appliances such as refrigerator, air conditioner, cellular phone, personal stereos and consumer electronics to computers. |

* Organization for Economic Cooperation and Development

*Table 4. E -waste producer and characterization in Indian scenario E-waste in India (source: research unit, 2011)*

| Material | Large household appliance | Small household appliance | ICT and consumer electronics | Lamp |
|---|---|---|---|---|
| Ferrous metal | 43 | 29 | 36 | - |
| Non-ferrous metal | 27.67 | 26.19 | 9 | 14.3 |
| Plastic | 19.31 | 37.75 | 30 | 3.7 |
| Glass | 0.02 | 0.16 | 19.3 | 77 |
| Other | 10 | 6.9 | 5.7 | 5 |

*Table 5. Overview of the composition of the appliances in the three categories is given in table (Source: Jalal Uddin, 2012)*

| Appliances | Ferrous wt. (%) | Non - ferrous wt. (%) | Glass wt. (%) | Plastic wt. (%) | Electronic component wt. (%) | Other wt. (%) |
|---|---|---|---|---|---|---|
| Refrigerator and Freezers | 64.4 | 6 | 1.4 | 13 | 0.2 | 15 |
| Personal Computers | 20 | 24 | 15 | 23 | 17.3 | 0.7 |
| TV sets | 5.3 | 5.4 | 6.2 | 22.9 | 0.9 | 3.5 |

Data from Table 5 indicates that the refrigerators and freezers have more percentage of ferrous metal (64.4%) then television and computers while televisions have more percentage of plastics (22.9%) than refrigerators and computers. But in today's time, manufacturers and companies tend to move towards plastics than ferrous metal. Similarly if televisions converted to recent TV sets, it is observed that the usage of more plastics has been taken place to make the product lighter and also to reduce the manufacturing cost. The overview composition of appliances enter plastic and metals are used is given below from Table 5.

The recoverable materials in the refrigerators are ferrous metals 46.61%, compressors 23.80%, and plastics 13.84% by weight. All these materials are refrigerators can be reused. The data of recoverable materials in refrigerators are given in Table 6. The reuse of these materials will effect the less usage of virgin materials and hence the manufacturing costs can be reduced.

## 4. E –WASTE (MANAGEMENT) RULES, 2016: WHAT'S NEW

Manufacturer, dealer, refurbisher and Producer Responsibility Organization (PRO) have been introduced as additional stakeholders in the rules. Compact Fluorescent Lamp (CFL) and other mercury containing lamp brought under the purview of rules. Collection mechanism based approach has been adopted to include collection centre, collection point, take back system etc for collection of e - waste by Producers under Extended Producer Responsibility (EPR). Option has been given for setting up of PRO, e - waste exchange, e - retailer, Deposit Refund Scheme as additional channel for implementation of EPR by Producers to ensure efficient channelization of e - waste. Provision for Pan India EPR Authorization by CPCB has been introduced replacing the state wise EPR authorization. Deposit Refund Scheme has been introduced as an additional economic instrument wherein the producer charges an additional amount as a deposit at the time of sale of the electrical and electronic equipment and returns it to the consumer along with interest when the end - of - life electrical and electronic equipment is returned. The e - waste exchange as an option has been provided in the rules as an independent market instrument offering as-

*Table 6. Recoverable Quantity of Materials in a Refrigerators (Source: Jalal Uddin,2012)*

| Material Type | % (by weight) |
|---|---|
| CFCs | 0.20 |
| Oil | 0.32 |
| Ferrous Metals | 46.61 |
| Non- Ferrous Metal | 4.97 |
| Plastics | 13.84 |
| Compressors | 23.80 |
| Cables / Plugs | 0.55 |
| Spent Foam | 7.60 |
| Glass | 0.81 |
| Mixed Waste | 1.30 |
| Total | 100.00 |

sistance or independent electronic systems offering services for sale and purchase of e - waste generated from end - of - life electrical and electronic equipment between agencies or organizations authorised under these rules. The manufacturer is also now responsible to collect e - waste generated during the manufacture of any electrical and electronic equipment and channelize it for recycling or disposal and seek authorization from SPCB. The dealer, if has been given the responsibility of collection on behalf of the producer, need to collect the e - waste by providing the consumer a box and channelize it to Producer. Dealer or retailer or e - retailer shall refund the amount as per take back system or De posit Refund Scheme of the producer to the depositor of e - waste. Refurbisher need collect e - waste generated during the process of refurbishing and channelise the waste to authorised dismantler or recycler through its collection centre and seek one time authorization from SPCB. The roles of the State Government has been also introduced in the Rules in order to ensure safety, health and skill development of the workers involved in the dismantling and recycling operations. State Government to prepare integrated plan for effective implementation of these provisions, and to submit annual report to Ministry of Environment, Forest and Climate Change .he transportation of e - waste shall be carried out as per the manifest system whereby the transporter shall be required to carry a document (three copies) prepared by the sender, giving the details. Liability for damages caused to the environment or third party due to improper management of e - waste including provision for levying financial penalty for violation of provisions of the Rules has also been introduced. Urban Local Bodies (Municipal Committee/Council/Corporation) has been assign the duty to collect and channelized the orphan products to authorized dismantler or recycler.

## 5. ROLE OF NON-FORMAL SECTOR

Ninety-five percentage of the e-waste in India is being recycled in non-formal sector and five percentage of the e-waste volume are handled in formal unit. In and around of metropolitan cities in India, there are over 3000 units engaged in non-formal sector for e-waste recycling. Non-formal units of e-waste recyclers are distributed all over India. A large cluster of industries exists in Delhi, Tamil Nadu, Uttar Pradesh, Karnataka, Maharashtra, Gujarat, Kerala, Andhra Pradesh, West Bengal, Rajasthan, etc. Non-formal units generally follow the steps such as collection of the e-waste from the rag pickers, disassembly of the products for their useable parts, components, modules, which are having resell value. The rest of the material is chemically treated to recover precious metals. Due to inadequate means, it may cause leaching of hazardous substances to the air, soil, and water. This recycling method has low efficiency and recovery is carried out only for valuable metals like gold, silver, aluminium, copper, etc. Other materials such as tantalum, cadmium, zinc, palladium etc. could not be recovered (Chatterjee,2012). Scrap collection is carried out by two categories of workers, waste pickers and the itinerant buyers. Waste pickers retrieve paper, plastic, metal and glass scrap from garbage bins or Receptacles that are provided by the municipalities for the disposal of garbage on the street, and from landfill sites where the collected garbage is transported and dumped. Itinerant waste buyers purchase small quantities of scrap from households, offices, shops and other small commercial establishments. All categories of scrap collectors rudimentarily sort and then sell the collected scrap commodities to retail scrap establishments on the basis of weight or unit. The factors that structure the scrap labour market are found to be age, gender, religion, ethnicity and 12 kinship (caste), geographical origin (migratory status), place of living (spatial distribution and locational specialisation) and education (deshmukh and mahajan,2015). Informal recycling systems can bring significant economic benefits to developing countries. From a macroeconomic perspective, they are well adapted to the prevailing conditions, namely abundant supply of working force, but scarce capital: they minimise capital expenditures and maximise hand (and animal) power. They are able to provide a steady, reliable supply of secondary raw materials for local manufacturing industry which can replace more expensive imported raw materials. This also stimulates the manufacture of low-cost, affordable products made from recycled materials. Recycling is always placed above disposal in the waste management 'hierarchy'. The informal waste recycling systems that already exist in many developing countries reduce the cost of formal waste management systems as they reduce the quantity of waste for collection, resulting in less money and time spent on collection and transport. Void space at disposal sites is preserved and only used for wastes with no potential value as recycled materials are diverted for reuse. For the most part, these economic benefits are achieved at no direct cost to the tax-payers. There are also social benefits associated with informal recycling. It provides employment and a livelihood for impoverished, marginalised and vulnerable individuals or social groups. Despite the particularly adverse working conditions associated with informal recyclers, it is important to recognise that it does allow those involved to survive and be employed in regions that often have high unemployment. Informal recycling has traditionally been practised by outcasts and marginal groups in developing countries such as gypsies, rural migrants, immigrants and members of religious minorities. In India, the Harijans are a cast of untouchables that deal with waste collection and recycling. In Cairo, the Zabbaleen informal recyclers belong to a Christian minority. Isolation as a part of social exclusion leads societies of scavengers to develop their own habits, customs, beliefs and values. As a result of their marginalisation, they are often a subject of harassment by the authorities and police and female scavengers in particular may be considered easy sexual targets (Wilson et al,2006)

## 6. ROLE OF FORMAL SECTOR

Few formal recyclers are operating in India. The processes followed informal sector is mainly limited to the segregation, dismantling of e-waste till the size reduction stage of printed circuit boards. A shredder is employed for PCBs size reduction. The pre-processed PCB is exported to smelting refineries in developed countries for further recovery of precious metals like copper, silver, gold, aluminium, palladium, tantalum, ruthenium, platinum etc. and also treating the slag by-product in an eco-friendly manner. The recycling of valuables substances by units in formal sector is carried out in the protected environment and with due care to minimize any damage to the environment or society. The use of advanced processes and technologies leads to efficient recovery of metals. Recovery technology by units in formal sector will be economically viable as the high cost of capital equipment and needed techniques could be shared by the volume of products. The efficiency of recovery in the formal recycling is high and metals at the trace level can also be recovered. Some technology works with a zero-landfill approach. Most of the e-waste in India is channelized to a non-formal sector, whereas, the formal sector is facing a problem of not having sufficient input materials (Chatterjee, 2012). Cost is another big issue for e-waste management. Cost of logistics and transportation is a challenge faced by most recyclers, preventing the flow of waste volumes in the country. Places where e-waste ends-up landfill where toxic chemicals in electronics products can leach into the land over time or be released into the atmosphere, impacting nearby communities and the environment.

Incineration process releases heavy metals such as lead, cadmium, and mercury into the atmosphere and which can bio-accumulate in the food chain, particularly in fish, which is the major source of exposure for the general public. Reuse is a good way to increase a product's lifespan. Many old products are exported to developing counties. Although the benefits of reusing electronics in this way is clear, the practice is causing serious problems because the old products are dumped after a short period of use in areas that are unlikely to have hazardous waste facilities. Although recycling can be a good way to reuse the raw materials in a product, the hazardous chemicals in e-waste mean that electronics can harm workers in the recycling yards, as well as their neighboring communities and the environment.

E-Waste is routinely exported by developed countries, often in violation of the international law. Inspections of 18 European seaports in 2005 found that as much as 47 percent of waste destined for export, including e-waste, was illegal. At least 23,000 metric tons of undeclared or "grey" market electronic waste was illegally shipped in 2003 to the Far East, India, Africa, and China. In the USA, it is estimated that 50–80 percent of the waste collected for recycling is being exported in this way. This practice is legal because the USA has not ratified the Basel Convention (Khurrum et al.,2011). Recycling of electronic waste is another important subject not only from the point of waste treatment but also from the recovery of valuable metals. The value distribution for personal computer shows that, the precious metal make up more than 70% of the value. This indicates that the major economic driver for recycling of electronic waste is from the recovery of precious metals and their industrial application (Jatindra and sudhir,2009).

The data summarized in the Table 7 is based on the environmental effect due to incineration, land filling and recycling of e-waste. In case of incineration, brominated flame retardants when burn at high temperature generate highly toxic dioxins and furans which are very harmful for human health as well as environment.

## 7. ENVIRONMENTAL AND HEALTH EFFECTS OF E-WASTE

Environmental effects of hazardous waste arise due to primary, secondary and tertiary emissions of hazardous waste. Primary emissions include hazardous waste present in e-waste including heavy metals like lead, arsenic, mercury and PCB whereas secondary emissions are generally due to incomplete treatment of e-waste which leads to generation of dioxins and furans. Tertiary emissions occur due to harmful chemicals used for recycling of hazardous waste. Often electronic goods can be classified into three main categories, white goods that consist of household appliances, brown goods including televisions, and cameras and grey goods include computers, printers and scanners. Literature review

*Table 7. Hazard due to improper disposal of e-waste end products (source: http://ewasteguide.info/e-waste-composition)*

| Disposal Type | Hazard |
|---|---|
| Incineration | Brominated flame retardants at a high temperature of 600-800$^0$C generate extremely toxic polybrominated dioxins (PBDDs) and polybrominated furans (PBDFs);PVCs generate toxic flue gas; |
| Land-filling | Hazardous metal (e.g. lead, chromium, mercury, cadmium etc.), PCBs, PBDEs leach into soil and groundwater. |
| Recycling | It often causes hazardous emission, due to recycling of plastics, halogenated substances and heavy metals like lead, cadmium, etc. shredding without proper disassembly causes hazardous substances dispersed into environment. |

suggests that grey goods are more hazardous than white and brown goods. Some of the major associated health impacts include affecting reproductive systems of humans both male and female, damaging lymphocytes and affecting growth retardation. Toxic chemicals present in e-waste can damage central nervous system and vital organs including liver and kidney. Continuous handling of such materials without use of protective gear can cause skin cancer, anaemia and carcinogenic tumors and may cause hormonal problems. Other problems include hypertension and psychological problems. Both scientific and non-scientific literature show growing incidence of several lethal or debilitating health conditions, which includes cancer, neurological, respiratory disorders, and birth defects. This impact is found to be worse in developing countries like India where people engaged in recycling e-waste are mostly in the unorganized sector, living in close proximity to dumps or landfills of untreated e-waste and often working without any protection or safeguards. Many workers engaged in these recycling operations are the urban poor and unaware of the hazards associated with them. For example, such recycling activities can lead to contamination of local drinking water sources causing severe health impacts (Ganguly, 2016).

## 8. ENVIRONMENTAL AND HEALTH HAZARDS OF E-WASTE

Arsenic is widely used in the semiconductors, diodes, microwaves, LEDs (light emitting diodes), solar cells which cause chronic exposure to arsenic can lead to various diseases of the skin and decrease nerve conductivity velocity. Chronic exposure to arsenic can also cause lung cancer and can often be fatal. Barium is used in electron tubes, filler for plastic and rubber, lubricant additives causes short-term exposure to barium could lead to brain swelling, muscle weakness, damage to the heart, liver, and spleen. Animal studies reveal increased blood pressure and changes in the heart from ingesting barium over a long period of time. Brominated flame retardants cause casing, circuit boards (plastic), cables, PVC cables can lead to combustion of halogenated case material and printed wiring board at lower temperature releases toxic emissions including dioxins which can lead to severe hormonal disorders.The management of e-waste in Indian context is quite difficult because the source of income is less and more and more people are engaged in this business. Recycling and reuse is the most adopted method for proper disposal of e-waste. As several private companies are also working in the e-waste recycling business. In Figure 2, proper management of e-waste is shown without affecting our environment is shown through Figure 1.

## 9. MANAGEMENT OF E-WASTE BY INDUSTRIES

E-Parisaraa Pvt. Ltd, India's first government authorised electronic waste recycler started operations from September 2005, is engaged in handling recycling and reusing of waste electrical and electronic equipment in eco-friendly way. The initiative is to aim at reducing the accumulation of used and discarded electronic and electrical equipment's, which most end up in landfills or partly recycled in a hygienic conditions by backyard recyclers and then partly thrown into waste streams damaging the environment(Source:http://ewasteindia.com).

Attero's recycling, is the only unit which does the complete e-waste management process with its end to end e-cycling plant, zero dumping technology. Attero's recycling technology aims to take today's waste and turn it into conflict-free, sustainable resources for tomorrow. Mobile phone recycling: Scanned and updated in the system, separating components which can be used again in different industries, such as electronics, plating, jewellery, automotive and art foundries. Display unit recycling: cathode ray tubes (CRT) contain lead, glass which can be recovered and reused. The process involves glass cutting, heating & air blow. Using magnetic separator, ferrous metals removed. Battery recycling: furnace smelting, treatment with alloys are the technologies involved. Attero recycles all types of batteries classified as e-waste by the EPA. PCB recycling: Printed circuit boards Component Removal Machine which separates components. I.T Goods recycling: Includes desktops, laptops, servers to printers, scanners, copiers etc (Source: http://www.attero.in). Earth e-waste management private limited industries in collection of e-waste material from premises and safe transportation to our factory premises. Sorting, Identify, Usefulness, Identify hazardousness, Dismantling, Segregation, Treatment/Disposal(http://www.eemplindia.com). Sims partners with local, national, and global businesses in the responsible reuse and recycling of end-of-life electronics and computers. We provide environmentally responsible solutions that ensure 100% digital data destruction. Our worldwide network of certified processing facilities, with no export mandates, reinforce our overall sustainable business model, and closed loop approach to responsible electronic recycling (http://www.simsrecycling.com). The organized sector uses WEEE recycling, a latest & environment friendly technology which not only ensures good health of the stakeholders but also recovers most of the materials used while making the product. Any equipment which uses either a plug or battery can be recycled using WEEE recycling. All types of obsolete electrical equipment including lighting, fridges and IT are treated safely by WEEEE recycling. WEEE recycling storage containers are also used to keep the waste safe. Since, most of these components are made of commodities which are re-saleable and re-usable, it makes WEEE recycling industry economically viable with the supplies of the required quantity of WEEE. While modern technologies carry on nearly hazard-free WEEE recycling of e-waste, precaution must be taken to control harmful emissions and toxics from causing detrimental impacts on health and the environment. The five step process is sourcing and collection, sorting and grading, dismantling, shredding and disposal (http://ecoreco.com). Hi-Tech Recycling India Pvt. Ltd. aims at becoming a trusted IT asset management (in full range of logistic management, data erasure, refurbishing, remarketing and recycling services) company for retiring IT asset equipments. To accomplish our mission, Hi-Tech Recycling commits its human, financial, and managerial resources to continue to

*Figure 2. Management of e-waste*
*(Source: E-waste in India Research Unit, Rajya Sabha Secretariat, New Delhi, June, 2011)*

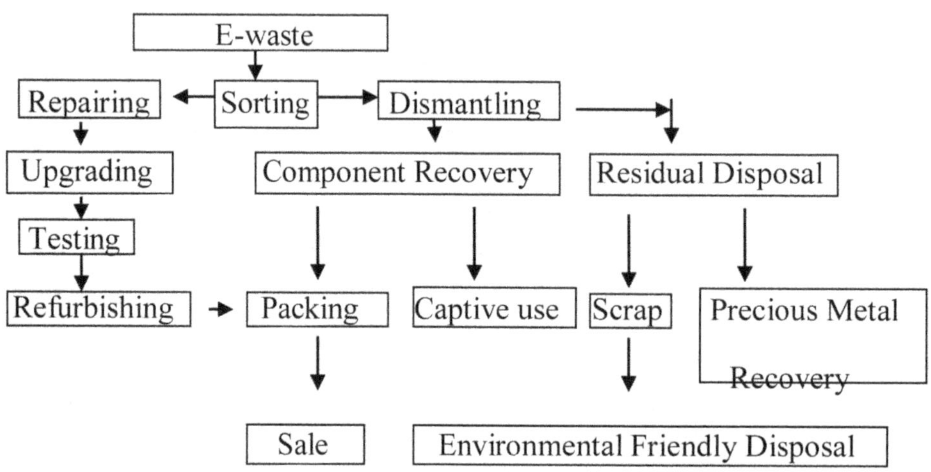

conduct a responsible in-house de-manufacturing service for non-reusable and/or non-resalable products in order to achieve the environmental obligations of zero landfill and zero export of electronic wastes to third world countries (http://www.hitechrecycling.in).

## 10. CONCLUSION

Due to the presence of the toxic elements, it is all the more dangerous for the society to stock them without carrying out appropriate disposal. There exist many hurdles to e-waste management in India. The major one is the dominance of the informal sector. The major challenges are to reduce e-waste through reuse, recycle, recovery and reduced use of toxic substances, to invent labor intensive intermediate technology to recycle. It is confirmed that the public awareness and cooperation of manufactures are essential for the advancement of e-waste management system. And also it is the responsibility of governments to allocate sufficient grants and protecting the internationally agreed environmental legislation within their borders. Licensing of certification like e-stewardship may ensure the security to prevent illegal smugglers and handlers of e-waste.

## REFERENCES


Baldé, C.P., Wang, F., Kuehr, R., & Huisman, J. (2014). *The Global E-Waste Monitor 2014 Quantities, Flows and Resources*. Academic Press.

Bhat, Rao, & Patil. (2012). Development of an integrated model to recover precious metals from electronic scrap - A novel strategy for e-waste management. *International Conference on Emerging Economics-Prospects and Challenges (ICEE-2012)*, 397-406.

Chatterjee. (2012). Sustainable electronic waste management and recycling process. *American Journal of Environmental Engineering, 2*(1), 23-33.

Deshmukh, S., & Mahajan, K. (2015, April 22). Role of Informal Sector in E-waste Management in Pune Region. *International Journal of Computers and Applications, 116*(5), 22–27. doi:10.5120/20333-2063

Ganguly. (2016). E-Waste Management in India – An Overview. *International Journal of Earth Science and Engineering, 9*(2), 574-588.

Jadhav. (2013). Electronic waste: A growing concern in today's environment sustainability. *International Journal of Social Science & Interdisciplinary Research, 2*(2), 139-147.

Jatindra, P. K., & Sudhir, K. (2009). E-Waste Management: A Case Study of Bangalore, India. *Research Journal Environmental and Earth Sciences, 1*(2), 111-115.

Karishma. (2016). Current Scenario of e-waste management in India: Issues and strategies. *International Journal of Scientific and Research Publications, 6*(1), 424-430.

Khurrum & Bhutta. (2011). *Electronic Waste: A Growing Concern in Today's Environment. Hindawi Publishing Corporation.*

Mohan, D., & Bhamawat, P. M. K. (2008). E-waste management-global scenario: A Review. *Journal of Environmental Research and Development*, *2*(4), 817–823.

Shagun, A. K., & Arora, A. (2013, October). Proposed Solution of e-waste management. *International Journal of Future Computer and Communication*, *2*(5), 490–493.

Sivaramanan. (2013). E-waste management, disposal and its impacts on the environment. *Universal Journal of Environmental Research and Technology, 3*(5), 531-537.

Uddin. (2012). Journal and Conference Paper On (Enviornment) e-waste management. *IOSR Journal of Mechanical and Civil Engineering, 2*(1), 25-45.

Wilson, Velis, & Cheeseman, . (2006). *Role of informal sector recycling in waste management in developing countries. In Habitat international* (pp. 797–808). Elsevier.

# Compilation of References


Abbasi, A. (2014). Agricultural Pollution: An Emerging Issue. In *Improvement of Crops in the Era of Climatic Changes*. Springer.

Abdallh, M. N., Abdelhalim, W. S., & Abdelhalim, H. S. (2016). *Industrial Wastewater Treatment of Food Industry Using Best Techniques*. Academic Press.

Abdel Ghafar, H. H., Ali, G. A. M., Fouad, O. A., & Makhlouf, S. A. (2015). Enhancement of adsorption efficiency of methylene blue on $Co_3O_{4/}SiO_2$ nanocomposite. *Desalination and Water Treatment*, *53*(11), 2980–2989. doi:10.1080/19443994.2013.871343

Abdel-Shafy, H., Hegemann, W., & Guldner, C. (1996). Fate of heavy metals *via* chemical–biological upgrading of sewage sludge treatment plants. *Environmental Management and Health*, *713*(3), 28–36. doi:10.1108/09566169610117895

Abhilash, P. C., & Nandita, S. (2009). Pesticides use and application: An Indian scenario. *Journal of Hazardous Materials*, *165*(1-3), 1–12. doi:10.1016/j.jhazmat.2008.10.061 PMID:19081675

Abhyankar, L. N., Jones, M. R., Guallar, E., & Acien, A. N. (2012). Arsenic exposure and hypertension: A systematic review. *Environmental Health Perspectives*, *120*(4), 494–500. doi:10.1289/ehp.1103988 PMID:22138666

Abo-Farha, S. A., Abdel-Aal, A. Y., Ashourb, I. A., & Garamon, S. E. (2009). Removal of some heavy metal cations by synthetic resin purolite. C100. *Journal of Hazardous Materials*, *169*(1-3), 190–194. doi:10.1016/j.jhazmat.2009.03.086 PMID:19403237

Achieng, G. O., Ongeri, D. M. K., & Omwoyo, W. N. (2013). Impact of anthropogenic activities on copper and lead levels in Kisumu city soils. *Sky Journal of Soil Science and Environmental Management*, *2*(5), 47–52.

Achyut, K. P., Singh, R. K., & Mishra, D. K. (2010). Thermolysis of waste plastics to liquid fuel A suitable method for plastic waste management and manufacture of value added products—A world prospective. Renewable and Sustainable Energy Reviews, 233–248.

Adamsand, B. A., & Holmes, E. L. (1935). Adsorptive properties of synthetic resins. *Journal of the Chemical Society*, *54*, 1–6.

Adams, L., Lyon, D., & Alvarez, P. (2006). Comparative eco-toxicity of nanoscale TiO2, SiO2, and ZnO water suspensions. *Water Resources*, *40*(19), 3527–3532. PMID:17011015

Agrafioti, E., Kalderis, D., & Diamadopoulos, E. (2013). Ca and Fe modified biochars as adsorbents of arsenic and chromium in aqueous solutions. *Journal of Environmental Management*, *146*, 444–450. doi:10.1016/j.jenvman.2014.07.029 PMID:25199600

Agrawal, G. D., Lunkad, S. K., & Malkhed, T. (1999). Diffuse agricultural nitrate pollution of groundwaters in India. *Water Science and Technology*, *39*(3), 67–75. doi:10.1016/S0273-1223(99)00033-5

Agrawal, M. (2005). Effects of air pollution on agriculture: An issue of national concern. *National Academy Science Letters (India)*, *28*, 93–106.

Ahmad, M. M., Getso, B. U., Ahmad, U. A., & Abdullahi, I. I. (2014). Impact of climate change on the distribution of tropical parasitic and other infectious diseases. *Toxicology and Food Technology*, *8*, 19–26.

Ahmad, M., Lee, S. S., Dou, X., Mohan, D., Sung, J. K., Yang, J. E., & Ok, Y. S. (2012). Effects of pyrolysis temperature on soybean stover-and peanut shell-derived biochar properties and TCE adsorption in water. *Bioresource Technology*, *118*, 536–544. doi:10.1016/j.biortech.2012.05.042 PMID:22721877

Ahmad, M., Lee, S. S., Rajapaksha, A. U., Vithanage, M., Zhang, M., Cho, J. S., ... Ok, Y. S. (2013). Trichloroethylene adsorption by pine needle biochars produced at various pyrolysis temperatures. *Bioresource Technology*, *143*, 615–622. doi:10.1016/j.biortech.2013.06.033 PMID:23838320

Ahmad, M., Rajapaksha, A., Lim, J., Zhang, M., Bolan, N., Mohan, D., ... Ok, Y. (2014). Biochar as a sorbent for contaminant management in soil and water: A review. *Chemosphere*, *99*, 19–33. doi:10.1016/j.chemosphere.2013.10.071 PMID:24289982

Ahmaruzzaman, M. (2009). Role of fly ash in the removal of organic pollutants from wastewater. *Energy & Fuels*, *23*(3), 1494–1511. doi:10.1021/ef8002697

Ahmed, M. J. K., & Ahmaruzzaman, M. (2016). A review on potential usage of industrial waste materials for binding heavy metal ions from aqueous solutions. *Journal of Water Process Engineering*, *10*, 39–47. doi:10.1016/j.jwpe.2016.01.014

Ainsworth, E. A., & McGrath, J. M. (2010). Direct effects of rising atmospheric carbon dioxide and ozone on crop yields. *Climate Change and Food Security*, *37*, 109–130. doi:10.1007/978-90-481-2953-9_7

Al Othman, Z. A., Naushad, M., & Nilchi, A. (2011). Development, characterization and ion exchange thermodynamics for a new crystalline composite cation exchange material: Application for the removal of $Pb^{2+}$ ion from a standard sample (rompin hematite). *Journal of Inorganic and Organometallic Polymers and Materials*, *21*(3), 547–559. doi:10.100710904-011-9491-9

Alam, M. M., Alothman, Z. A., & Naushad, M. (2013). Analytical and environmental applications of polyaniline Sn(IV) tungstoarsenate and polypyrrolepolyantimonic acid composite cation-exchangers. *Journal of Industrial and Engineering Chemistry*, *19*(6), 1973–1980. doi:10.1016/j.jiec.2013.03.006

Alam, S. S., & Mahbub, M. N. (2007). Karyotype comparison in two varieties of Vigna mungo after staining with orcein and CMA. *Bangladesh Journal of Botany*, *36*(2), 167–170.

Alam, Z., Inamuddin, & Nabi, S. A. (2010). Synthesis and characterization of a thermally stable strongly acidic Cd(II) ion selective composite cation-exchanger: PolyanilineCe(IV) molybdate. *Desalination*, *250*(2), 515–522. doi:10.1016/j.desal.2008.09.008

Alexander, (1999). Bioaccumulation, bioconcentration, biomagnification. *Environmental Geology*. doi:10.1007/1-4020-4494-1_31

Ali, G. A. M., Lih Teo, E. Y., Aboelazm, E. A. A., Sadegh, H., Memar, A. O. H., Shahryari-Ghoshekandi, R., & Chong, K. F. (2017). Capacitive performance of cysteamine functionalized carbon nanotubes. *Materials Chemistry and Physics*, *197*, 100–104. doi:10.1016/j.matchemphys.2017.05.019

Ali, G. A. M., Makhlouf, S. A., Yusoff, M. M., & Chong, K. F. (2015). Structural and electrochemical characteristics of graphene nanosheets as supercapacitor electrodes. *Reviews on Advanced Materials Science*, *40*(1), 35–43.

Ali, H., Khan, E., & Sajad, M. A. (2013). Phytoremediation of heavy metals—concepts and applications. *Chemosphere*, *91*(7), 869–881. doi:10.1016/j.chemosphere.2013.01.075 PMID:23466085

Alizadeh, T., Ganjali, M. R., Nourozi, P., Zare, M., & Hoseini, M. (2011). A carbon paste electrode impregnated with $Cd^{2+}$ imprinted polymer as a new and high selective electrochemical sensor for determination of ultra-trace $Cd^{2+}$ in water samples. *Journal of Electroanalytical Chemistry*, *657*(1-2), 98–106. doi:10.1016/j.jelechem.2011.03.029

Al-Tarazi, E., Abu Rajab, J., Al-Naqa, A., & El-Waheidi, M. (2008). Detecting leachate plumes and groundwater pollution at Ruseifa municipal landfill utilizing VLF-EM method. *Journal of Applied Geophysics*, *65*(3), 121–131. doi:10.1016/j.jappgeo.2008.06.005

Ambashta, R. D., & Sillanpää, M. (2010). Water purification using magnetic assistance: A review. *Journal of Hazardous Materials*, *180*(1-3), 38–49. doi:10.1016/j.jhazmat.2010.04.105 PMID:20488616

Ambec, S., & Coria, J. (2017). Policy spillovers in the regulation of multiple pollutants. *Journal of Environmental Economics and Management*. doi:10.1016/j.jeem.2017.05.011

American Academy of Pediatrics. (2012). Policy statement: Pesticide exposure in children. pediatrics. *The Official Journal of the American Academy of Pediatrics*, *131*, 1757–1763.

Amin, M. T., Alazba, A. A., & Manzoor, U. (2014). A review of removal of pollutants from water/wastewater using different types of nanomaterials. *Advances in Materials Science and Engineering*.

Amuda, O. S., & Amoo, I. A. (2007). Coagulation/flocculation process and sludge conditioning in beverage industrial wastewater treatment. *Journal of Hazardous Materials*, *141*(3), 778–783. doi:10.1016/j.jhazmat.2006.07.044 PMID:16959404

Anawar, H. M., Akai, J., & Sakugawa, H. (2004). Mobilization of arsenic from subsurface sediments by effect of bicarbonate ions in groundwater. *Chemosphere*, *54*(6), 753–762. doi:10.1016/j.chemosphere.2003.08.030 PMID:14602108

Andersson, E., Hagberg, S., Nilsson, T., Persson, B., Wingren, G., & Toren, K. (2001). A case-referent study of cancer mortality among sulfate mill workers in Sweden. *Occupational and Environmental Medicine*, *58*(5), 321–324. doi:10.1136/oem.58.5.321 PMID:11303081

Andersson, E., Nilsson, R., & Toren, K. (2002). Gliomas among men employed in the Swedish pulp and paper industry. *Scandinavian Journal of Work, Environment & Health*, *28*(5), 333–340. doi:10.5271jweh.683 PMID:12432987

Andrade, A. I. A. S. S., & Stigter, T. Y. (2009). Multi-method assessment of nitrate and pesticide contamination in shallow alluvial groundwater as a function of hydrogeological setting and land use. *Agricultural Water Management*, *96*(12), 1751–1765. doi:10.1016/j.agwat.2009.07.014

Aneja, V. P., Schlesinger, W. H., & Erisman, J. M. (2009). Effects of agriculture upon the air quality and climate: Research, policy, and regulations. *Environmental Science & Technology*, *43*(12), 4234–4240. doi:10.1021/es8024403 PMID:19603628

Angin, D. (2013). Effect of pyrolysis temperature and heating rate on biochar obtained from pyrolysis of safflower seeds press cake. *Bioresource Technology*, *128*, 593–597. doi:10.1016/j.biortech.2012.10.150 PMID:23211485

Anonymous. (2004). NATP – MM project report on 'Use of Urban and Industrial Effluent in Agriculture. CSSRI, Karnal, India.

APHA-AWWA-WPCF. (1998). Standard methods for the Examination of Water and Wasewater. American Public Health Association.

Araneda, M., Pavez, J., Luza, B., & Jeison, D. (2017). Use of activated sludge biomass as an agent for advanced primary separation. *Journal of Environmental Management*, *192*, 156–162. doi:10.1016/j.jenvman.2017.01.030 PMID:28160643

Archaya, S., Gopinath, L. R., & Bhuvaneswari, R. (2014). A Review on Advances in Phytoremediation of Pesticide Polluted Soils. *International Journal of Advances in Interdisciplinary Research*, *1*, 1–6.

Argun, M. E., Dursun, S., Ozdemir, C., & Karatas, M. (2007). Heavy metal adsorption by modified oak sawdust: Thermodynamics and kinetics. *Journal of Hazardous Materials*, *141*(1), 77–85. doi:10.1016/j.jhazmat.2006.06.095 PMID:16879919

Ashbolt, N. J. (2004). Microbial contamination of drinking water and disease outcomes in developing regions. *Toxicology*, *198*(1-3), 229–238. doi:10.1016/j.tox.2004.01.030 PMID:15138046

Augustijn-Beckers, P. W. M., Hornsby, A. G., & Wauchope, R. D. (1994). SCS/ARS/CES Pesticide Properties Database for Environmental Decisionmaking II. *Additional Properties Reviews of Environmental Contamination and Toxicology*, 137.

Azubuike, C. C., Chikere, C. B., & Okpokwasili, G. C. (2016). Bioremediation techniques–classification based on site of application: Principles, advantages, limitations and prospects. *World Journal of Microbiology & Biotechnology*, *32*(11), 180. doi:10.100711274-016-2137-x PMID:27638318

Bachmann, C., Groskreutz, R., Awad, J., & Lozier, J. (2005). Implementing a multi-barrier approach using ultrafiltration and UV disinfection. Proceedings of the Water Environment Federation, 2005(1), 183-195. doi:10.2175/193864705783978339

Bajpai P (2014). 14 – Water Reuse, Waste water Treatment and Closed-Cycle Operation. *Recycling and Deinking of Recovered Paper,* 251-269.

Baldé, C.P., Wang, F., Kuehr, R., & Huisman, J. (2014). *The Global E-Waste Monitor 2014 Quantities, Flows and Resources*. Academic Press.

Band, P.R., Le, N.D., Fang, R., Astrakianakis, G., Bert, J., Keefe, A., Krewski, D. (2001). Cohort cancer incidence among pulp and paper mill workers in British Columbia. *Scand J Work Environ Health, 27,* 113-119.

Bani, A., Echevarria, G., Sulce, C., & Morel, J. L. (2015). Improving the agronomy of *Alyssum murale* for extensive phytomining: A five-year field study. *International Journal of Phytoremediation, 17*(2), 117–127. doi:10.1080/15226514.2013.862204 PMID:25237722

Bansode, R. R., Losso, J. N., Marshall, W. E., Rao, R. M., & Portier, R. J. (2004). Pecan shell-based granular activated carbon for treatment of chemical oxygen demand (COD) in municipal wastewater. *Bioresource Technology*, *94*(2), 129–135. doi:10.1016/j.biortech.2003.12.009 PubMed

Barakat, M. A. (2011). New trends in removing heavy metals from industrial wastewater. *Arabian Journal of Chemistry*, *4*(4), 361–377. doi:10.1016/j.arabjc.2010.07.019

Barnes, K. K., Christenson, S. C., Kolpin, D. W., Focazio, M. J., Furlong, E. T., Zaugg, S. D., ... Barber, L. B. (2004). Pharmaceuticals and Other Organic Waste Water Contaminants Within a Leachate Plume Downgradient of a Municipal Landfill. *Ground Water Monitoring and Remediation*, *24*(2), 119–126. doi:10.1111/j.1745-6592.2004.tb00720.x

Barnes, K. K., Kolpin, D. W., Furlong, E. T., Zaugg, S. D., Meyer, M. T., & Barber, L. B. (2008). A national reconnaissance of pharmaceuticals and other organic wastewater contaminants in the United States--I) groundwater. *The Science of the Total Environment*, *402*(2-3), 192–200. doi:10.1016/j.scitotenv.2008.04.028 PMID:18556047

Bazrafshan, E., Moein, H., Kord Mostafapour, F. & Nakhaie, S. (2013). Application of electrocoagulation process for dairy wastewater treatment. *Journal of Chemisry,* 1-8.

Becerra-Castro, C., Kidd, P. S., Rodriguez-Garrido, B., Monterroso, C., Santos-Ucha, P., & Prieto-Fernandez, A. (2013). Phytoremediation of hexachlorocyclohexane (HCH)-contaminated soils using Cytisus striatus and bacterial inoculants in soils with distinct organic matter content. *Environmental Pollution*, *178*, 202–210. doi:10.1016/j.envpol.2013.03.027 PMID:23583940

Benavente, V., Calabuig, E., & Fullanan, A. (2014). (in press). Upgrading of moist agro-industrial wastes by hydrothermal carbonization. *Journal of Analytical and Applied Pyrolysis.*

Berny, P. (2007). Pesticides and the intoxication of wild animals. *Journal of Veterinary Pharmacology and Therapeutics*, *30*(2), 93–100. doi:10.1111/j.1365-2885.2007.00836.x PMID:17348893

Bhat, Rao, & Patil. (2012). Development of an integrated model to recover precious metals from electronic scrap - A novel strategy for e-waste management. *International Conference on Emerging Economics-Prospects and Challenges (ICEE-2012)*, 397-406.

Bingham, S. (2007). *Pesticides in rivers and groundwater*. Environment Agency.

Bitton, G. (2005). *Wastewater Microbiology* (3rd ed.). Hanaken, NJ: Wiley-Liss. doi:10.1002/0471717967

Borsari, M. (1994). *Encyclopedia of Inorganic Chemistry* (1st ed.). New York: Wiley.

Bouman, B. A. M., Castañeda, A. R., & Bhuiyan, S. I. (2002). Nitrate and pesticide contamination of groundwater under rice-based cropping systems: Past and current evidence from the Philippines. *Agriculture, Ecosystems & Environment*, *92*(2), 185–199. doi:10.1016/S0167-8809(01)00297-3

Bowker, R. P. (2000). Biological odour control by diffusion into activated sludge basins. *Water Science and Technology*, *41*(6), 127–132.

Boyd, C. E. (2015). *Water Pollution. In Water Quality: An Introduction* (pp. 313–338). Cham: Springer International Publishing.

Braden, J. B., & Shortle, J. S. (2013). Agricultural Sources of Water Pollution. *Reference Module in Earth System and Environmental Sciences*, *3*, 81–85.

Brewer, C. E., Schmidt-Rohr, K., Satrio, J. A., & Brown, R. C. (2009). Characterization of biochar from fast pyrolysis and gasification systems. *Environmental Progress & Sustainable Energy*, *28*(3), 386–396. doi:10.1002/ep.10378

Brown, P. L., Guerin, M., Hankin, S. I., & Lowson, R. T. (1998). Uranium and other contaminant migration in groundwater at a tropical Australian Uranium Mine. *Journal of Contaminant Hydrology*, *35*(1), 295–303. doi:10.1016/S0169-7722(98)00129-6

Bruner, M. A., Rao, M., Dumont, J. N., Hull, M., Jones, T., & Bantle, J. A. (1998). Ground and Surface Water Developmental Toxicity at a Municipal Landfill: Description and Weather-Related Variation. *Ecotoxicology and Environmental Safety*, *39*(3), 215–226. doi:10.1006/eesa.1998.9999 PMID:9570913

Buenrostro, O., & Bocco, G. (2003). Solid waste management in municipalities in Mexico: Goals and perspectives. *Resources, Conservation and Recycling*, *39*(3), 251–263. doi:10.1016/S0921-3449(03)00031-4

Buerge, I. J., Keller, M., Buser, H. R., Muller, M. D., & Poiger, T. (2011). Saccharin and other artificial sweeteners in soils: Estimated inputs from agriculture and households, degradation, and leaching to groundwater. *Environmental Science & Technology*, *45*(2), 615–621. doi:10.1021/es1031272 PMID:21142066

Bushra, R., Shahadat, M., Ahmad, A., Nabi, S. A., Umar, K., Oves, M., ... Muneer, M. (2014). Synthesis, characterization, antimicrobial activity and applications of polyanilineTi (IV) arsenophosphate adsorbent for the analysis of organic and inorganic pollutants. *Journal of Hazardous Materials*, *264*, 481–489. doi:10.1016/j.jhazmat.2013.09.044 PMID:24238807

Bushra, R., Shahadat, M., Khan, M. A., Inamuddin, Adnan, R., & Rafatullah, M. (2014). Optimization of polyaniline supported Ti(IV) arsenophosphate composite cation exchanger based ion-selective membrane electrode for the determination of lead. *Industrial & Engineering Chemistry Research*, *53*(50), 19387–19391. doi:10.1021/ie5034655

Bushra, R., Shahadat, M., Raeissi, A. S., & Nabi, S. A. (2012). Development of nano-composite adsorbent for removal of heavy metals from industrial effluent and synthetic mixtures; its conducting behaviour. *Desalination*, *289*, 1–11. doi:10.1016/j.desal.2011.12.013

Buszka, P. M., Yeskis, D. J., Kolpin, D. W., Furlong, E. T., Zaugg, S. D., & Meyer, M. T. (2009). Waste-Indicator and Pharmaceutical Compounds in Landfill-Leachate-Affected Ground Water near Elkhart, Indiana, 2000–2002. *Bulletin of Environmental Contamination and Toxicology*, *82*(6), 653–659. doi:10.100700128-009-9702-z PMID:19290448

Buyukkamaci, N., & Koken, E. (2010). Economic evaluation of alternative wastewater treatment plant options for pulp and paper industry. *The Science of the Total Environment*, *408*(24), 6070–6078. doi:10.1016/j.scitotenv.2010.08.045 PubMed

Canter, L. W., Knox, R. C., & Fairchild, D. M. (1988). Groundwater quality protection. M. Lewis Publishers Inc.

Carabias Martinez, R., Rodriguez Gonzalo, E., Fernandez Laespada, M., & Sanchez San Roman, F. (2000). Evaluation of surface- and ground-water pollution due to herbicides in agricultural areas of Zamora and Salamanca (Spain). *Journal of Chromatography. A*, *869*(1-2), 471–480. doi:10.1016/S0021-9673(99)01188-7 PMID:10720260

Carletti, G., Fatone, F., Bolzonella, D., & Cecchi, F. (2008). Occurrence and fate of heavy metals in large wastewater treatment plants treating municipal and industrial wastewaters. *Water Science and Technology*, *57*(9), 1329–1336. doi:10.2166/wst.2008.230 PMID:18495995

Carrara, C., Ptacek, C. J., Robertson, W. D., Blowes, D. W., Moncur, M. C., Sverko, E., & Backus, S. (2008). Fate of pharmaceutical and trace organic compounds in three septic system plumes, Ontario, Canada. *Environmental Science & Technology*, *42*(8), 2805–2811. doi:10.1021/es070344q PMID:18497127

Castro, P., & Huber, M. E. (2010). *Marine Biology* (8th ed.). New York: McGraw-Hill Companies Inc.

Catterina, F., Yan, Y., Jacqueline, C., Hugo, B., Roxana, P. L., Johanna, A., ... Craig, S. (2013). Arsenic, tobacco smoke, and occupation: Associations of multiple agents with lung and bladder cancer. *Epidemiology (Cambridge, Mass.)*, *24*(6), 898–905. doi:10.1097/EDE.0b013e31829e3e03 PMID:24036609

Central Pollution Control Board. (2015). *Assessment & Quantification of Plastics Waste Generation in Major cities –CPCB Report, 2015*. Retrieved from http://www.indiaenvironmentportal.org.in/files/file/Assessment%20and%20Quantification%20of%20Plastics%20Waste.pdf

Chandrasekharam, D. (2005). Arsenic pollution in groundwater of West Bengal, India: Where we stand? *Remediation and Management*, 25-29.

Chandra, V., Park, J., Chun, Y., Lee, J. W., Hwang, I.-C., & Kim, K. S. (2010). Water-dispersible magnetite-reduced graphene oxide composites for arsenic removal. *ACS Nano*, *4*(7), 3979–3986. doi:10.1021/nn1008897 PMID:20552997

Chang, Q., & Wang, G. (2007). Study on the macromolecular coagulant PEX which traps heavy metals. *Chemical Engineering Science*, *62*(17), 4636–4643. doi:10.1016/j.ces.2007.05.002

Chang, Q., Zhang, M., & Wang, J. X. (2009). Removal of $Cu^{2+}$ and turbidity from wastewater by mercaptoacetyl chitosan. *Journal of Hazardous Materials*, *169*(1-3), 621–625. doi:10.1016/j.jhazmat.2009.03.144 PMID:19414213

Chan, Y. J., Chong, M. F., Law, C. L., & Hassell, D. G. (2009). A review on anaerobic–aerobic treatment of industrial and municipal wastewater. *Chemical Engineering Journal*, *155*(1-2), 1–18. doi:10.1016/j.cej.2009.06.041

Chatterjee. (2012). Sustainable electronic waste management and recycling process. *American Journal of Environmental Engineering, 2*(1), 23-33.

Chatterjee, A., Das, D., & Chakraborti, D. (1993). A study of ground water contamination by arsenic in the residential area of behala, calcutta due to industrial pollution. *Environmental Pollution*, *80*(1), 57–65. doi:10.1016/0269-7491(93)90010-L PMID:15091873

Chemlal, R., Abdi, N., Lounici, H., Drouiche, N., Pauss, A., & Mameri, N. (2013). Modeling and qualitative study of diesel biodegradation using biopile process in sandy soil. *International Biodeterioration & Biodegradation*, *78*, 43–48. doi:10.1016/j.ibiod.2012.12.014

Chen, B., Zhou, D., & Zhu, L. (2008). Transitional adsorption and partition of nonpolar and polar aromatic contaminants by biochars of pine needles with different pyrolytic temperatures. *Environmental Science & Technology*, *42*(14), 5137–5143. doi:10.1021/es8002684 PMID:18754360

Chen, C., Hu, J., Shao, D., Li, J., & Wang, X. (2009). Adsorption behavior of multiwall carbon nanotube/iron oxide magnetic composites for Ni (II) and Sr (II). *Journal of Hazardous Materials*, *164*(2), 923–928. doi:10.1016/j.jhazmat.2008.08.089 PMID:18842337

Chen, C., Zhou, W., & Lin, D. (2015). Sorption characteristics of N-nitrosodimethylamine onto biochar from aqueous solution. *Bioresource Technology*, *179*, 359–366. doi:10.1016/j.biortech.2014.12.059 PMID:25553566

Chen, H., Arocena, J. M., Li, J., Thring, R. W., & Zhou, J. (2014). Assessments of chromium (and other metals) in vegetables and potential bio-accumulations in humans living in areas affected by tannery wastes. *Chemosphere*, *112*, 412–419. doi:10.1016/j.chemosphere.2014.04.091 PMID:25048934

Chen, S., Wu, W., Hu, K., & Li, W. (2010). The effects of land use change and irrigation water resource on nitrate contamination in shallow groundwater at county scale. *Ecological Complexity*, *7*(2), 131–138. doi:10.1016/j.ecocom.2010.03.003

Chen, X., Chen, G., Chen, L., Chen, Y., Lehmann, J., McBride, M. B., & Hay, A. G. (2011). Adsorption of copper and zinc by biochars produced from pyrolysis of hardwood and corn straw in aqueous solution. *Bioresource Technology*, *102*(19), 8877–8884. doi:10.1016/j.biortech.2011.06.078 PMID:21764299

Chen, Y., Litvak, P. F., Howe, G. R., Graziano, J. H., Rauf, P. B., & Parvez, F. (2007). Arsenic exposure from drinking water, dietary intakes of B vitamins and folate, and risk of high blood pressure in Bangladesh: A population-based, cross-sectional study. *American Journal of Epidemiology*, *165*(5), 541–552. doi:10.1093/aje/kwk037 PMID:17164464

Chen, Y., Wen, X., Wang, B., & Nie, P. (2017). Agricultural Pollution and Regulation: How to subsidize agriculture? *Journal of Cleaner Production*, *164*, 258–264. doi:10.1016/j.jclepro.2017.06.216

Chen, Z., Chen, B., & Chiou, C. T. (2012b). Fast and slow rates of naphthalene sorption to biochars produced at different temperatures. *Environmental Science & Technology*, *46*(20), 11104–11111. doi:10.1021/es302345e PMID:22970831

Chen, Z., Chen, B., Zhou, D., & Chen, W. (2012a). Bisolute sorption and thermodynamic behavior of organic pollutants to biomass-derived biochars at two pyrolytic temperatures. *Environmental Science & Technology*, *46*(22), 12476–12483. doi:10.1021/es303351e PMID:23121559

Cheremisinoff, N. P. (2001). *Handbook of water and wastewater treatment technologies*. Butterworth-Heinemann.

Chicone, C. (2017). Chapter 3 – An Environmental Pollutant. An Invitation to Applied Mathematics, 31-37.

Chilton, P. J., Lawrence, A. R., & Stuart, M. E. (1994). *The impact of tropical agriculture on groundwater quality in groundwater quality*. London: Chapman & Hall.

Chipasa, K. B. (2003). Accumulation and fate of selected heavy metals in a biological wastewater treatment system. *Waste Management (New York, N.Y.)*, *23*(2), 135–143. doi:10.1016/S0956-053X(02)00065-X PMID:12623088

Chithra, P. G., Raveendran, R., & Beena, B. (2008). Parachlorophenol anchored tin antimonite-An inorgano-organic ion-exchanger selective towards heavy metals like Bi(III) and Cu(II). *Desalination*, *232*(1-3), 20–25. doi:10.1016/j.desal.2008.01.006

Clarke, B. O., & Smith, S. R. (2011). Review of 'emerging' organic contaminants in biosolids and assessment of international research priorities for the agricultural use of biosolids. *Environment International*, *37*(1), 226–247. doi:10.1016/j.envint.2010.06.004 PMID:20797791

Coca-Cola. (n.d.). *Sustainability Update 2016/17. Message from our leadership*. Retrieved from https://www.coca-colaindia.com/sustainability-report/sustainability-message

Connell, D. W. O., Birkinshaw, C., & Dwyer, T. F. O. (2008). Heavy metal adsorbents prepared from the modification of cellulose: A review. *Bioresource Technology*, *99*(15), 6709–6724. doi:10.1016/j.biortech.2008.01.036 PMID:18334292

Cook, S. M., Khan, Z. R., & Pickett, J. A. (2007). The use of push-pull strategies in integrated pest management. *Annual Review of Entomology*, *52*(1), 375–400. doi:10.1146/annurev.ento.52.110405.091407 PMID:16968206

Corada-Fernandez, C., Candela, L., Torres-Fuentes, N., Pintado-Herrera, M. G., Paniw, M., & Gonzalez-Mazo, E. (2017). Effects of extreme rainfall events on the distribution of selected emerging contaminants in surface and groundwater: The Guadalete River basin (SW, Spain). *The Science of the Total Environment*, *605-606*, 770–783. doi:10.1016/j.scitotenv.2017.06.049 PMID:28679121

Cornell University. (2007). *Pesticides in the environment, Pesticide fact sheets and tutorial*. Pesticide Safety Education Program.

Council Directive 86/278/EEC on the protection of the environment, and in particular of the soil, when sewage sludge is used in agriculture, *Official Journal*, No. L181, 04/07/1986, p. 6.

CPCB. (2009). *Status of water supply, wastewater generation and treatment in Class I cities and Class II towns of India. Series: CUPS/70/2009-10*. Central Pollution Control Board.

CPCB. (2013). *Overview of plastic waste management*. Retrieved from http://www.cpcb.nic.in/waste/plasticwaste/ Plastic_waste-1.pdf

CPCB. (2016). Final Document on revised classification of industrial sectors under red, orange, green and white categories. New Delhi: Central Pollution Control Board.

Damalas, C. A., & Ilias, G. (2011). Pesticide Exposure, Safety Issues, and Risk Assessment Indicators. *International Journal of Environmental Research and Public Health*, *8*(5), 1402–1419. doi:10.3390/ijerph8051402 PMID:21655127

Dan'Azumi, S., & Bichi, M. H. (2010). Industrial Pollution and Heavy Metals Profile of Challawa River in Kanao, Nigeria. *Journal of Applied Sciences in Environmental Sanitation*, *5*, 23–29.

Danish Environmental Protection Agency. (1993). *Oversigt over revurderingen 1988, 1989, 1990, 1991 of 1992* [Summary of the reassessment of already approved pesticides in 1988, 1989, 1990, 1991 and 1992 based on stricter standards]. Author.

Datta, P. S., Deb, D. L., & Tyagi, S. K. (1997). Assessment of groundwater contamination from fertilizers in the Delhi area based on 180, N03– and K+ composition. *Journal of Contaminant Hydrology*, *27*(3), 249–262. doi:10.1016/S0169-7722(96)00099-X

Debendra, N., Mazumder, G., Haque, R., Ghosh, N., Binay, K. D., Santra, A., ... Smith, A. H. (1998). Arsenic levels in drinking water and the prevalence of skin lesionsin West Bengal, India. *International Journal of Epidemiology*, *27*(5), 871–877. doi:10.1093/ije/27.5.871 PMID:9839746

Deblonde, T., Leguille, C. C., & Hartemann, P. (2011). Emerging pollutants in wastewater: A review of the literature. *International Journal of Hygiene and Environmental Health*, *214*(6), 442–448. doi:10.1016/j.ijheh.2011.08.002 PMID:21885335

Deepali, & Gangwar, K.K. (2010). Metals concentration in textile and tannery effluents, Distribution of estrogens, 17beta-estradiol andestrone, in Canadian municipal wastewater treatment plants. *Science Total Environment, 336*, 155-170.

Demirbas, A. (2004). Effects of temperature and particle size on bio-char yield from pyrolysis of agricultural residues. *Journal of Analytical and Applied Pyrolysis*, *72*(2), 243–248. doi:10.1016/j.jaap.2004.07.003

Deshmukh, S., & Mahajan, K. (2015, April 22). Role of Informal Sector in E-waste Management in Pune Region. *International Journal of Computers and Applications*, *116*(5), 22–27. doi:10.5120/20333-2063

Dey, A., Bhowmick, A., & Chakraborty, D. (2003). Contamination through landfills in NCT, Delhi. In *Recent Trends in Hydrogeochemistry* (pp. 153–159). New Delhi: Capital Publ. Co.

Dias, J. M., Alvim-Ferraz, M. C. M., Almeida, M. F., Rivera-Utrilla, J., & Sanchez-Polob, M. (2007). Waste materials for activated carbon preparation and its use in aqueous phase treatment: A review. *Journal of Environmental Management*, *85*(4), 833–846. doi:10.1016/j.jenvman.2007.07.031 PMID:17884280

Ding, G. K. C. (2017). Waste water Treatment and Reuse – The Future Scope of Water Supply. Encyclopedia of Sustainable Technologies, 43-52.

Dong, X., Ma, L. Q., Zhu, Y., Li, Y., & Gu, B. (2013). Mechanistic investigation of mercury sorption by Brazilian pepper biochars of different pyrolytic temperatures based on X-ray photoelectron spectroscopy and flow calorimetry. *Environmental Science & Technology*, *47*(21), 12156–12164. doi:10.1021/es4017816 PMID:24040905

Downie, A., Crosky, A., & Munroe, P. (2009). Physical properties of biochar. In J. Lehmann & S. Joseph (Eds.), Biochar for environmental management: Science and technology (pp. 13–32). Academic Press.

Eaton, A. D., Clesceri, L. S., & Greenberg, A. E. (1995). *Standard Methods for the Examination of Water and Wastewater*. Washington, DC: American Public Health Association.

Edition, F. (2011). Guidelines for drinking-water quality. *WHO Chronicle*, *38*(4), 104–108. PMID:6485306

Edokpayi, J. N., Odiyo, J. O., Msagati, T. A. M., & Popoola, E. O. (2015). Removal efficiency of faecal indicator organisms, nutrients and heavy metals from a Peri-Urban wastewater treatment plant in thohoyandou, Limpopo Province, South Africa. *International Journal of Environmental Research and Public Health*, *12*(7), 7300–7320. doi:10.3390/ijerph120707300 PMID:26132481

El Samrani, A. G., Lartiges, B. S., & Villiéras, F. (2008). Chemical coagulation of combined sewer overflow: Heavy metal removal and treatment optimization. *Water Resources*, *42*, 951–960. PMID:17961629

El-Dessouky, H., & Ettouney, H. (2002). Teaching desalination-a multidiscipline engineering science. *Heat Transfer Engineering*, *23*(5), 1–3. doi:10.1080/01457630290090590

Emmanuel, D., Sekyere, E. O., & Owusu, V. (2016). Impact of agriculture extensive service on adoption of chemical fertilizers: Implications for rice productivity and development in Ghana. *NJAS Wageningen Journal of Life Sciences*, *79*, 41–49. doi:10.1016/j.njas.2016.10.002

Englande, A. J., Krenkel, P., & Shamas, J. (2003). Waste water Treatment & Water Reclamation. Encyclopedia of Physical Science and Technology, 639-670.

Environmental Management & Coordination (Water Quality) Regulations. (2006). *Legal Notice 120.* Third Schedule, Standards for Discharge of Effluent into the Environment.

EPA. (1994). *Land application of sewage sludge. A guide for land applies on the requirement of the Federal standards for the use or disposal of sewage sludge, 40 CFR 503*. EPA.

Espana, P. S., & Moran, B. S. (2010). Pesticide degradation in water. In H. S. Rathore & L. M. L. Nollet (Eds.), *Pesticides: Evaluation of environmental pollution*. CRC Press.

Estévez, E., Cabrera, M. C., Molina-Díaz, A., Robles-Molina, J., & Palacios-Díaz, M. P. (2012). Screening of emerging contaminants and priority substances (2008/105/EC) in reclaimed water for irrigation and groundwater in a volcanic aquifer (Gran Canaria, Canary Islands, Spain). *The Science of the Total Environment*, *433*(Supplement C), 538–546. doi:10.1016/j.scitotenv.2012.06.031 PMID:22858460

Eyssen, G. E. M., & Ruedy, J. (1983). Methyl mercury exposure in northern quebec I. neurologic findings in adults. *American Journal of Epidemiology*, *118*(4), 461–469. doi:10.1093/oxfordjournals.aje.a113651 PMID:6637973

Falsanisi, D., Liberti, L., & Notarnicola, M. (2010). Ultrafiltration (UF) pilot plant for municipal wastewater reuse in agriculture: Impact of the operation mode on process performance. *Water (Basel)*, *2*(4), 872–885. doi:10.3390/w2040872

FAO, Network on Erosion-Induced Loss in Soil Productivity. (1991). *Report of a Workshop, Bogor, Indonesia*. Land and Water Development Division, FAO.

Farooq, U., Kozinski, J. A., Khan, M. A., & Athar, M. (2010). Biosorption of heavy metal ions using wheat based biosorbents – A review of the recent literature. *Bioresource Technology*, *101*(14), 5043–5053. doi:10.1016/j.biortech.2010.02.030 PMID:20223652

Fatima, S. (2014). Chemical Recycle of Plastics. *American Journal of Engineering Research*, 3(11), 93-108.

Fattore, E., Muller, L., Davoli, E., Castelli, D., & Benfenati, E. (1998). Industrial pollutants in ground waters from northern Milan. *Chemosphere*, *36*(9), 2007–2017. doi:10.1016/S0045-6535(97)10085-6 PMID:9532728

Fawell, J., & Nieuwenhuijsen, M. J. (2003). Contaminants in drinking water. *British Medical Bulletin*, *68*(1), 199–208. doi:10.1093/bmb/ldg027 PMID:14757718

Ferri, R., Hashim, D., Smith, D. R., Guazzetti, S., Donna, F., Ferretti, E., ... Lucchini, R. G. (2015). Metal contamination of home garden soils and cultivated vegetables in the province of Brescia, Italy. *Implications for human exposure. The Science of the Total Environment*, *518-519*, 507–517. doi:10.1016/j.scitotenv.2015.02.072 PMID:25777956

Fishel, F. M. (2013). Pest Management and Pesticides: A Historical Perspective. University of Florida/ IFAS Extension, PI 219.

Fishel, F. M. (2015). *Pesticides labelling*. Signal Words.

Fleischli, M. A., Franson, J. C., Thomas, N. J., Finley, D. L., & Riley, W. (2004). Avian Mortality Events in the United States Caused by Anticholinesterase Pesticides: A Retrospective Summary of National Wildlife Health Center Records from 1980 to 2000. *Archives of Environmental Contamination and Toxicology*, 46. PMID:15253053

Fram, M. S., & Belitz, K. (2011). Occurrence and concentrations of pharmaceutical compounds in groundwater used for public drinking-water supply in California. *The Science of the Total Environment*, *409*(18), 3409–3417. doi:10.1016/j.scitotenv.2011.05.053 PMID:21684580

Frankland, A. (1869). *Franklands Bericht "über Wasserversorgung an die Royal Institution of Great Britain", Dtsch Vierteljahresschr.* öffentliche Gesundheitspfl.

Freddo, A., Cai, C., & Reid, J. (2012). Environmental contextualisation of potential toxic elements and polycyclic aromatic hydrocarbons in biochar. *Environmental Pollution*, *171*, 18–24. doi:10.1016/j.envpol.2012.07.009 PMID:22863991

Freemark, K. (1995). Impacts of agricultural herbicide use on terrestrial wildlife in temperate landscapes: A review with special reference to North America. *Agriculture, Ecosystems & Environment*, *52*(2-3), 67–91. doi:10.1016/0167-8809(94)00534-L

Frutos, F. J. G., Perez, R., Escolano, O., Rubio, A., Gimeno, A., Fernandez, M. D., ... Laguna, J. (2012). Remediation trials for hydrocarbon-contaminated sludge from a soil washing process: Evaluation of bioremediation technologies. *Journal of Hazardous Materials*, *199–200*, 262–271. doi:10.1016/j.jhazmat.2011.11.017 PMID:22118850

Fulekar, M. H., Geetha, M., & Sharma, J. (2009). Bioremediation of Trichlorpyr Butoxyethyl Ester (TBEE) in bioreactor using adapted Pseudomonas aeruginosa in scale up process technique. *Biology and Medicine*, *1*(3), 1–6.

Galloway, T. S., & Depledge, M. H. (2001). Immunotoxicity in Invertebrates: Measurement and Ecotoxicological Relevance. *Ecotoxicology (London, England)*, *10*(1), 5–23. doi:10.1023/A:1008939520263 PMID:11227817

Galloway, T., & Handy, R. (2003). Immunotoxicity of Organophosphorous Pesticides. *Ecotoxicology (London, England)*, *12*(1/4), 345–363. doi:10.1023/A:1022579416322 PMID:12739880

Ganesh, R., Rajinikanth, R., Thanikal, J. V., Ramanujam, R. A., & Torrijos, M. (2010). Anaerobic treatment of winery wastewater in fixed bed reactors. *Bioprocess and Biosystems Engineering*, *33*(5), 619–628. doi:10.100700449-009-0387-9 PMID:19876651

Ganguly. (2016). E-Waste Management in India – An Overview. *International Journal of Earth Science and Engineering, 9*(2), 574-588.

Gans, R. (1905). Jahrb Preuss Geol. *Landesanstalt Berlin*, *26*, 179.

Garg, S., & Chaudhry, S. (2017). Treatment of Wastewater of Food Industry by Membrane Bioreactor.

Gautam, R., Bhardwaj, N., & Saini, Y. (2010). Fluoride accumulation by vegetables and crops grown in Nawa Tehsil of Nagaur district (Rajasthan, India). *Journal of Phytology*, *2*, 80–85.

Ghosh, P., Banerjee, M., Chaudhuri, S. D., Das, J. K., Sarma, N., Basu, A., & Giri, A. K. (2007). Increased chromosome aberration frequencies in the Bowen's patients compared to non-cancerous skin lesions individuals exposed to arsenic. *Mutation Research/Genetic Toxicology and Environmental Mutagenesis*, *632*(1-2), 104–110. doi:10.1016/j.mrgentox.2007.05.005 PMID:17600756

Gillion, R. J., Barbash, J. E., Crawford, G. G., Hamilton, P. A., Martin, J. D., Nakagaki, N., … Wolock, D. M. (2006). *Overview of Findings and Implications. Pesticides in the Nation's Streams and Ground Water (1992–2001)*. Academic Press.

Glance, A. (2016). Local and global environment. *TERI Energy & Environment Data Diary and Yearbook (TEDDY) 2015/16 (Updated Edition)*, 339.

Gode, F., & Pehlivan, E. (2006). Removal of chromium (III) from aqueous solutions using Lewatit S 100: The effect of pH, time, metal concentration and temperature. *Journal of Hazardous Materials*, *136*(2), 330–337. doi:10.1016/j.jhazmat.2005.12.021 PMID:16439060

Godfray, H. C. J., Beddington, J. R., Crute, I. R., Haddad, L., Lawrence, D., Muir, J. F., ... Toulmin, C. (2010). Food Security: The Challenge of Feeding 9 Billion People. *Science*, *327*(5967), 812–818. doi:10.1126cience.1185383 PMID:20110467

Godfrey, E., Woessner, W. W., & Benotti, M. J. (2007). Pharmaceuticals in on-site sewage effluent and ground water, Western Montana. *Ground Water*, *45*(3), 263–271. doi:10.1111/j.1745-6584.2006.00288.x PMID:17470115

González-Naranjo, V., Boltes, K., & Biel, M. (2013). Mobility of ibuprofen, a persistent active drug, in soils irrigated with reclaimed water. *Plant, Soil and Environment*, *59*(2), 68–73. doi:10.17221/590/2012-PSE

Gopalakannan, V., & Viswanathan, N. (2016). One pot synthesis of metal ion anchored alginate–gelatin binary biocomposite for efficient Cr(VI) removal. *International Journal of Biological Macromolecules*, *83*, 450–459. doi:10.1016/j.ijbiomac.2015.10.010 PMID:26456290

Government of India. (2008). *Eleventh Five-Year Plan: 2008–2012*. New Delhi: Planning Commission of India.

Government of Western Australia, Department of Water. (2010). *Aerial spraying of crops with pesticides*. Water quality protection note 104. Author.

Grassi, S., Amadori, M., Pennisi, M., & Cortecci, G. (2014). Identifying sources of B and As contamination in surface water and groundwater downstream of the Larderello geothermal – industrial area (Tuscany–Central Italy). *Journal of Hydrology (Amsterdam)*, *509*(Supplement C), 66–82. doi:10.1016/j.jhydrol.2013.11.003

Grimmeisen, F., Lehmann, M. F., Liesch, T., Goeppert, N., Klinger, J., Zopfi, J., & Goldscheider, N. (2017). Isotopic constraints on water source mixing, network leakage and contamination in an urban groundwater system. *The Science of the Total Environment*, *583*(Supplement C), 202–213. doi:10.1016/j.scitotenv.2017.01.054 PMID:28104328

Grobelak, A., Napora, A., & Kacprzak, M. (2015). Using plant growth-promoting rhizobacteria (PGPR) to improve plant growth. *Ecological Engineering*, *84*, 22–28. doi:10.1016/j.ecoleng.2015.07.019

Gulyas, G., Fazekas, B., Varga, R., & Karpati, A. (2010). Biological purification of chemically pre-treated dairy industry. *Current World Environment*, *5*(2), 373–378.

Gupta, P. K. (2004). Pesticide exposure-Indian scene. *Toxicology*, *198*(1-3), 83–90. doi:10.1016/j.tox.2004.01.021 PMID:15138033

Gupta, V. K., Agarwal, S., Bharti, A. K., & Sadegh, H. (2017a). Adsorption mechanism of functionalized multi-walled carbon nanotubes for advanced Cu (II) removal. *Journal of Molecular Liquids*, *230*, 667–673. doi:10.1016/j.molliq.2017.01.083

Gupta, V. K., Agarwal, S., Sadegh, H., Ali, G. A. M., Bharti, A. K., & Hamdy, A. S. (2017b). Facile route synthesis of novel graphene oxide-β-cyclodextrin nanocomposite and its application as adsorbent for removal of toxic bisphenol A from the aqueous phase. *Journal of Molecular Liquids*, *237*, 466–472. doi:10.1016/j.molliq.2017.04.113

Gupta, V. K., Agarwal, S., & Saleh, T. A. (2011). Synthesis and characterization of alumina-coated carbon nanotubes and their application for lead removal. *Journal of Hazardous Materials*, *185*(1), 17–23. doi:10.1016/j.jhazmat.2010.08.053 PMID:20888691

Gupta, V. K., Jain, R., & Varshney, S. (2007). Removal of Reactofix golden yellow 3 RFN from aqueous solution using wheat husk—an agricultural waste. *Journal of Hazardous Materials*, *142*(1-2), 443–448. doi:10.1016/j.jhazmat.2006.08.048 PMID:17010514

Gupta, V. K., Tyagi, I., Agarwal, S., Sadegh, H., Shahryari-ghoshekandi, R., Yari, M., & Yousefi-nejat, O. (2015b). Experimental study of surfaces of hydrogel polymers HEMA, HEMA–EEMA–MA, and PVA as adsorbent for removal of azo dyes from liquid phase. *Journal of Molecular Liquids*, *206*, 129–136. doi:10.1016/j.molliq.2015.02.015

Gupta, V., Moradi, O., Tyagi, I., Agarwal, S., Sadegh, H., Shahryari-Ghoshekandi, R., ... Garshasbi, A. (2016). Study on the removal of heavy metal ions from industry waste by carbon nanotubes: effect of the surface modification: a review. *Critical Reviews in Environmental Science and Technology*, *46*(2), 93–118. doi:10.1080/10643389.2015.1061874

Gupta, V., Sadegh, H., Yari, M., Ghoshekandi, R. S., Maazinejad, B., & Chahardori, M. (2015a). Removal of ammonium ions from wastewater: A short review in development of efficient methods. *Global Journal of Environmental Science and Management*, *1*(2), 149.

Habeeb, O. A., Kanthasamy, R., Ali, G. A. M., Sethupathi, S., & Yunus, R. B. M. (2017a). Hydrogen sulfide emission sources, regulations, and removal techniques: A review. *Reviews in Chemical Engineering*, *0*(0). doi:10.1515/revce-2017-0004

Habeeb, O. A., Ramesh, K., Ali, G. A. M., & Yunus, R. M. (2017b). Experimental design technique on removal of hydrogen sulfide using CaO-eggshells dispersed onto palm kernel shell activated carbon: Experiment, optimization, equilibrium and kinetic studies. Journal of Wuhan University of Technology-Mater. *Science Editor*, *32*(2), 305–320. doi:10.100711595-017-1597-7

Habeeb, O. A., Ramesh, K., Ali, G. A. M., & Yunus, R. M. (2017c). Low-cost and eco-friendly activated carbon from modified palm kernel shell for hydrogen sulfide removal from wastewater: Adsorption and kinetic studies. *Desalination and Water Treatment*, *84*, 205–214. doi:10.5004/dwt.2017.21196

Hajek, E. A. (2010). *Natural Enemies: An Introduction to Biological control*. Cambridge University Press.

Hameed, B. H., & El-Khaiary, M. I. (2008). Kinetics and equilibrium studies of malachite green adsorption on rice straw-derived char. *Journal of Hazardous Materials*, *153*(1-2), 701–708. doi:10.1016/j.jhazmat.2007.09.019 PMID:17942219

Hamoda, M. F., Ghusain, I. A. I., & Mutairi, N. Z. A. (2004). Sand Filtration of wastewater for tertiary treatment and water reuse. *Desalination*, *164*(3), 203–211. doi:10.1016/S0011-9164(04)00189-4

Hao, F., Zhao, X., Ouyang, W., Lin, C., Chen, S., Shan, Y., & Lai, X. (2013). Molecular structure of corncob-derived biochars and the mechanism of atrazine sorption. *Journal of Agronomy*, *105*(3), 773–782. doi:10.2134/agronj2012.0311

Heberer, T. (2002). Occurrence, fate, and removal of pharmaceutical residues in the aquatic environment: A review of recent research data. *Toxicology Letters*, *131*(1-2), 5–17. doi:10.1016/S0378-4274(02)00041-3 PMID:11988354

Henneberger, P. K., & Lax, M. B. (1998). Lung cancer mortality in a cohort ofolder pulp and paper workers. *International Journal of Occupational and Environmental Health*, *4*(3), 147–154. doi:10.1179/oeh.1998.4.3.147 PMID:10026475

Herbert, R. B. Jr. (2006). Seasonal variations in the composition of mine drainage-contaminated groundwater in Dalarna, Sweden. *Journal of Geochemical Exploration*, *90*(3), 197–214. doi:10.1016/j.gexplo.2005.12.002

Hoddle, M. S., & Parra, J. R. P. (2013). Potential Lepidoteran Pests Associated with Avocado Fruit in Parts of Home Range of Persea Americana. In J. E. Pena (Ed.), Potential Invasive parts of Agricultural Crop. Academic Press. doi:10.1079/9781845938291.0086

Hogstedt, C. (1990). Cancer epidemiology in the paper and pulp industry. *IARC Scientific Publications*, 382–389. PMID:2228137

Hohenblum, P., Gans, O., Moche, W., Scharf, S., & Lorbeer, G. (2004). Monitoring of selected estrogenic hormones and industrial chemicals in groundwaters and surface waters in Austria. *The Science of the Total Environment*, *333*(1-3), 185–193. doi:10.1016/j.scitotenv.2004.05.009 PMID:15364528

Hohener, P., & Ponsin, V. (2014). In situ vadose zone bioremediation. *Current Opinion in Biotechnology*, *27*, 1–7. doi:10.1016/j.copbio.2013.08.018 PMID:24863890

Hu, X., Zhou, Q., & Luo, Y. (2010). Occurrence and source analysis of typical veterinary antibiotics in manure, soil, vegetables and groundwater from organic vegetable bases, northern China. *Environmental Pollution, 158*(9), 2992-2998. doi: 10.1016/j.envpol.2010.05.023

Hua, M., Yang, B., Shan, C., Zhang, W., He, S., Lv, L., & Pan, B. (2017). Simultaneous removal of As(V) and Cr(VI) from water by macroporous anion exchanger supported nanoscale hydrous ferric oxide composite. *Chemosphere, 171*, 126–133. doi:10.1016/j.chemosphere.2016.12.051 PMID:28012384

Hu, B., Wang, K., Wu, L., Yu, S.-H., Antonietti, M., & Titirici, M.-M. (2010). Engineering carbon materials from the hydrothermal carbonization process of biomass. *Advanced Materials, 22*(7), 813–828. doi:10.1002/adma.200902812 PMID:20217791

IBI. (2013). *Standardized product definition and product testing guidelines for biochar that is used in soil.* International Biochar Initiative.

Ihsanullah, A., Abbas, A., Al-Amer, A. M., Laoui, T., Al-Marri, M. J., Nasser, M. S., ... Atieh, M. A. (2016). Heavy metal removal from aqueous solution by advanced carbon nanotubes: Critical review of adsorption applications. *Separation and Purification Technology, 157*, 141–161. doi:10.1016/j.seppur.2015.11.039

Iijima, S. (1991). Helical microtubules of graphitic carbon. *Nature, 354*(6348), 56–58. doi:10.1038/354056a0

Inamuddin, & Ismail, Y.A. (2010). Synthesis and characterization of electrically conducting poly-o-methoxyanilineZr(1V) molybdate Cd(II) selective composite cationexchanger. *Desalination, 250*, 523-529.

Inamuddin, N., Naushad, M., Rangreez, T. A., & ALOthman, Z. A. (2015). Ion-selective potentiometric determination of Pb(II) ions using PVC-based carboxymethyl cellulose Sn(IV) phosphate composite membrane electrode. *Desalination and Water Treatment, 56*(3), 806–813. doi:10.1080/19443994.2014.941307

Inamuddin, R. T.A., & Khan, A. (2015). Synthesis of single-walled carbon nanotubes cerium(IV) phosphate composite cation exchanger: Ion exchange studies and its application as ion-selective membrane electrode for determination of Cd(II) ions. *Polymer Composites*. doi:10.1002/pc

Inyang, M., Gao, B., Zimmerman, A., Zhang, M., & Chen, H. (2014). Synthesis, characterization, and dye sorption ability of carbon nanotube–biochar nanocomposites. *Chemical Engineering Journal, 236*, 39–46. doi:10.1016/j.cej.2013.09.074

Islam, M., & Patel, R. (2008). Polyacrylamide thorium (IV) phosphate as an importantlead selective fibrousion exchanger: Synthesis, characterization and removal study. *Journal of Hazardous Materials, 156*(1-3), 509–520. doi:10.1016/j.jhazmat.2007.12.046 PMID:18242841

Jadhav. (2013). Electronic waste: A growing concern in today's environment sustainability. *International Journal of Social Science & Interdisciplinary Research, 2*(2), 139-147.

Jain, A. K., Sondhi, S. M., & Sharma, V. K. (2000). Synthesis, characterization and Hg(II) ion selectivity of 1-(2-Nitro-4-methyl phenyl)-6-methyl-6-methoxy-1,4,5,6-tetrahydropyrimidine- 2-(3H) thione (TPT). *Electroanalysis, 12*(4), 301–305. doi:10.1002/(SICI)1521-4109(20000301)12:4<301::AID-ELAN3013.0.CO;2-2

Jambhulkar, P. P., & Sharma, P. (2014). Development of Bioformulation and Delivery System of *Pseudomonas fluorescens* Against Bacterial Leaf Blight of Rice (*Xanthomonas oryzae pv.oryzae*). *Journal of Environmental Biology*, *35*, 843–849. PMID:25204057

James, G. W., David, P. K., Gary, H. H., & Anton, M. S. (2002). *Handbook of Ecotoxicology* (2nd ed.). CRC Press.

Jatindra, P. K., & Sudhir, K. (2009). E-Waste Management: A Case Study of Bangalore, India. *Research Journal Environmental and Earth Sciences, 1*(2), 111-115.

Javaid, M. K., Ashiq, M., & Tahir, M. (2016). Potential of Biological Agents in Decontamination of Agricultural Soil. *Scientifica*, *2016*, 1–9. doi:10.1155/2016/1598325 PMID:27293964

Javed, M., & Usmani, N. (2013). Assessment of heavy metal (Cu, Ni, Fe, Co, Mn, Cr, Zn) pollution in effluent dominated rivulet water and their effect on glycogen metabolism and histology of Mastacembelusarmatus. *SpringerPlus*, *2*(1), 390–402. doi:10.1186/2193-1801-2-390 PMID:24133639

Jensen, D. L., Ledin, A., & Christensen, T. H. (1999). Speciation of heavy metals in landfill-leachate polluted groundwater. *Water Research*, *33*(11), 2642–2650. doi:10.1016/S0043-1354(98)00486-2

Jeyaratnam, J. (1990). Acute pesticide poisoning: A major global health problem. *World Health Statistics Quarterly*, *43*, 139–144. PMID:2238694

Jimenez, B., Barrios, J. A., Mendez, J. M., & Diaz, J. (2004). Sustainable sludge management in developing countries. *Water Science and Technology*, *49*(10), 251–258. PMID:15259962

Jin, Y., Wu, S., Zeng, Z., & Fu, Z. (2017). Effects of environmental pollutants on gut microbiota. *Environmental Pollution*, *222*, 1–9. doi:10.1016/j.envpol.2016.11.045 PMID:28086130

Johnson, C. D., & Juengst, D. (1997). *Polluted urban runoff: A source of concern. No. I-02-97-5M-20-S.* Madison, WI: University of Wisconsin Extension.

Johnson, P. C., Johnson, R. L., Bruce, C. L., & Leeson, A. (2010). Advances in In Situ Air Sparging/Biosparging. *Bioremediation Journal*, *5*(4), 251–266. doi:10.1080/20018891079311

Johnston, A. E. (1986). Soil organic-matter, effects on soils and crops. *Soil Use and Management*, *2*(3), 97–105. doi:10.1111/j.1475-2743.1986.tb00690.x

Jones, D. L., Freeman, C., & Rodriguez, A. R. S. (2017). Waste Water Treatment. *Encyclopedia of applied. Plant Science*, *3*, 352–362.

Jones, J. W., Antle, J. M., Basso, B., Boote, K. J., Conant, R. T., Foster, I., ... Wheeler, T. R. (2017). Brief history of agricultural system modelling. *Agricultural Systems*, *155*, 240–254. doi:10.1016/j.agsy.2016.05.014 PMID:28701816

Judd, S., & Jefferson, B. (Eds.). (2003). *Membranes for industrial wastewater recovery and re-use*. Elsevier.

Kadirvelu, K., Thamaraiselvi, K., & Namasivayam, C. (2001). Removal of heavy metals from industrial wastewaters by adsorption onto activated carbon prepared from an agricultural solid waste. *Bioresource Technology*, *76*(1), 63–65. doi:10.1016/S0960-8524(00)00072-9 PMID:11315812

Kakar, Y. (1990). Groundwater pollution due to industrial effluents. *Central Ground Water Board, Govt. of Indian Archives*, *2*, 1–12.

Kalfa, O. M., Yalcinkaya, O., & Turker, A. R. (2009). Synthesis of nano $B_2O_3/TiO_2$ composite material and its application to preconcentration and separation of cadmium. *Journal of Hazardous Materials*, *166*, 455–461. doi:10.1016/j.jhazmat.2008.11.112 PMID:19135793

Kang, S. Y., Lee, J. U., Moon, S. H., & Kim, K. W. (2004). Competitive adsorption characteristics of $Co^{2+}$, $Ni^{2+}$, and $Cr^{3+}$ by IRN 77 cation exchange resin in synthesized, wastewater. *Chemosphere*, *56*(2), 141–147. doi:10.1016/j.chemosphere.2004.02.004 PMID:15120560

Kao, C. M., Chen, C. Y., Chen, S. C., Chien, H. Y., & Chen, Y. L. (2008). Application of in situ biosparging to remediate a petroleum-hydrocarbon spill site: Field and microbial evaluation. *Chemosphere*, *70*(8), 1492–1499. doi:10.1016/j.chemosphere.2007.08.029 PMID:17950413

Karishma. (2016). Current Scenario of e-waste management in India: Issues and strategies. *International Journal of Scientific and Research Publications, 6*(1), 424-430.

Karvelas, M., Katsoyiannis, A., & Samara, C. (2003). Occurrence and fate of heavy metals in the wastewater treatment process. *Chemosphere*, *53*(10), 1201–1210. doi:10.1016/S0045-6535(03)00591-5 PMID:14550351

Kasozi, G. N., Zimmerman, A. R., Nkedi-Kizza, P., & Gao, B. (2010). Catechol and humic acid sorption onto a range of laboratory-produced black carbons (Biochars). *Environmental Science & Technology*, *44*(16), 6189–6195. doi:10.1021/es1014423 PMID:20669904

Kaushal, S. S., Likens, G. E., Utz, R. M., Pace, M. L., Grese, M., & Yepsen, M. (2013). Increased river alkalization in the eastern U.S. *Environmental Science & Technology*, *47*(18), 10302–10311. PMID:23883395

Kehew, A. (2001). *Applied Chemical Hydrogeology*. Prentice Hall.

Kellogg, R. L., Nehring, R., Grube, A., Goss, D. W., & Plotkin, S. (2000). *Environmental indicators of pesticide leaching and runoff from farm fields Archived June 18, 2002, at the Way back Machine*. United States Department of Agriculture Natural Resources Conservation Service.

Khan, A., Asiri, A.M., Rub, M.A., Azum, N., Khan, A.A.P., Khan, S.B., Rahman, M.M., & Khan, I. (2013). Synthesis, characterization of silver nanoparticle embedded polyaniline tungstophosphate-nanocomposite cation exchanger and its application for heavy metal selective membrane. *Composite B Engineering, 45*, 1486-1492.

Khan, A.A., & Inamuddin. (2006). Preparation, physico-chemical characterization, analytical applications and electrical conductivity measurement studies of an 'organic– inorganic' composite cation-exchanger: PolyanilineSn(IV) phosphate. *Reactive and Functional Polymer, 6*6, 1649-1663.

Khan, B., & Ullah, Z. (1991). *Aquatic Pollution Due to industries in river Kabul at Aan Garh (Nowshera)* (M.Sc thesis report). Department of environmental planning and management, Uni. of Peshawar, NWFP, Pakistan.

Khan, A. A., & Akhtar, T. (2008). Preparation, physico-chemical characterization and electrical conductivity measurement studies of an organic–inorganic nanocomposite cation-exchanger: Poly-o-toluidine Zr(IV) phosphate. *Electrochimica Acta*, *53*(17), 5540–5548. doi:10.1016/j.electacta.2008.03.002

Khan, A. A., & Akhtar, T. (2009). Synthesis, characterization and ion-exchange properties of a fibrous type 'polymeric-inorganic' composite cation-exchanger Nylon-6,6 Sn(IV) phosphate: Its application in making Hg(II) selective membrane electrode. *Electrochimica Acta*, *54*(12), 3320–3329. doi:10.1016/j.electacta.2008.12.038

Khan, A. A., & Akhtar, T. (2011). Synthesis, characterization and analytical application of nano-composite cation-exchange material, poly-o-toluidine Ce(IV) phosphate: Its application in making Cd(II) ion selective membrane electrode. *Solid State Sciences*, *13*(3), 559–568. doi:10.1016/j.solidstatesciences.2010.12.026

Khan, A. A., & Baig, U. (2012). Electrically conductive membrane of polyaniline–titanium(IV)phosphate cation exchange nanocomposite: Applicable for detection of Pb(II) using its ion-selective electrode. *Journal of Industrial and Engineering Chemistry*, *18*(6), 1937–1944. doi:10.1016/j.jiec.2012.05.008

Khan, A. A., Habiba, U., & Khan, A. (2009). Synthesis and characterization of organicinorganic nanocomposite poly-o-anisidinesn(IV) arsenophosphate: Its analytical applications as Pb(II) ion-selective membrane electrode. *International Journal of Analytical Chemistry. Article*, *ID659215*, 10.

Khan, A. A., & Inamuddin. (2006). Applications of Hg(II) sensitive polyanilineSn(IV) phosphate composite cation-exchange material in determination of $Hg^{2+}$ from aqueous solutions and in making ion-selective membrane electrode. *Sensors andActuators. B, Chemical*, *120*(1), 10–18. doi:10.1016/j.snb.2006.01.033

Khan, A. A., & Inamuddin. (2006). Preparation, physico-chemical characterization, analytical applications and electrical conductivity measurement studies of an 'organic–inorganic' composite cation-exchanger: Polyaniline Sn(IV) phosphate. *Reactive & Functional Polymers*, *66*(12), 1649–1663. doi:10.1016/j.reactfunctpolym.2006.06.007

Khan, A. A., Inamuddin, & Alam, M. M. (2005). Preparation, characterization and analytical applications of a new and novel electrically conducting fibrous type polymeric–inorganic composite material: Polypyrrole Th(IV) phosphate used as a cation-exchanger and Pb(II) ion-selective membrane electrode. *Materials Research Bulletin*, *40*(2), 289–305. doi:10.1016/j.materresbull.2004.10.014

Khan, A. A., & Khan, A. (2009). Synthesis, characterization and electrical conductivity measurement studies of poly-o-anisidine Sn(IV) phosphate nano-composite cation-exchange material. *Materials Science and Engineering B*, *158*(1-3), 92–97. doi:10.1016/j.mseb.2009.01.024

Khan, A. A., & Khan, A. (2010). Ion-exchange studies on poly-o-anisidineSn(IV) phosphate nano composite and its application as Cd(II) ion-selective membrane electrode. *Central European Journal of Chemistry*, *8*, 396–408.

Khan, A. A., Khan, A., & Inamuddin. (2007). Preparation and characterization of a new organic–inorganic nano-composite poly-o-toluidine Th(IV) phosphate: Its analytical applications as cation-exchanger and in making ion-selective electrode. *Talanta*, *72*(2), 699–710. doi:10.1016/j.talanta.2006.11.044 PMID:19071675

Khan, A. A., & Paquiza, L. (2011). Characterization and ion-exchange behavior of thermally stable nano-composite polyaniline zirconium titanium phosphate: Its analytical application in separation of toxic metals. *Desalination*, *265*(1-3), 242–254. doi:10.1016/j.desal.2010.07.058

Khan, A. A., Paquiza, L., & Khan, A. (2010). An advanced nano-composite cation-exchanger polypyrrole zirconium titanium phosphate as a Th(IV)-selective potentiometric sensor: Preparation, characterization and its analytical application. *Journal of Materials Science*, *45*(13), 3610–3625. doi:10.100710853-010-4407-6

Khan, A. A., & Shaheen, S. (2013). Synthesis and characterization of a novel hybrid nano composite cation exchanger poly-o-toluidine Sn(IV) tungstate: Its analytical applications as ion-selective electrode. *Solid State Sciences*, *16*, 158–167. doi:10.1016/j.solidstatesciences.2012.10.024

Khan, A. M., Ganai, S. A., & Nabi, S. A. (2009). Synthesis of a crystalline organic-inorganic composite exchanger, acrylamide stannic silicomolybdate: Binary and quantitative separation of metal ions. *Colloids and Surfaces. A, Physicochemical and Engineering Aspects*, *337*(1-3), 141–145. doi:10.1016/j.colsurfa.2008.12.012

Khan, M. A., Ahmad, A., Umar, K., & Nabi, S. A. (2015). Synthesis, characterization, and biological applications of nanocomposites for the removal of heavy metals and dyes. *Industrial & Engineering Chemistry Research*, *54*(1), 76–82. doi:10.1021/ie504148k

Khan, M. M. A., Rafiuddin, & Inamuddin. (2013). PVC based polyvinyl alcohol zinc oxide composite membrane: Synthesis and electrochemical characterization for heavy metal ions. *Journal of Industrial and Engineering Chemistry*, *19*(4), 1365–1370. doi:10.1016/j.jiec.2012.12.041

Khan, M. N., Mobin, M., & Abbas, Z. K. (2017). *Fertilizers and their contaminants in Soil, Surface and Groundwater*. Reference Module in Earth System and Environmental Sciences. doi:10.1016/B978-0-12-409548-9.09888-2

Khan, T. A. (2015). Heavy metal ion-exchange kinetic studies over cellulose acetate Zr(IV) molybdophosphate composite cation-exchanger. *Desalination and Water Treatment*, *53*(6), 1675–1682. doi:10.1080/19443994.2013.855676

Khan, W., Rayirath, U. P., Subramanian, S., Jithesh, M. N., Rayorath, P., Hodges, D. M., ... Prithiviraj, B. (2009). Seaweed extracts as biostimulants of plant growth and development. *Journal of Plant Growth Regulation*, *28*(4), 386–399. doi:10.100700344-009-9103-x

Khurrum & Bhutta. (2011). *Electronic Waste: A Growing Concern in Today's Environment. Hindawi Publishing Corporation.*

Kim, H., Kaown, D., Mayer, B., Lee, J. Y., Hyun, Y., & Lee, K. K. (2015). Identifying the sources of nitrate contamination of groundwater in an agricultural area (Haean basin, Korea) using isotope and microbial community analyses. *The Science of the Total Environment*, *533*, 566–575. doi:10.1016/j.scitotenv.2015.06.080 PMID:26204420

Kimosop, S. J., Getenga, Z. M., Orata, F., Okello, V. A., & Cheruiyot, J. K. (2016). Residue levels and discharge loads of antibiotics in wastewater treatment plants (WWTPs), hospital lagoons, and rivers within Lake Victoria Basin, Kenya. *Environmental Monitoring and Assessment*, *188*(9), 532. doi:10.100710661-016-5534-6 PMID:27553946

Kimosop, S. J., Okello, V. A., Orata, F., Getenga, Z. M., & Shikuku, V. O. (2017). Green Remediation of Carbamazepine from Water Using Novel Magnetic Iron Modified Carbonized Baggasse: Kinetics, Equilibrium and Mechanistic Studies. *Chemical Science International Journal*, *18*(3), 1–9. doi:10.9734/CSJI/2017/32444

Kim, S., Krajmalnik-Brown, R., Kim, J. O., & Chung, J. (2014). Remediation of petroleum hydrocarbon-contaminated sites by DNA diagnosis-based bioslurping technology. *The Science of the Total Environment*, *497-498*, 250–259. doi:10.1016/j.scitotenv.2014.08.002 PMID:25129160

Knapp, J. S. (2001). Use of wood-rotting fungi for the decolorization of dyes and industrial. *Fungi in Bioremediation*, (23), 242.

Ko, C. H., & Fan, C. (2010). Enhanced chemical oxygen demand removal and flux reduction in pulp and paper wastewater treatment using laccase-polymerized membrane filtration. *Journal of Hazardous Materials*, *181*(1-3), 763–770. doi:10.1016/j.jhazmat.2010.05.079 PMID:20638964

Kohler, H. R., & Triebskorn, R. (2013). Wildlife Ecotoxicology of Pesticides: Can We Track Effects to the Population Level and Beyond? *Science*, *341*(6147), 759–765. doi:10.1126cience.1237591 PMID:23950533

Kolodziej, E. P., Harter, T., & Sedlak, D. L. (2004). Dairy wastewater, aquaculture, and spawning fish as sources of steroid hormones in the aquatic environment. *Environmental Science & Technology*, *38*(23), 6377–6384. doi:10.1021/es049585d PMID:15597895

Kresic, N. (2007). *Hydrogeology and Groundwater Modeling* (2nd ed.). Boca Raton, FL: CRC Press.

Kumar, P. J. (2014). Evolution of groundwater chemistry in and around Vaniyambadi Industrial Area: Differentiating the natural and anthropogenic sources of contamination. *Chemie der Erde - Geochemistry, 74*(4), 641-651. doi: 10.1016/j.chemer.2014.02.002

Kumar, M., Ramanathan, A. L., Tripathi, R., Farswan, S., Kumar, D., & Bhattacharya, P. (2017). A study of trace element contamination using multivariate statistical techniques and health risk assessment in groundwater of Chhaprola Industrial Area, Gautam Buddha Nagar, Uttar Pradesh, India. *Chemosphere*, *166*(Supplement C), 135–145. doi:10.1016/j.chemosphere.2016.09.086 PMID:27693874

Kumar, P., & Chandra, R. (2004). Detoxification of distillery effluent through Bacillus thuringiensis (MTCC 4714) enhanced phytoremediation potential of Spirodela polyrrhiza (L.) Schliden. *Bulletin of Environmental Contamination and Toxicology*, *73*(5), 903–910. doi:10.100700128-004-0512-z PMID:15669736

Kumar, S., Loganathan, V. A., Gupta, R. B., & Barnett, M. O. (2011). An assessment of U (VI) removal from groundwater using biochar produced from hydrothermal carbonization. *Journal of Environmental Management*, *92*(10), 2504–2512. doi:10.1016/j.jenvman.2011.05.013 PMID:21665352

Kurniawan, T. A., Chan, G. Y. S., Lo, W. H., & Babel, S. (2006). Physicoechemical treatment techniques for wastewater laden with heavy metals. *Chemical Engineering Journal*, *118*(1-2), 83–98. doi:10.1016/j.cej.2006.01.015

Ku, Y., & Jung, I. L. (2001). Photocatalytic reduction of Cr(VI) in aqueous solutions by UV irradiation with the presence of titanium dioxide. *Water Resources*, *35*, 135–142. PMID:11257867

Lampman, C. (1995). The relationship between experience and attitudes concerning epilepsy. *Journal of Applied Social Psychology*, *25*(7), 619–631. doi:10.1111/j.1559-1816.1995.tb01602.x

Lamprecht, H. O. (1988). Bau und Materialtechnik bei antiken Wasserversorgungsanlagen. *Geschichte der Wasserversorgung*, *3*, 127–155.

Lapworth, D. J., Baran, N., Stuart, M. E., and Ward, R. S. (2012). Emerging organic contaminants in groundwater: A review of sources, fate and occurrence. *Environmental Pollution 163*, 287-303. doi: 10.1016/j.envpol.2011.12.034

Lapworth, D. J., Baran, N., Stuart, M. E., Manamsa, K., and Talbot, J. (2015). Persistent and emerging micro-organic contaminants in Chalk groundwater of England and France. *Environmental Pollution, 203*, 214-225. doi: 10.1016/j.envpol.2015.02.030

Latifossglu, A. A., Surucu, G., & Evirgen, M. (1997). Water Pollut. IV: Model. In *Meas., Predict., 4th Int. Conf* (pp. 733-742). Academic Press.

Lazzari, L., Sperni, L., Bertin, P., & Pavoni, B. (2000). Correlation between inorganic (heavy metals) and organic (PCBs and PAHs) micropollutant concentrations during sewage sludge composting processes. *Chemosphere*, *41*(3), 427–435. doi:10.1016/S0045-6535(99)00289-1 PMID:11057605

Lee, C. S., Robinson, J., & Chong, M. F. (2014). A review on application of flocculants in wastewater treatment. *Process Safety and Environmental Protection*, *92*(6), 489–508. doi:10.1016/j.psep.2014.04.010

Lee, J. H. (2013). An overview of phytoremediation as a potentially promising technology for environmental pollution control. *Biotechnology and Bioprocess Engineering; BBE*, *18*(3), 431–439. doi:10.100712257-013-0193-8

Lee, Y.-C., & Yang, J.-W. (2012). Self-assembled flower-like $TiO_2$ on exfoliated graphite oxide for heavy metal removal. *Journal of Industrial and Engineering Chemistry*, *18*(3), 1178–1185. doi:10.1016/j.jiec.2012.01.005

LeGrand, H. (1970). *Movement of agricultural pollutants with groundwater. In Agricultural practices and water quality* (pp. 303–313). Ames, IA: Iowa State University Press.

Lehmann, J. (2007). A handful of carbon. *Nature*, *447*(7141), 143–144. doi:10.1038/447143a PMID:17495905

Lehmann, J., Rillig, M. C., Thies, J., Masiello, C. A., Hockaday, W. C., & Crowley, D. (2011). Biochar effects on soil biota—a review. *Soil Biology & Biochemistry*, *43*(9), 1812–1836. doi:10.1016/j.soilbio.2011.04.022

Lei, L., Song, C., Xie, X., Li, Y., & Wang, F. (2010). Acid mine drainage and heavy metal contamination in groundwater of metal sulfide mine at arid territory (BS mine, Western Australia). *Transactions of Nonferrous Metals Society of China*, *20*(8), 1488–1493. doi:10.1016/S1003-6326(09)60326-5

Lemke, R., McConkey, B., & Janzen, H. (2004). Impacts of agriculture on air quality (with emphasis on $CO_2$ and $N_2O$). *Agriculture and Agri-Food Canada,* 23–36.

Leonard, P., Hearty, S., Brennan, J., Dunne, L., Quinn, J., Chakraborty, T., & Kennedy, R. O. (2003). Advances in biosensors for detection of pathogens in food and water. *Enzyme and Microbial Technology*, *32*(1), 3–13. doi:10.1016/S0141-0229(02)00232-6

Leung, A., & Sinha, D. (1998). *Brewery Industry China Hong Kong Management Case Study*. Hong Kong: Management Development Center of Hong Kong.

Liao, M.-H., Wu, K.-Y., & Chen, D.-H. (2003). Fast Removal of Basic Dyes by a Novel Magnetic Nano-adsorbent. *Chemistry Letters*, *32*(6), 488–489. doi:10.1246/cl.2003.488

Li, C.-T., Zhuang, H.-K., Hsieh, L.-T., Lee, W.-J., & Tsao, M.-C. (2001). PAH emission from the incineration of three plastic wastes. *Environment International*, *27*(1), 61–67. doi:10.1016/S0160-4120(01)00056-3 PMID:11488391

Li, K., Wang, Y. W., Huang, M., Yan, H., Yang, H., Xiao, S. J., & Li, A. M. (2015). Preparation of chitosan-graft-polyacrylamide magnetic composite microspheres for enhanced selective removal of mercury ions from water. *Journal of Colloid and Interface Science*, *455*, 261–270. doi:10.1016/j.jcis.2015.05.043 PMID:26073848

Li, N., Kang, Y., Pan, W., Zeng, L., Zhang, Q., & Luo, J. (2015). Concentration and transportation of heavy metals in vegetables and risk assessment of human exposure to bio accessible heavy metals in soil near a waste-incinerator site. *South China. The Science of the Total Environment*, *521-522*, 144–151. doi:10.1016/j.scitotenv.2015.03.081 PMID:25829292

Li, N., Zheng, M., Chang, X., Ji, G., Lu, H., Xue, L., ... Cao, J. (2011). Preparation of magnetic $CoFe_2O_4$-functionalized graphene sheets via a facile hydrothermal method and their adsorption properties. *Journal of Solid State Chemistry*, *184*(4), 953–958. doi:10.1016/j.jssc.2011.01.014

Lingamdinne, L. P., Koduru, J. R., Roh, H., Choi, Y.-L., Chang, Y.-Y., & Yang, J.-K. (2016). Adsorption removal of Co (II) from waste-water using graphene oxide. *Hydrometallurgy*, *165*, 90–96. doi:10.1016/j.hydromet.2015.10.021

Lin, P. C., Lin, H. J., Liao, Y. Y., Guo, H. R., & Chen, K. T. (2013). Acute poisoning with neonicotinoid insecticides: A case report and literature review. *Basic & Clinical Pharmacology & Toxicology*, *112*(4), 282–286. doi:10.1111/bcpt.12027 PMID:23078648

Lin, T. C., Pan, P. T., & Cheng, S. S. (2010). Ex situ bioremediation of oil-contaminated soil. *Journal of Hazardous Materials*, *176*(1-3), 27–34. doi:10.1016/j.jhazmat.2009.10.080 PMID:20053499

Li, Q., Mahendra, S., Lyon, D., Brunet, L., Liga, M., Li, D., & Alvarez, P. (2008). Antimicrobial nanomaterials for water disinfection and microbial control: Potential applications and implications. *Water Resources*, *42*(18), 4591–4602. PMID:18804836

Liu, G. D., Wu, W. L., & Zhang, J. (2005). Regional differentiation of non-point source pollution of agriculture-derived nitrate nitrogen in groundwater in northern China. *Agriculture, Ecosystems & Environment*, *107*(2), 211–220. doi:10.1016/j.agee.2004.11.010

Liu, Z., Hamuti, A., Abdulla, H., Zhang, F., & Mao, X. (2016). Accumulation of metallic elements by native species thriving in two mine tailings in Aletai, China. *Environmental Earth Sciences*, *75*(9), 781. doi:10.100712665-016-5594-5

Liu, Z., Zhang, F.-S., & Wu, J. (2010). Characterization and application of chars produced from pinewood pyrolysis and hydrothermal treatment. *Fuel*, *89*(2), 510–514. doi:10.1016/j.fuel.2009.08.042

Loos, R., Locoro, G., Comero, S., Contini, S., Schwesig, D., Werres, F., ... Gawlik, B. M. (2010). Pan-European survey on the occurrence of selected polar organic persistent pollutants in ground water. *Water Research*, *44*(14), 4115–4126. doi:10.1016/j.watres.2010.05.032 PMID:20554303

Lopes, D. D., Silva, S. M. C. P., Fernandes, F., Teixeira, R. S., Celligoi, A., & Dall'Antônia, L. H. (2012). Geophysical technique and groundwater monitoring to detect leachate contamination in the surrounding area of a landfill – Londrina (PR – Brazil). *Journal of Environmental Management*, *113*(Supplement C), 481–487. doi:10.1016/j.jenvman.2012.05.028 PMID:22749799

Lors, C., Damidot, D., Ponge, J. F., & Perie, F. (2012). Comparison of a bioremediation process of PAHs in a PAH-contaminated soil at field and laboratory scales. *Environmental Pollution*, *165*, 11–17. doi:10.1016/j.envpol.2012.02.004 PMID:22390976

Lotter, D. W., Seidel, R., & Liebhardt, W. (2003). The performance of organic and conventional cropping systems in an extreme climate year. *American Journal of Alternative Agriculture*, *18*(03), 146–154. doi:10.1079/AJAA200345

Low, K. S., & Lee, C. K. (1991). Cadmium uptake by the Moss, Calymperesdelessertii, Besch. *Bioresource Technology*, *38*(1), 1–6. doi:10.1016/0960-8524(91)90214-5

Loyon, L. (2017). Overview of manure treatment in France. *Waste Management (New York, N.Y.)*, *61*, 516–520. doi:10.1016/j.wasman.2016.11.040 PMID:27955906

Lucar, G. B., Campbell, C. L., & Lucas, L. T. (1992). *Introduction to plant disease identification and management* (2nd ed.). Chapman & Hall.

Lu, H., Zhang, W., Yang, Y., Huang, X., Wang, S., & Qiu, R. (2012). Relative distribution of $Pb^{2+}$ sorption mechanisms by sludge-derived biochar. *Water Resources*, *46*, 854–862. PMID:22189294

Lu, Y., Song, S., Wang, R., Liu, Z., Meng, J., Sweetman, A. J., ... Wang, T. (2015). Impacts of soil and water pollution on food safety and health risk in China. *Environment International*, *77*, 5–15. doi:10.1016/j.envint.2014.12.010 PMID:25603422

Madsen, K. H., & Streibig, I. C. (2003). Benefits and risks of the use of herbicide resistant crops. In R. Labrada (Ed.), Weed management for developing countries. Academic Press.

Mahmoud, D. K., Salleh, M. A. M., Karim, W. A. W. A., Idris, A., & Abidin, Z. Z. (2012). Batch adsorption of basic dye using acid treated kenaf fibre char: Equilibrium, kinetic and thermodynamic studies. *Chemical Engineering Journal*, *181–182*(1), 449–457. doi:10.1016/j.cej.2011.11.116

Marina, P. E., Ali, G. A. M., See, L. M., Teo, E. Y. L., Ng, E.-P., & Chong, K. F. (2016). In situ growth of redox-active iron-centered nanoparticles on graphene sheets for specific capacitance enhancement. *Arabian Journal of Chemistry*. doi:10.1016/j.arabjc.2016.02.006

Maroni, M., Fanetti, A. C., & Metruccio, F. (2006). Risk assessment and management of occupational exposure to pesticides in agriculture. *La Medicina del Lavoro*, *9*, 430–437. PMID:17017381

Marwas, A. G. (2010). *Water & water treatment in India, Market opportunities for Swiss companies*. Academic Press.

Mateo-Sagasta, J., Raschid-Sally, L., & Thebo, A. (2015). Global wastewater and sludge production, treatment and use. In *Wastewater* (pp. 15–38). Dordrecht: Springer. doi:10.1007/978-94-017-9545-6_2

Matsubara, M., Morimoto, S., Sase, H., Ohizumi, T., Sumida, H., Nakata, M., & Ueda, H. (2009). Long-term declining trends of river water pH in Central Japan. *Water, Air, and Soil Pollution*, *200*(1-4), 253–265. doi:10.100711270-008-9909-3

Matsumura, F., & Krishna Murti, C.R. (1982). *Biodegradation of Pesticides*. Academic Press.

Ma, X., Tan, Z., Pang, L., & Liu, J. (2013). Determination of Vocs in groundwater at an industrial contamination site using a homemade low-density polyethylene passive diffusion sampler. *Journal of Environmental Sciences (China)*, *25*(11), 2338–2343. doi:10.1016/S1001-0742(12)60322-2 PMID:24552064

Mckinlay, R., Dassyne, J., Djamgoz, M. B. A., Plant, J. A., & Voulvoulis, N. (2012). Agricultural pesticides and chemical fertilizers In Pollutants, Human health and the environment: A risk based approach (pp. 147-179). Wiley Blackwell.

McSorley, R., & Gallaher, R. N. (1996). Effect of Yard Waste Compost on Nematode Densities and Maize Yield. *Journal of Nematology*, *2*, 655–660. PMID:19277191

Media, D. (n.d.). *Dabur Joins EPR Connect: Towards a Sustainable Future Project*. Retrieved from https://www.dabur.com/in/en-us/media/dabur-joins-epr-connect-towards-a-sustainable-future-project

Medina, M. (2004). *Scavenger cooperatives in Asia and Latin America*. Global Development Network. Retrieved from http://www.gdnet.org/pdf/medina.pdf

Medina, M. (1997). The effect of income on municipal solid waste generation rates for countries of varying levels of economic development: A model. *Journal of Solid Waste Technology Management*, *24*(3), 149–155.

Medina, M. (2000). Scavenger co-operatives in Asia and Latin America. *Resources, Conservation and Recycling*, *31*(1), 51–69. doi:10.1016/S0921-3449(00)00071-9

Meffe, R., & de Bustamante, I. (2014). Emerging organic contaminants in surface water and groundwater: A first overview of the situation in Italy. *The Science of the Total Environment*, *481*, 280–295. doi:10.1016/j.scitotenv.2014.02.053 PMID:24602913

Meneghetti, L. R. R., Thome, A., Schnaid, F., Prietto, P. D. M., & Cavelhao, G. (2012). Natural attenuation and biostimulation of biodiesel contaminated soils from Southern Brazil with different particle sizes. *Journal of Environmental Science & Engineering*, *1*(2B), 155–162.

Mesa, J., Rodriguez-Llorente, I. D., Pajuelo, E., Piedras, J. M., Caviedes, M. A., Redondo-Gomez, S., & Mateos-Naranjo, E. (2015). Moving closer towards restoration of contaminated estuaries: Bioaugmentation with autochthonous rhizobacteria improves metal rhizoaccumulation in native Spartina maritima. *Journal of Hazardous Materials*, *300*, 263–271. doi:10.1016/j.jhazmat.2015.07.006 PMID:26188869

Michael, I., Rizzo, L., McArdell, C. S., Manaia, C. M., Merlin, C., Schwartz, T., ... Fatta-Kassinos, D. (2013). Urban wastewater treatment plants as hotspots for the release of antibiotics in the environment: A review. *Water Resources*, *47*, 957–995. PMID:23266388

Miller, G. T. (2004). Sustaining the Earth (6th ed.). Thompson Learning, Inc.

MoE Japan. (2006). *Annual report on the environment in Japan.* Retrieved from https://www.env.go.jp/en/wpaper/2006/fulltext.pdf

Mohammad, A., Inamuddin, & Hussain, S. (2014). Poly (3,4-ethylenedioxythiophene): polystyrene sulfonate (PEDOT:PSS) Zr(IV) phosphate composite cation exchanger: sol-gel synthesis and physicochemical characterization. *Ionics*, *21*(4), 1063–1071. doi:10.100711581-014-1247-4

Mohana, S., Desai, C., & Madamwar, D. (2007). Biodegradation and decolourization of anaerobically treated distillery spent wash by a novel bacterial consortium. *Bioresource Technology*, *98*(2), 333–339. doi:10.1016/j.biortech.2005.12.024 PMID:16473005

Mohana, S., Shrivastava, S., Divecha, J., & Madamwar, D. (2008). Response surface methodology for optimization of medium for decolorization of textile dye Direct Black 22 by a novel bacterial consortium. *Bioresource Technology*, *99*(3), 562–569. doi:10.1016/j.biortech.2006.12.033 PMID:17804220

Mohan, D., & Bhamawat, P. M. K. (2008). E-waste management-global scenario: A Review. *Journal of Environmental Research and Development*, *2*(4), 817–823.

Mohan, D., Pittman, C. U., & Steele, P. H. (2006). Pyrolysis of wood/biomass for bio-oil, a critical review. *Energy & Fuels*, *20*(3), 848–889. doi:10.1021/ef0502397

Mohebi-Fard, E., Reyahi-Khoram, M. & Sobhan-Ardakani, S. (2015). Performance evaluation of the wastewater treatment plant of Pelareh Dairy Industry, Iran. *Journal of Advances in Environmental Health Research,* 3(4), 250-257.

Monte, M. C., Fuente, E., Blanco, A., & Negro, C. (2009). Waste management from pulp and paper production in the European Union. *Waste Management (New York, N.Y.)*, *29*(1), 293–308. doi:10.1016/j.wasman.2008.02.002 PMID:18406123

Moore, M., Gould, P., & Keary, B. S. (2003). Global urbanization and impact on health. *International Journal of Hygiene and Environmental Health*, *206*(4-5), 269–278. doi:10.1078/1438-4639-00223 PMID:12971682

Moradi, O., Gupta, V. K., Agarwal, S., Tyagi, I., Asif, M., Makhlouf, A. S. H., ... Shahryari-ghoshekandi, R. (2015a). Characteristics and electrical conductivity of graphene and graphene oxide for adsorption of cationic dyes from liquids: Kinetic and thermodynamic study. *Journal of Industrial and Engineering Chemistry*, *28*, 294–301. doi:10.1016/j.jiec.2015.03.005

Moradi, O., Sadegh, H., Shahryari-Ghoshekandi, R., & Norouzi, M. (2015b). Application of carbon nanotubes in nanomedicine: New medical approach for tomorrow. *Handbook of Research on Diverse Applications of Nanotechnology in Biomedicine. Chemical Engineering (Albany, N.Y.)*, 90–128.

Morman, S. A., & Plumlee, G. S. (2013). The role of airborne mineral dusts in human disease. *Aeolian Research*, *9*, 203–212. doi:10.1016/j.aeolia.2012.12.001

Mostafa, A. A. (2013). Treatment of cheese processing wastewater by physicochemical and biological methods. *International Journal of Microbiology*, *4*(3), 321–332.

Mozaz, S. R., de Alda, M. J. L., & Barcelo, D. (2004). Monitoring of estrogens, pesticides and bisphenol A in natural waters and drinking water treatment plants by solid phase extraction-liquid chromatography-mass spectrometry. *Journal of Chromatography. A*, *1045*(1-2), 85–92. doi:10.1016/j.chroma.2004.06.040 PMID:15378882

Mui, E. L. K., Cheung, W. H., Valix, M., & McKay, G. (2010). Dye adsorption onto char from bamboo. *Journal of Hazardous Materials*, *177*(1–3), 1001–1005. doi:10.1016/j.jhazmat.2010.01.018 PMID:20097002

Mukherjee, A., Zimmerman, A., & Harris, W. (2011). Surface chemistry variations among a series of laboratory-produced biochars. *Geoderma*, *163*(3–4), 247–255. doi:10.1016/j.geoderma.2011.04.021

Mulbry, W., Westhead, E. K., Pizarro, C., & Sikora, L. (2005). Recycling of Manure nutrients: Use of algal biomass from dairy manure treatment as a slow release fertilizer. *Bioresource Technology*, *96*(4), 451–458. doi:10.1016/j.biortech.2004.05.026 PMID:15491826

Munoz, J. I., Roca, J. A. M., & Clar, A. I. (2008). Study of different alternatives of tertiary treatments for wastewater reclamation to optimize the water quality for irrigation reuse. *Desalination*, *222*(1-3), 222–229. doi:10.1016/j.desal.2007.01.157

Murphy, E., Zachara, J., Smith, S., Phillips, J., & Wietsma, T. (1994). Interaction of hydrophobic organic compounds with mineral-bound humic substances. *Environmental Science & Technology*, *28*(7), 1291–1299. doi:10.1021/es00056a017 PMID:22176321

Musungu, C. P., Jondiko, I. O., Lalah, J., Ongeri, D., Chepkui, R., & Kiema, F. (2013). The extent of nutrient removal by wastewater treatment plants along the Nyalenda Wigwa Stream and the River Kisat (Kenya). *Ecohydrology & Hydrobiology*, *13*(4), 236–240. doi:10.1016/j.ecohyd.2013.10.002

Muth, P., & Premnath. (2006). Plastics materials flow analysis for India. *Resources, Conservation and Recycling, 47*, 222–244.

Nabi, S. A., Akhtar, A., Khan, M. D. A., & Khan, M. A. (2014). Synthesis, characterization and electrical conductivity of Polyaniline-Sn(IV)tung- stophosphate hybrid cation exchanger: Analytical application for removal of heavy metal ions from wastewater. *Desalination, 340*, 73–83. doi:10.1016/j.desal.2014.02.020

Nabi, S. A., Bushra, R., Naushad, M., & Khan, A. M. (2010). Synthesis, characterization and analytical applications of a new composite cation exchange material poly-otoluidine stannic molybdate for the separation of toxic metal ions. *Chemical Engineering Journal, 165*(2), 529–536. doi:10.1016/j.cej.2010.09.064

Nabi, S. A., Bushra, R., & Shahadat, M. (2012). Removal of toxic metal ions by using compositecation-exchange material. *Journal of Applied Polymer Science, 125*(5), 3438–3446. doi:10.1002/app.36325

Nabi, S. A., Ganai, A. S., & Shalla, A. H. (2008). New organic-inorganic type acrylamide aluminumtungstate: Preparation, characterization and analytical applications as a cation exchange material. *Separation Science and Technology, 43*(14), 3695–3711. doi:10.1080/01496390802212674

Nabi, S. A., Ganai, S. A., & Khan, A. M. (2011). Synthesis, characterization and ion exchange behavior of polyaniline stannic silicomolybdate, an organic–inorganic composite material: Quantitative separation of $Pb^{2+}$ ions from industrial effluents. *Journal of Inorganic and Organometallic Polymers, 21*(1), 25–35. doi:10.100710904-010-9423-0

Nabi, S. A., & Naushad, M. (2008). Synthesis, characterization and analytical applications of a new composite cation exchanger cellulose acetate-Zr (IV) molybdophosphate. *Colloids and Surfaces. A, Physicochemical and Engineering Aspects, 316*(1-3), 217–225. doi:10.1016/j.colsurfa.2007.09.005

Nabi, S. A., Naushad, M., & Bushra, R. (2009). Synthesis and characterization of a new organic–inorganic $Pb^{2+}$ selective composite cation exchanger acrylonitrile stannic(IV) tungstate and its analytical applications. *Chemical Engineering Journal, 152*(1), 80–87. doi:10.1016/j.cej.2009.03.033

Nabi, S. A., & Shalla, A. H. (2009). EDTA-stannic (IV) iodate: Preparation, characterization and its analytical applications for metal content determination in real and synthetic samples. *Journal of Porous Materials, 16*(5), 587–597. doi:10.100710934-008-9236-5

Nabi, S. A., & Shalla, A. H. (2009). Synthesis, characterization and analytical application of hybrid; Acrylamide zirconium (IV) arsenate a cation exchanger, effect of dielectric constant on distribution coefficient of metal ions. *Journal of Porous Materials, 163*, 657–664. PMID:18722710

Naja, G. M., & Volesky, B. (2009). Treatment of metal-bearing effluents: removal and recovery. In Handbook on Heavy Metals in the Environment. Taylor and Francis.

Namasivayam, C., & Kadirvelu, K. (1999). Uptake of mercury (II) from wastewater by activated carbon from unwanted agricultural solid by-product: Coirpith. *Carbon, 37*(1), 79–84. doi:10.1016/S0008-6223(98)00189-4

Narayana, T. (2009). Municipal solid waste management in India: From waste disposal to recovery of resources. Waste Management, 29, 1163–1166.

Narvaez, O. M. R., Hernandez, J. M. P., & Goonetilleke, A. (2017). Treatment technologies for emerging contaminants in water: A review. *Chemical Engineering Journal*, *323*, 361–380. doi:10.1016/j.cej.2017.04.106

Naseem, R., & Tahir, S. S. (2001). Removal of Pb(II) from aqueous solution by using bentonite as an adsorbent. *Water Resources*, *35*, 3982–3986. PMID:12230182

Nath, B., Chakraborty, S., & Jana, J. (2006). Arsenic in the groundwater of the Bengal delta plain: hydrogeochemical studies from the districts of Nadia and South 24 Parganas, West Bengal, India. *Conference on Groundwater for Sustainable Dev*.

National Park Service. (2006). US Department of the Interior, Sequoia & Kings Canyon National Park: Air quality-Airborne synthetic chemicals. Author.

Naushad, M., AL-Othman, Z. A., & Islam, M. (2013). Adsorption of cadmium ion using a new composite cation-exchanger polyaniline Sn(IV) silicate: Kinetics, thermodynamic and isotherm studies. *International Journal of Environmental Science and Technology*, *10*(3), 567–578. doi:10.100713762-013-0189-0

Navarro, A., & Carbonell, M. (2007). Evaluation of groundwater contamination beneath an urban environment: The Besos river basin (Barcelona, Spain). *Journal of Environmental Management*, *85*(2), 259–269. doi:10.1016/j.jenvman.2006.08.021 PMID:17223248

Navarro, A., & Font, X. (1993). Discriminating different sources of groundwater contamination caused by industrial wastes in the Besós River Basin, Barcelona, Spain. *Applied Geochemistry*, *8*(Supplement 2), 277–279. doi:10.1016/S0883-2927(09)80050-4

Naymik, T. G. (1993). Contaminant Hydrogeology: C. W. Fetter. Macmillan, 1993, xvii + 458p., $61.25 (ISBN 0-02-337135-8). *Geochimica et Cosmochimica Acta*, *57*(15), 3820. doi:10.1016/0016-7037(93)90162-P

Ng'eno, E., Orata, F., Lilechi, D. B., Shikuku, V. O., & Kimosop, S. (2016). Adsorption of caffeine and ciprofloxacin onto pyrolytically derived water hyacinth biochar: Isothermal, kinetics and thermodynamics. *Journal of Chemistry and Chemical Engineering*, *10*, 185194.

Nguyen, T. A. H., Ngo, H. H., Guo, W. S., Zhang, J., Liang, S., Yue, Q. Y., ... Nguyen, T. V. (2013). Applicability of agricultural waste and by-products for adsorptive removal of heavy metals from wastewater. *Bioresource Technology*, *148*, 574–585. doi:10.1016/j.biortech.2013.08.124 PMID:24045220

Nguyen, T., Cho, H., Poster, D., & Ball, W. (2007). Evidence for a pore-filling mechanism in the adsorption of aromatic hydrocarbons to a natural wood char. *Environmental Science & Technology*, *41*(4), 1212–1217. doi:10.1021/es0617845 PMID:17593721

Nikakhtari, H., Kumar, P., Nemati, M., & Hill, G. A. (2009). Biodegradation of diesel oil in a baffled roller bioreactor. *Journal of Chemical Technology and Biotechnology (Oxford, Oxfordshire)*, *84*(4), 525–532. doi:10.1002/jctb.2074

Nikolaou, A., Meric, S., & Fatta, D. (2007). Occurrence patterns of pharmaceuticals in water and wastewater environments. *Analytical and Bioanalytical Chemistry*, *387*(4), 1225–1234. doi:10.100700216-006-1035-8 PMID:17205270

Noonan, D. C., & Curtij, J. T. (1990). Gronnwater remediation and petroleum: A guide for underground storage tanks. M. L. P. Chelsea.

Nouri, J., Khorasani, N., Lorestani, B., Karami, M., Hassani, A. H., & Yousefi, N. (2009). Accumulation of heavy metals in soil and uptake by plant species with phytoremediation potential. *Environmental Earth Sciences*, *59*(2), 315–323. doi:10.100712665-009-0028-2

Novoselov, K. S. (2004). Electric Field Effect in Atomically Thin Carbon Films. *Science*, *306*(5696), 666–669. doi:10.1126cience.1102896 PMID:15499015

Nriagu, J. O., & Pacyna, J. M. (1988). Quantitative assessment of worldwide contamination of air, water and soils by trace metals. *Nature*, *333*(6169), 134–139. doi:10.1038/333134a0 PMID:3285219

Ntaliani, M., Costopoulou, C., & Karetsos, S. (2008). Mobile Government: A challenge for agriculture. *Government Information Quarterly*, *25*(4), 699–716. doi:10.1016/j.giq.2007.04.010

Obasi, M. N. (1999). Environmental pollution with particular reference to chemical pollutant in Nigeria. White and White Publishers Ltd.

Ogola, J. S., Chimuka, L., & Maina, D. (2009). Occurrence and fate of trace metals in and around treatment and disposal facilities in Limpopo Province, South Africa (a case of two areas). *Proceedings of the IASTED International Conference on Environmental Management and Engineering*.

Ojovan, M.I., & Lee, W.E. (2014). 14 – Treatment of Radioactive Waste. *An Introduction to Nuclear Waste Immobilisation*, 171-203.

Olaniran, N. S. (1995). Environment and Health: An Introduction. In Environment and Health. Academic Press.

Olanrewaju, O. O., Olalekan, S. F., Odendaal, J., Adegbenro, P. D., & Ogheneochuko, O. (2012). Preliminary Investigation into Occurrence and Removal of Arsenic, Cadmium, Mercury and Zinc in Wastewater Treatment Plants in Cape Town and Stellenbosch. *Polish Journal of Environmental Studies*, *21*, 1755–1765.

Oliveira, A., Bocio, A., Trevilato, T., Takayanagui, A., Domingo, J., & Segura-munoz, S. (2007). Heavy metals in untreated/treated urban effluent and sludge from a biological wastewater treatment plant. *Environmental Science and Pollution Research International*, *14*(7), 483–489. doi:10.1065/espr2006.10.355 PMID:18062480

Onay, O. (2007). Influence of pyrolysis temperature and heating rate on the production of bio-oil and char from sunflower seed by pyrolysis, using a well-swept fixed-bed reaction. *Fuel Processing Technology*, *88*(5), 523–553. doi:10.1016/j.fuproc.2007.01.001

Ongley, E. D. (1996). *Control of water pollution from agriculture*. Academic Press.

Ongley, E. D. (1997). Pesticides, as water contaminants. In *Combating Agricultural Contamination of Water Resources*. Academic Press.

Onsdorff, K. A. (1996). What the Weitzenhoff Court got wrong. *Journal of Water Environment Law and Practice*, *4*(1), 14–18.

Osenbruck, K., Glaser, H. R., Knoller, K., Weise, S. M., Moder, M., Wennrich, R., ... Strauch, G. (2007). Sources and transport of selected organic micropollutants in urban groundwater underlying the city of Halle (Saale), Germany. *Water Research*, *41*(15), 3259–3270. doi:10.1016/j.watres.2007.05.014 PMID:17575997

Oyaro, N., Juddy, O., Murago, E. N. M., & Gitonga, E. (2007). The contents of Pb, Cu, Zn and Cd in meat in Nairobi, Kenya. *International Journal* of *Food. Agriculture & Environment*, *5*, 119–121.

Ozcimen, D., & Ersoy-Mericboyu, A. (2010). Characterization of biochar and bio-oil samples obtained from carbonization of various biomass materials. *Renewable Energy*, *35*(6), 1319–1324. doi:10.1016/j.renene.2009.11.042

Palmer, W. E., Bromley, P. T., & Brandenburg, R. L. (2007). *Wildlife & pesticides -Peanuts*. North Carolina Cooperative Extension Service.

Panda, A., Singh, R. K., & Mishra, D. K. (2010). *Thermolysus of Waste Plastics to Liquid Fuel: A Suitable Method for Plastic Waste Management and Manufacture of Value Added Products—A World Prospective*. Retrieved from https://www.researchgate.net/publication/227421226_Thermolysis_of_waste_plastics_to_liquid_fuel_A_suitable_method_for_plastic_waste_management_and_manufacture_of_value_added_products--A_world_prospective

Pappu, A., Saxena, M., & Asolekar, S. R. (2007). Solid waste generation in India and their recycling potentials in building materials. *Building and Environment*, *42*(6), 2311–2320. doi:10.1016/j.buildenv.2006.04.015

Park, J. H., Choppola, G. K., Bolan, N. S., Chung, J. W., & Chuasavathi, T. (2011). Biochar reduces the bioavailability and phytotoxicity of heavy metals. *Plant and Soil*, *348*(1-2), 439–451. doi:10.100711104-011-0948-y

Parshetti, G. K., Chowdhury, S., & Balasubramanian, R. (2014). Hydrothermal conversion of urban food waste to chars for removal of textile dyes from contaminated waters. *Bioresource Technology*, *161*, 310–319. doi:10.1016/j.biortech.2014.03.087 PMID:24727353

Pathak, U., Das, P., Banerjee, P., & Datta, S. (2016). Treatment of wastewater from a dairy industry using rice husk as adsorbent: Treatment efficiency, isotherm, thermodynamics, and kinetics modelling. *Journal of Thermodynamics*, *2016*, 1–7. doi:10.1155/2016/3746316

Patil, D. S., Chavan, S. M., & Oubagaranadin, J. U. K. (2016). A review of technologies for manganese removal from wastewaters. *Journal of Environmental Chemical Engineering*, *4*(1), 468–487. doi:10.1016/j.jece.2015.11.028

Patil, P. U., Kapadnis, B. P., & Dhamankar, V. S. (2003). Decolorisation of synthetic melanoidin and biogas effluent by immobilised fungal isolate of Aspergillus niger UM2. *International Sugar Journal*, *105*(1249), 8–13.

Paulino, A. T., Minasse, F. A. S., Guilherme, M. R., Reis, A. V., Muniz, E. C., & Nozaki, J. (2006). Novel adsorbent based on silkworm chrysalides for removal of heavy metals from wastewaters. *Journal of Colloid and Interface Science*, *301*(2), 479–487. doi:10.1016/j.jcis.2006.05.032 PMID:16780853

Pavani, P., & Rajeswari, T. R. (2014). Impact of Plastics on Environmental Pollution. *Journal of Chemical and Pharmaceutical Sciences*, (3).

Pearce, J. M. (2007). Burton's line in lead poisoning. *European Neurology*, *57*(2), 118–119. doi:10.1159/000098100 PMID:17179719

Pedersen, T. L. (1997). *Pesticide residues in drinking water*. Academic Press.

Penn State Extension. (2017). *Toxicity of Pesticides*. Author.

Penradee, C., Kyoung-Woong, K., & Suthipong, S. (2008). Metal Contents and Its Variation in Wastewater and Sewage Sludge: A Case Study of Bangkok Central Wastewater Treatment Plants. *Proceedings of the International Symposia on Geoscience Resources and Environments of Asian Terranes*, 489-492.

Perdigón-Melón, J. A., Carbajo, J. B., Petre, A. L., Rosal, R., & García-Calvo, E. (2010). Coagulation–Fenton coupled treatment for ecotoxicity reduction in highly polluted industrial wastewater. *Journal of Hazardous Materials*, *181*(1-3), 127–132. doi:10.1016/j.jhazmat.2010.04.104 PMID:20537462

*Pesticides in the Environment : Exposure Pathways*. (2003). Academic Press.

Phillips, P. J., Schubert, C., Argue, D., Fisher, I., Furlong, E. T., Foreman, W., ... Chalmers, A. (2015). Concentrations of hormones, pharmaceuticals and other micropollutants in groundwater affected by septic systems in New England and New York. *The Science of the Total Environment*, *512-513*, 43–54. doi:10.1016/j.scitotenv.2014.12.067 PMID:25613769

Pignatello, J., Kwon, S., & Lu, Y. (2006). Effect of natural organic substances on the surface and adsorptive properties of environmental black carbon (char): Attenuation of surface activity by humic and fulvic acids. *Environmental Science & Technology*, *40*(24), 7757–7763. doi:10.1021/es061307m PMID:17256524

Pimentel, D. (2005). Environmental and economic costs of the application of pesticides primarily in the United States. *Environment, Development and Sustainability*, *7*(2), 229–252. doi:10.100710668-005-7314-2

Pindihama, G. K., Gumbo, J. R., & Oberholster, P. J. (2011). Evaluation of a low-cost technology to manage algal toxins in rural water supplies. *African Journal of Biotechnology*, *10*(86), 19883–19889. doi:10.5897/AJBX11.024

Plangklang, P., & Reungsang, A. (2010). Bioaugmentation of carbofuran by *Burkholderia cepacia* PCL3 in a bioslurry phase sequencing batch reactor. *Process Biochemistry*, *45*(2), 230–238. doi:10.1016/j.procbio.2009.09.013

Planquart, P., Bonin, G., Prone, A., & Massiani, C. (1999). Distribution, movement, and plant availability of trace metals in soils amended with sewage sludge composts: Application to low loadings. *The Science of the Total Environment*, *241*(1-3), 161–179. doi:10.1016/S0048-9697(99)00338-1

Power, A. G. (2010). Ecosystem services and agriculture: Tradeoffs and synergies. *Philosophical Transaction of Royal Society B*, *365*(1554), 2959–2971. doi:10.1098/rstb.2010.0143 PMID:20713396

Press Trust of India. (2016). *15,342 tn plastic waste generated in India everyday: Dave*. Retrieved from http://www.business-standard.com/article/pti-stories/15-342-tn-plastic-waste-generated-in-india-everyday-dave-116080200866_1.html

Preston, T. R., & Leng, R. A. (1989). The greenhouse effect and its implications for world agriculture, the need for environmentally friendly development. *Livestock Research for Rural Development.*

Principi, P., Villa, F., Bernasconi, M., & Zanardini, E. (2006). Metal toxicity in municipal wastewater activated sludge investigated by multivariate analysis and in situ hybridization. *Water Research*, *40*(1), 99–106. doi:10.1016/j.watres.2005.10.028 PMID:16343589

Qian, K., Kumar, A., Patil, K., Bellmer, D., Wang, D., Yuan, W., & Huhnke, R. (2013). Effects of biomasss feedstocks and gasification conditions on the physicochemical properties of char. *Energies*, *6*(8), 3972–3986. doi:10.3390/en6083972

Qiu, H., Lv, L., Pan, B., Zhang, Q., Zhang, W., & Zhang, Q. (2009). Critical review in adsorption kinetic models. *Journal of Zhejiang University. Science A*, *10*(5), 716–724. doi:10.1631/jzus.A0820524

Qiu, Y., Zheng, Z., Zhou, Z., & Sheng, G. D. (2009). Effectiveness and mechanisms of dye adsorption on a straw-based biochar. *Bioresource Technology*, *100*(21), 5348–5351. doi:10.1016/j.biortech.2009.05.054 PMID:19540756

Quinn, L. & Amie. (2012). The impacts of agriculture and temperature on the physiological stress response in fish. Uleth. *University of Lethbridge.*

Ragavan, K. V., & Rastogi, N. K. (2017). β-Cyclodextrin capped graphene-magnetite nanocomposite for selective adsorption of Bisphenol-A. *Carbohydrate Polymers*, *168*, 129–137. doi:10.1016/j.carbpol.2017.03.045 PMID:28457432

Raghuvanshi, S., Bhakar, V., Sowmya, C., & Sangwan, K. S. (2017). Waste water treatment Plant lifecycle Assessment: Treatment Process to Reuse of water. *Procedia CIRP*, *61*, 761–766. doi:10.1016/j.procir.2016.11.170

Rahimi, A., & García, J.M. (2017). Chemical recycling of waste plastics for new materials production. *Nature Reviews Chemistry, 1.*

Rain, D. A., & Guillette, L. J. Jr. (1998). Reptiles as models of contaminant-induced endocrine disruption. *Animal Reproduction Science*, *53*(1-4), 77–86. doi:10.1016/S0378-4320(98)00128-6 PMID:9835368

Rajabi, M., Mirza, B., Mahanpoor, K., Mirjalili, M., Najafi, F., Moradi, O., ... Tyagi, I. (2016). Adsorption of malachite green from aqueous solution by carboxylate group functionalized multi-walled carbon nanotubes: Determination of equilibrium and kinetics parameters. *Journal of Industrial and Engineering Chemistry*, *34*, 130–138. doi:10.1016/j.jiec.2015.11.001

Rangreez, T. A., Inamuddin, Naushad, M., & Ali, H. (2015). Synthesis and characterisation of poly(3,4-ethylenedioxythiophene)-poly(styrenesulfonate) (PEDOT:PSS) Zr(IV) monothiophosphate composite cation exchanger: analytical application in the selective separation of lead metal ions. *International Journal of Environmental Analytical Chemistry*, *95*(6), 556–568. doi:10.1080/03067319.2015.1036863

Rao, M. N., & Datta, A. K. (1978). Waste water treatment (3rd ed.). Academic Press.

Rattner, B. A. (2009). History of wildlife toxicology. *Ecotoxicology (London, England)*, *18*(7), 773–783. doi:10.100710646-009-0354-x PMID:19533341

Ravanbakhsh, M., Ronaghi, A.-M., Taghavi, S. M., & Jousset, A. (2016). Screening for the next generation heavy metal hyperaccumulators for dryland decontamination. *Journal of Environmental Chemical Engineering*, *4*(2), 2350–2355. doi:10.1016/j.jece.2016.04.013

Ravva, S. V., & Sarreal, C. Z. (2014). Survival of Salmonella enterica in aerated and non-aerated wastewaters from dairy lagoons. *International Journal of Environmental Research and Public Health*, *11*(11), 11249–11260. doi:10.3390/ijerph111111249 PMID:25358096

Raza, A., Friedel, J. K., & Bodner, G. (2012). Improving water use efficiency for sustainable agriculture. In *Agroecology and strategies for climate change* (pp. 167–211). Dordrecht: Springer. doi:10.1007/978-94-007-1905-7_8

Reddy, D. H. K., & Lee, S. M. (2012). Water pollution and treatment technologies. *Journal of Environmental & Analytical Toxicology*, 2.

Rehmann, L., Prpich, G. P., & Daugulis, A. J. (2008). Remediation of PAH contaminated soils: Application of a solid–liquid two-phase partitioning bioreactor. *Chemosphere*, *73*(5), 798–804. doi:10.1016/j.chemosphere.2008.06.006 PMID:18640698

Ren, G., Hu, D., Cheng, E., Vargasreus, M., Reip, P., & Allaker, R. (2009). Characterisation of copper oxide nanoparticles for antimicrobial applications. *International Journal of Antimicrobial Agents*, *33*(6), 587–590. doi:10.1016/j.ijantimicag.2008.12.004 PMID:19195845

Richard, P., & Sittig's, P. (2015). *Handbook of Pesticides and Agricultural Chemicals* (2nd ed.). William Andrew Publishing.

Rickert, D. (1993). Water quality assessment to determine the nature and extent of water pollution by agriculture and related activities. In *Prevention of Water Pollution by Agriculture and Related Activities. Proceedings of the FAO Expert Consultation*. Santiago, Chile: FAO.

Rinkesh. (n.d.). *Agricultural Pollution*. Retrieved from https://www.conserve-energy-future.com/causes-and-effects-of-agricultural-pollution.php

Roberts, M. T. (2016). *Food law in the United States.* Cambridge University Press.

Rossel, S. A., & Jorge, M. F. (1999). Cuban strategy for management and control of waste. In *Proceedings of Recovery, recycling, re-integration (R '99) congress*. EMPA.

Roy, S., Beig, G., & Ghude, S. (2009). Exposure-plant response of ambient ozone over the tropical Indian region. *Atmospheric Chemistry and Physics*, *9*, 4141–4157.

Sacher, F., Lange, F. T., Brauch, H. J., & Blankenhorn, I. (2001). Pharmaceuticals in groundwaters analytical methods and results of a monitoring program in Baden-Wurttemberg, Germany. *Journal of Chromatography. A*, *938*(1-2), 199–210. doi:10.1016/S0021-9673(01)01266-3 PMID:11771839

Sadegh, H. (2017). Development of graphene oxide from graphite: A review on synthesis, characterization and its application in wastewater treatment. *Reviews on Advanced Materials Science*, *49*, 38–43.

Sadegh, H., Ali, G. A. M., Abbasi, Z., & Nadagoud, M. N. (2017a). Adsorption of Ammonium Ions onto Multi-Walled Carbon Nanotubes. *Studia Universitatis Babes-Bolyai. Chemia*, *62*(2), 233–245. doi:10.24193ubbchem.2017.2.18

Sadegh, H., Ali, G. A. M., Gupta, V. K., Makhlouf, A. S. H., Shahryari-Ghoshekandi, R., Nadagouda, M. N., ... Megiel, E. (2017b). The role of nanomaterials as effective adsorbents and their applications in wastewater treatment. *Journal of Nanostructure in Chemistry*, *7*(1), 1–14. doi:10.100740097-017-0219-4

Sadegh, H., Shahryari-ghoshekandi, R., Agarwal, S., Tyagi, I., Asif, M., & Gupta, V. K. (2015). Microwave-assisted removal of malachite green by carboxylate functionalized multi-walled carbon nanotubes: Kinetics and equilibrium study. *Journal of Molecular Liquids*, *206*, 151–158. doi:10.1016/j.molliq.2015.02.007

Sadegh, H., Zare, K., Maazinejad, B., Shahryari-Ghoshekandi, R., Tyagi, I., Agarwal, S., & Gupta, V. K. (2016). Synthesis of MWCNT-COOH-Cysteamine composite and its application for dye removal. *Journal of Molecular Liquids*, *215*, 221–228. doi:10.1016/j.molliq.2015.12.042

Sahu, O. P., & Chaudhari, P. K. (2013). Review on chemical treatment of industrial waste water. *Journal of Applied Science & Environmental Management*, *17*(2), 241–257.

Sakai, S., Yoshida, H., & Hirai, Y. (2011). International comparative study of 3R and waste management policy developments. *Journal of Material Sciences and Waste Management,* 86–102.

Sala-Serra, M., Sunyer, J., Kogevinas, M., McFarlane, D., & Anto, J. M. (1996). Cohort study on cancer mortality among workers in the pulp and paper industry in Catalonia, Spain. *American Journal of Industrial Medicine*, *30*, 87–92. doi:10.1002/(SICI)1097-0274(199607)30:1<87::AID-AJIM153.0.CO;2-4 PMID:8837689

Salem, S. M., Lettieri, P., & Baeyens, J. (2009). Recycling and recovery routes of plastic solid waste (PSW): A review. Waste Management, 29, 2625–2643.

Sanborn, M., Kerr, K. J., Sanin, L. H., Cole, D. C., Bassil, K. L., & Vakil, C. (2007). Non-cancer health effects of pesticides: Systematic review and implications for family doctors. *Canadian Family Physician Medecin de Famille Canadien*, *53*, 1712–1720. PMID:17934035

Sánchez Pérez, J. M., Antiguedad, I., Arrate, I., Garcialinares, C., & Morell, I. (2003). The influence of nitrate leaching through unsaturated soil on groundwater pollution in an agricultural area of the Basque country: A case study. *The Science of the Total Environment*, *317*(1), 173–187. doi:10.1016/S0048-9697(03)00262-6 PMID:14630420

Sanneh, E. S., Hu, H. A., Chang, Y. M., & Sanyang, E. (2011). Introduction of a recycling system for sustainable municipal solid waste management: A case study on the greater Banjul area of the Gambia. *Environment, Development and Sustainability*, *13*(6), 1065–1080. doi:10.100710668-011-9305-9

Sarmah, A. K., Meyer, M. T., & Boxall, A. B. (2006). A global perspective on the use, sales, exposure pathways, occurrence, fate and effects of veterinary antibiotics (VAs) in the environment. *Chemosphere*, *65*(5), 725–759. doi:10.1016/j.chemosphere.2006.03.026 PMID:16677683

Sato, T., Qadir, M., Yamamoto, S., Endo, T., & Zahoor, A. (2013). Global, regional, and country level need for data on wastewater generation, treatment, and use. *Agricultural Water Management*, *130*, 1–13. doi:10.1016/j.agwat.2013.08.007

Schimmelpfennig, S., & Glaser, B. (2012). One step forward toward characterization: Some important material properties to distinguish biochars. *Journal of Environmental Quality*, *41*(4), 1001–1013. doi:10.2134/jeq2011.0146 PMID:22751042

Schwarzenbach, R. P., Escher, B. I., & Fenner, K. (2006). The challenge of micropollutants in aquatic systems. *Science*, *313*(5790), 1072–1077. doi:10.1126cience.1127291 PMID:16931750

Selvakumar, S., Chandrasekar, N., & Kumar, G. (2017). Hydrogeochemical characteristics and groundwater contamination in the rapid urban development areas of Coimbatore, India. *Water Resources and Industry*, *17*(Supplement C), 26–33. doi:10.1016/j.wri.2017.02.002

Sethi, A. C. (2001). Key note address. *International Conference on Fresh Water*, Bonn, Germany.

Shagun, A. K., & Arora, A. (2013, October). Proposed Solution of e-waste management. *International Journal of Future Computer and Communication*, *2*(5), 490–493.

Shahadat, M., Bushra, R. (2015). Synthesis, characterization and significant applications of PANI-Zr(IV) sulphosalicylate nanocomposite. *Advances in Nanotechnology, 6*, 181-206.

Shahadat, M., Nabi, S. A., Bushra, R., Raeissi, A. S., Umar, K., & Ansari, M. O. (2012). Synthesis, characterization, photolytic degradation, electrical conductivity and applications of a nanocomposite adsorbent for the treatment of pollutants. *RSC Advances*, *2*(18), 7207–7220. doi:10.1039/c2ra20589k

Shahadat, M., Shalla, A. H., & Raeissi, A. S. (2012). Synthesis, characterization, and sorption behavior of a novel composite cation exchange adsorbent. *Industrial & Engineering Chemistry Research*, *51*(47), 15525–15529. doi:10.1021/ie3014555

Shamsipur, M., Dezaki, A. S., Akhond, M., Sharghi, H., Paziraee, Z., & Alizadeh, K. (2009). Novel PVC-membrane potentiometric sensors based on a recently synthesized sulfur-containing macrocyclicdiamide for $Cd^{2+}$ ion. Application to flow-injection Potentiometry. *Journal of Hazardous Materials*, *172*(2-3), 566–573. doi:10.1016/j.jhazmat.2009.07.003 PMID:19640638

Shannon, M. A., Bohn, P. W., Elimelech, M., Georgiadis, J. G., Marias, B. J., & Mayes, A. M. (2008). Science and technology for water purification in the coming decades. *Nature*, *452*(7185), 301–310. doi:10.1038/nature06599 PMID:18354474

Sharma, S. K., & Kennedy, M. D. (2017). Soil aquifer treatment for wastewater treatment and reuse. *International Biodeterioration & Biodegradation*, *119*, 671–677. doi:10.1016/j.ibiod.2016.09.013

Shestopalov, V. M., Rudenko, Y. F., Bohuslavsky, A. S., & Bublias, V. N. (2006). Chernobyl-born radionuclides: groundwater protectability withrespect to preferential flow zones. In H. Vereecken, A. Binley, G. Cassiani, A. Revil, & K. Titov (Eds.), *Applied Hydrogeophysics* (pp. 341–376). Dordrecht: Springer Netherlands. doi:10.1007/978-1-4020-4912-5_12

Shikuku, V.O., Zanella, R., Kowenje, C.O., Donato, F.F., Bandeira, N., & Prestes, O.D. (2017). Single and Competitive Removal of Sulfachloropyridazine and Sulfadimethoxine onto Natural Kaolinite Clay: Kinetics, Isotherms and Thermodynamics Studies. *South African Journal of Chemistry*, *70*.

Shirmardi, M., Mesdaghinia, A., Mahvi, A. H., Nasseri, S., & Nabizadeh, R. (2012). Kinetics and equilibrium studies on adsorption of acid red 18 (Azo-Dye) using multiwall carbon nanotubes (MWCNTs) from aqueous solution. *Journal of Chemistry*, *9*(4), 2371–2383.

Shukla, G., Kumar, A., Bhanti, M., Joseph, P. E., & Taneja, A. (2006). Organochlorine pesticide contamination of ground water in the city of Hyderabad. *Environment International*, *32*(2), 244–247. doi:10.1016/j.envint.2005.08.027 PMID:16183122

Siddiqui, W.A., Khan, S.A., & Inamuddin. (2007). Synthesis, characterization and ionexchange properties of a new and novel 'organic–inorganic' hybrid cation- exchanger: Poly(methyl methacrylate) Zr(IV) phosphate. *Colloids and Surfaces A: Physicochemical and Engineering Aspect, 295*, 193-199.

Silva-Castro, G. A., Rodelas, B., Perucha, C., Laguna, J., Gonzalez-Lopez, J., & Calvo, C. (2013). Bioremediation of diesel-polluted soil using biostimulation as post treatment after oxidation with Fenton-like reagents: Assays in a pilot plant. *The Science of the Total Environment*, *445–446*, 347–355. doi:10.1016/j.scitotenv.2012.12.081 PMID:23354375

Silva-Castro, G. A., Uad, I., Rodrıguez-Calvo, A., Gonzalez-Lopez, J., & Calvo, C. (2015). Response of autochthonous microbiota of diesel polluted soils to land- farming treatments. *Environmental Research*, *137*, 49–58. doi:10.1016/j.envres.2014.11.009 PMID:25486545

Sims, G. K., & Cupples, A. M. (1999). Factors controlling degradation of pesticides in soil. *Pesticide Science*, *55*(5), 598–601. doi:10.1002/(SICI)1096-9063(199905)55:5<598::AID-PS9623.0.CO;2-N

Sims, G. K., & Sommers, L. E. (1986). Biodegradation of pyridine derivatives in soil suspensions. *Environmental Toxicology and Chemistry*, *5*(6), 503–509. doi:10.1002/etc.5620050601

Singh, S., Barick, K., & Bahadur, D. (2011). Novel and efficient three dimensional mesoporous ZnO nanoassemblies for environmental remediation. *International Journal of Nanoscience, 10*(4-5), 1001-1005.

Singhal, B. B. S., & Gupta, R. P. (2010). Groundwater Contamination. In B. B. S. Singhal & R. P. Gupta (Eds.), *Applied Hydrogeology of Fractured Rocks* (2nd ed.; pp. 221–236). Dordrecht: Springer Netherlands. doi:10.1007/978-90-481-8799-7_12

Sivaramanan. (2013). E-waste management, disposal and its impacts on the environment. *Universal Journal of Environmental Research and Technology, 3*(5), 531-537.

Skylab. (2007). *Classroom in Space: Part III - Science Demonstrations, Chapter 17: Life Sciences*. Author.

Smith, L., Inman, A., & Zhang, H. (2017). Mitigation of diffuse water pollution in England and China, and the scope for policy transfer. *Land Use Policy, 61*, 208–219. doi:10.1016/j.landusepol.2016.09.028

Soares, W. L., & Porto, M. F. D. (2009). Estimating the social cost of pesticide use: An assessment from acute poisoning in Brazil. *Ecological Economics, 68*(10), 2721–2728. doi:10.1016/j.ecolecon.2009.05.008

Sohi, S. P. (2012). Carbon storage with benefits. *Science, 338*(6110), 1034–1035. doi:10.1126cience.1225987 PMID:23180849

Sorensen, J. P. R., Lapworth, D. J., Nkhuwa, D. C. W., Stuart, M. E., Gooddy, D. C., Bell, R. A., ... Pedley, S. (2015). Emerging contaminants in urban groundwater sources in Africa. *Water Research, 72*(Supplement C), 51–63. doi:10.1016/j.watres.2014.08.002 PMID:25172215

*Sources of common contaminants and their health effects*. (2008). Emergency Response Program, EPA.

Speight, J. G. (2017). Chapter 4 – Sources and types of organic pollutants. Environmental Organic Chemistry for Engineers, 153-201.

Spiegel, J., & Maystre, L. Y. (1998). International Labour Office. Environmental pollution control and prevention. In *Encyclopedia of Occupational Health and Safety* (4th ed.). Geneva: International Labour Office.

Srivastava, N. K., & Majumder, C. B. (2008). Novel biofiltration methods for the treatment of heavy metals from industrial wastewater. *Journal of Hazardous Materials, 151*(1), 1–8. doi:10.1016/j.jhazmat.2007.09.101 PMID:17997034

Standley, L. J., Rudel, R. A., Swartz, C. H., Attfield, K. R., Christian, J., Erickson, M., & Brody, J. G. (2008). Wastewater-contaminated groundwater as a source of endogenous hormones and pharmaceuticals to surface water ecosystems. *Environmental Toxicology and Chemistry, 27*(12), 2457–2468. doi:10.1897/07-604.1 PMID:18616377

Stenberg, A. J., Heil, M., Ahman, I., & Bjorkman, C. (2015). Optimizing Crops for Biocontrol of Pests and Disease. *Trends in Plant Science, 20*, 698–712.

Stephenson, G. A., & Solomon, K. R. (1993). *Pesticides and the Environment. Department of Environmental Biology*. Guelph, Ontario, Canada: University of Guelph.

Stoate, C., Boatman, N. D., Borralho, R. J., Carvalho, C. R., de Snoo, G. R., & Eden, P. (2001). Ecological impacts of arable intensification in Europe. *Journal of Environmental Management, 63*(4), 337–365. doi:10.1006/jema.2001.0473 PMID:11826719

Story, P., & Cox, M. (2001). Review of the effects of organophosphorus and carbamate insecticides on vertebrates. Are there implications for locust management in Australia? *Wildlife Research, 28*(2), 179. doi:10.1071/WR99060

Stuart, M., & Lapworth, D. (2013). Emerging Organic Contaminants in Groundwater. In S. C. Mukhopadhyay & A. Mason (Eds.), *Smart Sensors for Real-Time Water Quality Monitoring* (pp. 259–284). Berlin: Springer Berlin Heidelberg. doi:10.1007/978-3-642-37006-9_12

Stuart, M., Lapworth, D., Crane, E., & Hart, A. (2012). Review of risk from potential emerging contaminants in UK groundwater. *The Science of the Total Environment*, *416*, 1–21. doi:10.1016/j.scitotenv.2011.11.072 PMID:22209399

Suffet, I. H., Ho, J., Chou, D., Khiari, D., & Mallevialle, J. (1995). Taste-and- odor problems observed during drinking water treatment. In I. H. Suffet, J. Mallevialle, & E. Kawczynski (Eds.), *Advances in Taste-and-odor Treatment and Control*. American Water Works Association.

Suman, G., & Dikshit, A. K. (2010). Biopesticides: An ecofriendly approach for pest control. *Journal of Biopesticides*, *3*(1), 186–188.

Sumathi, S., & Hung, Y. T. (2006). Treatment of pulp and paper mill wastes. In L. K. Wang, Y. T. Hung, H. H. Lo, & C. Yapijakis (Eds.), *Waste treatment in the process industries* (pp. 453–497). Taylor & Francis.

Sun, J., Lian, F., Liu, Z., Zhu, L., & Song, Z. (2014). Biochars derived from various crop straws: Characterization and Cd (II) removal potential. *Ecotoxicology and Environmental Safety*, *106*, 226–231. doi:10.1016/j.ecoenv.2014.04.042 PMID:24859708

Sun, K., Keiluweit, M., Kleber, M., Pan, Z., & Xing, B. (2011). Sorption of fluorinated herbicides to plant biomass-derived biochars as a function of molecular structure. *Bioresource Technology*, *102*(21), 9897–9903. doi:10.1016/j.biortech.2011.08.036 PMID:21907572

Sun, L., Wan, S., & Luo, W. (2013). Biochars prepared from anaerobic digestion residue, palm bark, and eucalyptus for adsorption of cationic methylene blue dye: Characterization, equilibrium, and kinetic studies. *Bioresource Technology*, *140*, 406–413. doi:10.1016/j.biortech.2013.04.116 PMID:23714096

Swartz, C. H., Reddy, S., Benotti, M. J., Yin, H., Barber, L. B., Brownawell, B. J., & Rudel, R. A. (2006). Steroid estrogens, nonylphenol ethoxylate metabolites, and other wastewater contaminants in groundwater affected by a residential septic system on Cape Cod, MA. *Environmental Science & Technology*, *40*(16), 4894–4902. doi:10.1021/es052595+ PMID:16955883

Szekacs, A., Mortl, M., & Darvas, B. (2015). Monitoring Pesticide Residues in Surface and Ground Water in Hungary: Surveys in 1990-2015. *Journal of Chemistry*, *2015*, 1–15. doi:10.1155/2015/717948

Szewzyk, U., Szewzyk, R., Manz, W., & Schleifer, K. H. (2000). Microbiological Safety of Drinking Water. *Annual Review of Microbiology*, *54*(1), 81–127. doi:10.1146/annurev.micro.54.1.81 PMID:11018125

Szolnoki, Z., & Farsang, A. (2013). Evaluation of metal mobility and bioaccessibility in soils of urban vegetable gardens using sequential extraction. *Water, Air, and Soil Pollution*, *224*(10), 1737. doi:10.100711270-013-1737-4

Taha, A. W., Dakroury, A. M., Sayed, G. E., & El-Salam, S. A. (2007). Assessment removal of heavy metals ions from wastewater by cement kiln dust (CKD). *Eleventh International Water Technology Conference*, 879-893.

Taiwo, A. M., Gbadebo, A. M., Oyedepo, J. A., Ojekunle, Z. O., Alo, O. M., Oyeniran, A. A., ... Taiwo, O. T. (2016). Bioremediation of industrially contaminated soil using compost and plant technology. *Journal of Hazardous Materials*, *304*, 166–172. doi:10.1016/j.jhazmat.2015.10.061 PMID:26551220

Tang, L., Fang, Y., Pang, Y., Zeng, G., Wang, J., Zhou, Y., ... Chen, J. (2014). Synergistic adsorption and reduction of hexavalent chromium using highly uniform polyaniline–magnetic mesoporous silica composite. *Chemical Engineering Journal*, *254*, 302–312. doi:10.1016/j.cej.2014.05.119

Tang, Z., Zhang, L., Huang, Q., Yang, Y., Nie, Z., Cheng, J., ... Chai, M. (2015). Contamination and risk of heavy metals in soils and sediments from a typical plastic waste recycling area in North China. *Ecotoxicology and Environmental Safety*, *122*, 343–351. doi:10.1016/j.ecoenv.2015.08.006 PMID:26318969

Tan, X., Yunguo, L., Guangming, Z., Xin, W., Xinjiang, H., Zhongzhu, Y., & Yanling, G. (2015). Application of biochar for the removal of pollutants from aqueous solutions. *Chemosphere*, *125*, 70–85. doi:10.1016/j.chemosphere.2014.12.058 PMID:25618190

Teijon, G., Candela, L., Tamoh, K., Molina-Diaz, A., & Fernandez-Alba, A. R. (2010). Occurrence of emerging contaminants, priority substances (2008/105/CE) and heavy metals in treated wastewater and groundwater at Depurbaix facility (Barcelona, Spain). *The Science of the Total Environment*, *408*(17), 3584–3595. doi:10.1016/j.scitotenv.2010.04.041 PMID:20593552

Teschke, W. J. (2002). Mortality from lung cancer in workers exposed to sulfur dioxide in the pulp and paper industry. *Environmental Health Perspectives*, *110*(10), 991–995. doi:10.1289/ehp.02110991 PMID:12361923

Thakkar, A. P., Dhamankar, V. S., & Kapadnis, B. P. (2006). Biocatalytic decolourisation of molasses by Phanerochaete chrysosporium. *Bioresource Technology*, *97*(12), 1377–1381. doi:10.1016/j.biortech.2005.07.005 PMID:16115759

*The biological control of pests*. (2007). Biology Pages.

The Federation of Indian Chambers of Commerce and Industry (FICCI). (2014). *Potential Of Plastics Industry In Northern India With Special Focus On Plasticulture And Food Processing: A report on plastics industry.* Retrieved from http://ficci.in/spdocument/20396/Knowledge-Paper-ps.pdf

The Nation Karachi. (2012). Industrial discharge playing havoc with lives, ecosystem. *The Nation.* Retrieved from http://www.nation.com.pk/karachi/21-Nov-2012/industrialdischargeplaying-havoc-with-lives-ecosystem

Thijs, S., Sillen, W., Rineau, F., Weyens, N., & Vangronsveld, J. (2016). Towards an Enhanced Understanding of Plant–Microbiome Interactions to Improve Phytoremediation: Engineering the Metaorganism. *Frontiers in Microbiology*, *7*, 341. doi:10.3389/fmicb.2016.00341 PMID:27014254

Thompson, G., Swain, J., Kay, M., & Forster, C. (2001). The treatment of pulp and paper milleffluent: A review. *Bioresource Technology*, *77*(3), 275–286. doi:10.1016/S0960-8524(00)00060-2 PMID:11272013

Thompson, H. S. (1850). On the absorbent power of soils. *Journal of the Royal Agricultural Society of England*, *11*, 68–74.

Tian, Z., Niu, Y., & Fan, D. (2017). Maintaining rice production while mitigating methane and nitrous oxide emissions from paddy field in China: Evaluating tradeoffs by using coupled agricultural system models. *Agricultural Systems*. doi:10.1016/j.agsy.2017.04.006

Tiwari, K. K., Dwivedi, S., Mishra, S., Srivastava, S., Tripathi, R. D., Singh, N. K., & Chakraborty, S. (2008). Phytoremediation efficiency of Portulaca tuberose sox and Portulaca oleracea L. Naturally growing in an industrial effluent irrigated area in Vadodara, Gujarat. India. *Environmental Monitoring and Assessment*, *147*(1-3), 15–22. doi:10.100710661-007-0093-5 PMID:18193484

Tiwary, R. K., Dhakate, R., Ananda Rao, V., & Singh, V. S. (2005). Assessment and prediction of contaminant migration in ground water from chromite waste dump. *Environmental Geology*, *48*(4), 420–429. doi:10.100700254-005-1233-2

Tomei, M. C., Angelucci, D. M., Annesini, M. C., & Daugulis, A. J. (2013). Ex situ remediation of polluted soils by absorptive polymers, and a comparison of slurry and two-phase partitioning bioreactors for ultimate contaminant degradation. *Journal of Hazardous Materials*, *262*, 31–37. doi:10.1016/j.jhazmat.2013.08.026 PMID:24007996

Toren, K., Persson, B., Wingren, G. (1996). Health effects of working in pulp and paper mills: malignant diseases. *American Journal of Industrial Medicine*, *29*, 123-130.

Toxic Link. (2016). *WEEE plastic and Brominated Flame Retardants: A Report on WEEE Plastic Recycling*. Retrieved from http://toxicslink.org/docs/Report-WEEE-plastics.pdf

Trivedy, R. K., & Nakate, S. S. (2001). *Treatment of hospital waste and sewage in hyacinth ponds*. Jaipur, India: Low Cost Wastewater Treatment Technologies. ABD.

Troschinetz, A. M., & Mihelcic, J. R. (2009). Sustainable recycling of municipal solid waste in developing countries. *Waste Management (New York, N.Y.)*, *29*(2), 915–923. doi:10.1016/j.wasman.2008.04.016 PMID:18657963

Tsai, W. T., & Chen, H. R. (2013). Adsorption kinetics of herbicide paraquat in aqueous solution onto a low-cost adsorbent, swine-manure-derived biochar. *International Journal of Environmental Science and Technology*, *10*(6), 1349–1356. doi:10.100713762-012-0174-z

Turusov, V., Rakitsky, V., & Tomatis, L. (2002). Dichlorodiphenyltrichloroethane (DDT): Ubiquity, persistence, and risks. *Environmental Health Perspectives*, *110*(2), 125–128. doi:10.1289/ehp.02110125 PMID:11836138

U.S. Environmental Protection Agency. (2007). *Pesticides: Health and Safety*. National Assessment of the Worker Protection Workshop.

UC IPM Online (2007). *Statewide IPM Program, Agriculture and Natural Resources*. University of California. Retrieved from http://www.ipm.ucdavis.edu

Uddin. (2012). Journal and Conference Paper On (Enviornment) e-waste management. *IOSR Journal of Mechanical and Civil Engineering, 2*(1), 25-45.

Ulusoy, U., & Akkaya, R. (2009). Adsorptive features of polyacrylamide-apatite composite for $Pb^{2+}$, $UO_2^{2+}$ and $Th^{4+}$. *Journal of Hazardous Materials*, *163*(1), 98–108. doi:10.1016/j.jhazmat.2008.06.064 PMID:18656311

UN WWAP. (2003). *United Nations World Water Assessment Programme. The World Water Development Report 1: Water for People - Water for Life*. Paris, France: UNESCO.

United Nations-Water. (2009). *World Water Day brochure*. Retrieved from http://www.unwater.org/worldwaterday/downloads/wwd09 brochure en LOW.pdf

United States Environmental Protection Agency (2017). *Human health issues related to pesticides*. Author.

United States Public Health. (2012). *Drinking water standards*. Author.

UN-Water. (2009). *World Water Day brochure*. Available at: http://www.unwater.org/worldwaterday/downloads/wwd09brochureenLOW.pdf

Urase, T., & Kikuta, T. (2005). Separate estimation of adsorption and degradation of pharmaceutical substances and estrogens in the activated sludge process. *Water Research*, *39*(7), 1289–1300. doi:10.1016/j.watres.2005.01.015 PMID:15862328

USEPA. (1976). RCRA Subtitle I and Sections 7003, 9003(h), 9005 and 9006.

Üstün, G. E., Solmaz, S. K. A., & Birgül, A. (2007). Regeneration of industrial district wastewater using a combination of Fenton process and ion exchange—A case study. *Resources, Conservation and Recycling*, *52*(2), 425–440. doi:10.1016/j.resconrec.2007.05.006

Vamvuka, D. (2011). Bio-oil, solid and gaseous biofuels fro biomass pyrolysis processes- an overview. *International Journal of Energy Research*, *35*(10), 835–862. doi:10.1002/er.1804

Van der Werf, H. M. G. (1996). Assessing the impact of pesticides on the environment. *Agriculture, Ecosystems & Environment*, *60*(2-3), 81–96. doi:10.1016/S0167-8809(96)01096-1

Van Doesburg, M.J., Noordijk, H., & Van den Anker, I.M. (1993). *Evaluation of the smog situation in the summer period of 1992*. RIVM Report 722101005.

Varo-Arguello, W. E., Camacho-Perez, B., Rios-Leal, E., Vazquez-Landaverde, P. A., Ponce-Noyola, M. T., Barrera-Cortes, J., ... Poggi-Varaldo, H. M. (2012). Triphasic slurry bioreactors for the bioremediation of lindane-impacted soil under aerobic and anaerobic conditions. *Environmental Engineering and Management Journal*, *11*(10), 1811–1823.

Velimirovic, M., Tosco, T., Uyttebroek, M., Luna, M., Gastone, F., De Boer, C., ... Bastiaens, L. (2014). Field assessment of guar gum stabilized microscale zerovalent iron particles for in-situ remediation of 1,1,1-trichloroethane. *Journal of Contaminant Hydrology*, *164*, 88–99. doi:10.1016/j.jconhyd.2014.05.009 PMID:24963597

Venkataraman, C., Habib, G., & Kadamba, D. (2006). Emissions from open biomass burning in India, Integrating the inventory approach with high-resolution moderate resolution imaging spectroradiometer (MODIS) active and land cover data. *Global Biogeochemical Cycles*, 20.

Verma, R., & Suthar, S. (2015). Impact of density loads on performance of duckweed bioreactor: A potential system for synchronized wastewater treatment and energy biomass production. *American Institute of Chemical Engineers*, *34*, 1596–1604.

Vorosmarty, C. J., McIntyre, P. B., Gessner, M. O., Dudgeon, D., Prusevich, A., Green, P., ... Davies, P. M. (2010). Global threats to human water security and river biodiversity. *Nature*, *467*(7315), 555–561. doi:10.1038/nature09440 PMID:20882010

Voutsa, D., Zaxariadis, G., Gantidis, N., Samara, C., & Kouimtzis, T. (1996). Evaluation of municipal and industrial wastewater sludges for agricultural purposes. *Fresenius Environmental Bulletin*, *5*, 1–6.

Wan Ngah, W. S., Teong, L. C., & Hanafiah, M. A. K. M. (2011). Adsorption of dyes and heavy metal ions by chitosan composites: A review. *Carbohydrate Polymers*, *83*(4), 1446–1456. doi:10.1016/j.carbpol.2010.11.004

Wang, C., Wang, F., Wang, T., Bian, Y., Yang, X., & Jiang, X. (2010). PAHs biodegradation potential of indigenous consortia from agricultural soil and contaminated soil in two-liquid-phase bioreactor (TLPB). *Journal of Hazardous Materials*, *176*(1-3), 41–47. doi:10.1016/j.jhazmat.2009.10.123 PMID:19954884

Wang, H., & Nie, Y. (2001). Municipal solid waste characteristics and management in China. *Journal of the Air & Waste Management Association*, *51*(2), 250–263. doi:10.1080/10473289.2001.10464266 PMID:11256500

Wang, X., Cai, W., Liu, S., Wang, G., Wu, Z., & Zhao, H. (2013). ZnO hollow microspheres with exposed porous nanosheets surface: Structurally enhanced adsorption towards heavy metal ions. *Colloids and Surfaces. A, Physicochemical and Engineering Aspects*, *422*, 199–205. doi:10.1016/j.colsurfa.2013.01.031

Way, J. T. (1850). On the power of soils to absorb manure. *Journal of the Royal Agricultural Society of England*, *11*, 313–379.

Wells, M. (2007). *Vanishing bees threaten U.S. crops*. BBC News.

Whelan, M. J., Coulon, F., Hince, G., Rayner, J., McWatters, R., Spedding, T., & Snape, I. (2015). Fate and transport of petroleum hydrocarbons in engineered biopiles in polar regions. *Chemosphere*, *131*, 232–240. doi:10.1016/j.chemosphere.2014.10.088 PMID:25563162

White, C. (1972). *Handbook of chlorination*. Van Nostrand Reinhold Company.

White, D., Lapworth, D. J., Stuart, M. E., & Williams, P. J. (2016). Hydrochemical profiles in urban groundwater systems: New insights into contaminant sources and pathways in the subsurface from legacy and emerging contaminants. *The Science of the Total Environment*, *562*(Supplement C), 962–973. doi:10.1016/j.scitotenv.2016.04.054 PMID:27155350

WHO and UNICEF. (2013). *Progress on Sanitation and Drinking Water*. Geneva: World Health Organization.

Williams, P., Whitfield, M., Biggs, J., Bray, S., Fox, G., Nicolet, P., & Sear, D. (2004). Comparative diversity of rivers, streams, ditches and ponds in an agricultural landscape in southern England. *Biological Conservation*, *115*(2), 329–341. doi:10.1016/S0006-3207(03)00153-8

Wilson, Velis, & Cheeseman, . (2006). *Role of informal sector recycling in waste management in developing countries. In Habitat international* (pp. 797–808). Elsevier.

Wilson, C., & Tisdell, C. (2001). Why farmers continue to use pesticides despite environmental, health and sustainability costs. *Ecological Economics*, *39*(3), 449–462. doi:10.1016/S0921-8009(01)00238-5

Winde, F. (2006). Long-term Impacts of Gold and Uranium Mining on Water Quality in Dolomitic Regions — examples from the Wonderfonteinspruit catchment in South Africa. In B. J. Merkel & A. Hasche-Berger (Eds.), *Uranium in the Environment: Mining Impact and Consequences* (pp. 807–816). Berlin: Springer Berlin Heidelberg. doi:10.1007/3-540-28367-6_83

Wolt, J. D., Smith, J. K., Sims, J. K., & Duebelbeis, D. O. (1996). Products and kinetics of cloramsulam-methylaerobic soil metabolism. *Journal of Agricultural and Food Chemistry*, *44*(1), 324–332. doi:10.1021/jf9503570

Woodard, C. (2005). Industrial Waste Treatment Handbook (2nd ed.). Butterworth Heinemann.

Woo, S. H., Jeon, C. O., Yun, Y. S., Choi, H., Lee, C. S., & Lee, D. S. (2009). On-line estimation of key process variables based on kernel partial least squares in an industrial cokes wastewater treatment plant. *Journal of Hazardous Materials*, *161*(1), 538–544. doi:10.1016/j.jhazmat.2008.04.004 PMID:18486333

World Health Organisation. (2011). *Guidelines for Drinking Water Quality* (4th ed.).

World Health Organization. (1989). *Health guidelines for the use of wastewater in agriculture and aquaculture: report of a WHO scientific group*. WHO.

World Health Organization. (2003). *Assessing Microbial Safety of Drinking Water Improving Approaches and Methods: Improving Approaches and Methods*. OECD Publishing.

Wu, M., Tang, X., Li, Q., Yang, W., Jin, F., Tang, M., & Scholz, M. (2013). Review of ecological engineering solutions for rural non-point source water pollution control in Hubei Province, China. *Water, Air, and Soil Pollution*, *224*(5), 1561–1579. doi:10.100711270-013-1561-x

Wu, Z., Banuelos, G. S., Lin, Z.-Q., Liu, Y., Yuan, L., Yin, X., & Li, M. (2015). Biofortification and phytoremediation of Selenum in China. *Frontiers in Plant Science*, *6*, 1–8. doi:10.3389/fpls.2015.00136 PMID:25653664

Xu, R. K., Xiao, S. C., Yuan, J. H., & Zhao, A. Z. (2011). Adsorption of methyl violet from aqueous solutions by the biochars derived from crop residues. *Bioresource Technology*, *102*(22), 10293–10298. doi:10.1016/j.biortech.2011.08.089 PMID:21924897

Xu, T. (2005). Ion exchange membranes: State of their development and perspective. *Journal of Membrane Science*, *263*(1-2), 1–29. doi:10.1016/j.memsci.2005.05.002

Xu, X., Zhao, Y., Zhao, X., Wang, Y., & Deng, W. (2013). Sources of heavy metal pollution in agricultural soils of a rapidly industrializing area in the Yangtze Delta of China. *Ecotoxicology and Environmental Safety*, *108*, 161–167. doi:10.1016/j.ecoenv.2014.07.001 PMID:25063882

Yadav, R. D., Chaudhry, S., & Dhiman, S. S. (2009). Biopulping and its potential to reduce effluent loads from bleaching of hardwood kraft pulp. *BioResources*, *5*(1), 159–171.

Yang, Y., Lin, X., Wei, B., Zhao, Y., & Wang, J. (2013). Evaluation of adsorption potential of bamboo biochar for metal-complex dye: Equilibrium, kinetics and artificial neural network modeling. *International Journal of Environmental Science and Technology*. doi:10.100713762-013-0306-0

Yao, Y., Gao, B., Chen, H., Jiang, L., Inyang, M., Zimmerman, A. R., ... Li, H. (2012). Adsorption of sulfamethoxazole on biochar and its impact on reclaimed water irrigation. *Journal of Hazardous Materials*, *209*, 408–413. doi:10.1016/j.jhazmat.2012.01.046 PMID:22321858

Yu, L., Han, M., & He, F. (2017). A review of treating oily wastewater. *Arabian Journal of Chemistry*, *10*, S1913–S1922. doi:10.1016/j.arabjc.2013.07.020

Zahra, A., Hashmi, M. Z., Malik, R. N., & Ahmed, Z. (2014). Enrichment and geo-accumulation of heavy metals and risk assessment of sediments of the Kurang Nallah – feeding tributary of the Rawal Lake reservoir, Pakistan. *The Science of the Total Environment*, *470-471*, 925–933. doi:10.1016/j.scitotenv.2013.10.017 PMID:24239813

Zambon, P., Ricci, P., Bovo, E., Casula, A., Gattolin, M., Fiore, A. R., Chiosi, F., & Guzzinati, S. (2007). Sarcoma risk and dioxin emissions from incinerators and industrial plants: a population-based case-control study (Italy). *Environmental Health*, *6*, 19.

Zapata, F., & Nguyen, M. L. (2009). Chapter 7 Soil Erosion and Sedimentation Studies using Environmental Radionuclides. Radioactivity in the Environment, 16, 295-322.

Zarbin, A. J. G. (2007). Química de (nano)materiais. *Quimica Nova*, *30*(6), 1469–1479. doi:10.1590/S0100-40422007000600016

Zare, K., Gupta, V. K., Moradi, O., Makhlouf, A. S. H., Sillanpää, M., Nadagouda, M. N., ... Kazemi, M. (2015a). A comparative study on the basis of adsorption capacity between CNTs and activated carbon as adsorbents for removal of noxious synthetic dyes: A review. *Journal of Nanostructure in Chemistry*, *5*(2), 227–236. doi:10.100740097-015-0158-x

Zare, K., Sadegh, H., Shahryari-ghoshekandi, R., Asif, M., Tyagi, I., Agarwal, S., & Gupta, V. K. (2016). Equilibrium and kinetic study of ammonium ion adsorption by $Fe_3O_4$ nanoparticles from aqueous solutions. *Journal of Molecular Liquids*, *213*, 345–350. doi:10.1016/j.molliq.2015.08.045

Zare, K., Sadegh, H., Shahryari-ghoshekandi, R., Maazinejad, B., Ali, V., Tyagi, I., ... Gupta, V. K. (2015b). Enhanced removal of toxic Congo red dye using multi walled carbon nanotubes: Kinetic, equilibrium studies and its comparison with other adsorbents. *Journal of Molecular Liquids*, *212*, 266–271. doi:10.1016/j.molliq.2015.09.027

Zhai, Y., Zhao, X., Teng, Y., Li, X., Zhang, J., Wu, J., & Zuo, R. (2017). Groundwater nitrate pollution and human health risk assessment by using HHRA model in an agricultural area, NE China. *Ecotoxicology and Environmental Safety*, *137*(Supplement C), 130–142. doi:10.1016/j.ecoenv.2016.11.010 PMID:27918944

Zhang, J., Dai, J., Chen, H., Du, X., Wang, W., & Wang, R. (2012). Petroleum contamination in groundwater/air and its effects on farmland soil in the outskirt of an industrial city in China. *Journal of Geochemical Exploration*, *118*(Supplement C), 19–29. doi:10.1016/j.gexplo.2012.04.002

Zhang, J., Zhai, S., Li, S., Xiao, Z., Song, Y., An, Q., & Tian, G. (2013). Pb (II) removal of Fe3O4@SiO2–NH2 core–shell nanomaterials prepared via a controllable sol–gel process. *Chemical Engineering Journal*, *215*, 461–471. doi:10.1016/j.cej.2012.11.043

Zhang, P., Sun, H., Yu, L., & Sun, T. (2013). Adsorption and catalytic hydrolysis of carbaryl and atrazine on pig manure-derived biochars: Impact of structural properties of biochars. *Journal of Hazardous Materials*, *244*, 217–224. doi:10.1016/j.jhazmat.2012.11.046 PMID:23246958

Zhang, W., & Di Giano, F. A. (2002). Comparison of bacterial regrowth in distribution systems using free chlorine and chloramine: A statistical study of causative factors. *Water Research*, *36*(6), 1469–1482. doi:10.1016/S0043-1354(01)00361-X PMID:11996337

Zhang, Y., Li, Q., Sun, L., Tang, R., & Zhai, J. (2010). High efficient removal of mercury from aqueous solution by polyaniline/humic acid nanocomposite. *Journal of Hazardous Materials*, *175*(1-3), 404–409. doi:10.1016/j.jhazmat.2009.10.019 PMID:19896766

Zhao, G., Li, J., Ren, X., Chen, C., & Wang, X. (2011). Few-layered graphene oxide nanosheets as superior sorbents for heavy metal ion pollution management. *Environmental Science & Technology*, *45*(24), 10454–10462. doi:10.1021/es203439v PMID:22070750

Zhao, Y., Wei, Y., Zhang, Y., Wen, X., Xi, B., Zhao, X., ... Wei, Z. (2017). Roles of composts in soil based on the assessment of humification degree of fulvic acids. *Ecological Indicators*, *72*, 473–480. doi:10.1016/j.ecolind.2016.08.051

Zheng, H., Wang, Z., Zhao, J., Herbert, S., & Xing, B. (2013). Sorption of antibiotic sulfamethoxazole varies with biochars produced at different temperatures. *Environmental Pollution*, *181*, 60–67. doi:10.1016/j.envpol.2013.05.056 PMID:23811180

Zhou, Y., Gao, B., Zimmerman, A., & Cao, X. (2014). Biochar-supported zerovalent iron reclaims silver from aqueous solution to form antimicrobial nanocomposite. *Chemosphere*, *117*, 801–805. doi:10.1016/j.chemosphere.2014.10.057 PMID:25461951

Zhu, D., Kwon, S., & Pignatello, J. (2005). Adsorption of single-ring organic compounds to wood charcoals prepared under different thermochemical conditions. *International Journal of Environmental Science and Technology*, *39*(11), 3990–3998. PMID:15984774

Zouboulis, A. I., & Moussas, P. A. (2011). Groundwater and Soil Pollution: Bioremediation. Encyclopedia of Environmental Health, 1037–1044.

# About the Contributors

**George Achieng'** is a Researcher in Environmental Chemistry and serves as Quality Assurance Manager at CSI International Ltd, Kenya, a private laboratory services provider.

**Somesh Ajnavi** is presently working as Assistant Professor in the Department of Environmental Engineering at Hindustan College of Science & Technology, Mathura. Holding Master Degree (M.Tech.) in Environmental Science & Technology from Thapar University, Patiala. Having experience of industry (about Two and half years), Academic & Research (more than 7 years) activities. Published 8 papers in National/International journals. Also reviewed 9 papers for Cogent Engineering (Taylor & Francis Publications) in last 3 years. Guided more than 15 undergraduate (B.Tech.) projects of Civil Engineering and Environmental Engineering students.

**Gomaa A. M. Ali** works as Assistant Professor at Chemistry Department, Faculty of Science, Al-Azhar University Egypt. He obtained his B.Sc. and M.Sc in Physical Chemistry from Al-Azhar University, Faculty of Science. He was awarded his Ph.D. in Advanced Nanomaterials for Energy Storage from UMP, Malaysia. He has 40 published papers, 2 patents and 10 medals (gold, platinum, and diamond). He joined more than 7 intentional conferences and many workshops. His research interest includes: Materials Science, Nanocomposite, Graphene, Supercapacitors, Water treatment. His M.Sc. thesis title is: Preparation and Characterization of Co/SiO2 nanocomposites of Technological Applications His Ph.D. thesis title is: Preparation and Characterization of Manganese-based and Carbon-Based Nanomaterials for Supercapacitors Application.

**Tabassum Ara** is Head, Department of Chemistry, National Institute of Technology, Srinagar. She received her Master's and Ph. D. degree in Chemistry from the University of Kashmir. She has more than two decades of experience in teaching in the Department of Pharmaceutical Sciences, Kashmir University and Department of Chemistry, NIT Srinagar. Her research activities have been in organic and environmental Chemistry. She has published various papers in International Journals.

**Rafia Bashir** is working as a Faculty in the Department of Chemistry at the National Institute of Technology, Srinagar since 2011. She received her Graduate degree from the University of Kashmir and her Master's and Ph.D. degree in chemistry from Jamia Hamdard, New Delhi. Her research activities have been organic synthesis, natural products, polymer, and environmental chemistry. Dr. Rafia has published various research papers in international reputed journals (cited in Scopus and Elsevier) and has attended many national as well as international conferences and workshops. Dr. Rafia is a member of many science academies and reviewer of many reputed journals.

**Smita Chaudhry** is Professor & Director, Institute of Environmental Studies, Kurukshetra University, Kurukshetra, Haryana, India, actively involved in its establishment; 1991-2007, served as Assistant and Associate Professor in Department of Bioscience & Biotechnology, Banasthali Vidyapeeth University, Rajasthan. She received her post graduation and doctoral from Kumaun University, Nainital, India. She has done pioneer work in restoration of landslides of Himalayas, published in reputed Journals. She has 27 years research experience & completed various research projects and guided Ph.D. and also a large number of M.Sc and M. Tech students in various fields of Restoration Ecology, Bioremediation and Phytoremediation, Carbon Sequestration & Climate Change, Impact Analysis, Landuse/Land cover changes, Agroecology, Hydropower & Environmental Sustainability, Energy Auditing & Management. She has more than 60 international and national publications in addition to project reports. She is the member of various International and National Societies and has organized a number of seminars and workshops. She has also chaired sessions in National and International Seminars. She is Fellow of National Institute of Ecology, New Delhi.

**Hamida-Tun-Nisa Chisti** is an Assistant Professor in Department of Chemistry at National Institute of Technology, Srinagar; where she has been working as faculty since 2008. She received her B.Sc Hons', Master's and her doctorate in Inorganic Chemistry from Aligarh Muslim University. Her research activities have been inorganic synthesis and environmental chemistry. Dr. Hamida is a member of many science academies and international societies. Her work has been published in various international journals and has been cited by a number of scientific journals. Dr Hamida has attended many national as well as international conferences, seminars and workshops.

**Shashi Kant Dubey** is an Assistant Professor in Department of Environmental Engineering at Hindustan College of Science and Technology, Farah (Mathura). He holds M.Tech degree in Environmental Science and Technology from Thapar University, Patiala. Mr. Dubey has about 10 years of experience in teaching undergraduate classes and research. He has also guided more than 12 B.Tech student projects. He has published about 10 papers in international referred journals. Mr. Dubey has is also working as reviewer for International journal of Environmental Technology, Cogent Engineering and Cogent Environmental Science.

**Shivani Garg**, since 2015, Assistant Professor at Institute of Environmental Studies, Kurukshetra University, Kurukshetra, Haryana, India. She has completed M.Tech (Energy and Environmental Management) and M.Sc.(Botany) from Kurukshetra University. Her main area research is connected with Wastewater Treatment, wastewater, energy management, biogas studies, waste management and environmental pollution. She has also presented the papers in National and International Seminars and conferences.

**P. Senthil Kumar** is working as an Associate Professor in the Department of Chemical Engineering, SSN College of Engineering, India, was born on 08th May, 1982 in Madurai, Tamilnadu, India. He has 10 years of teaching experience, 9 year of research experience and 1 year of industrial experience. He is a recognized Anna University Research Supervisor to guide Ph.D./M.S. candidates. Under his guidance, 03 Ph.D Scholars have already completed and 10 Ph.D. scholars are currently doing their active research in the area of Waste Water Treatment and Alternative Fuels. He has published more than 120+ research papers in highly reputed International Journals and 70 research papers in National/International

Conferences. He has also published more than 25+ book chapters in the reputed publishers by Elsevier, Springer, etc. He has organized/attended many externally/internally funded conferences /workshops / seminars at national and international level.

**Chispin Kowenje** is an Associate Professor in Physical, Materials, and Environmental Chemistry and Head of Department of Chemistry at Maseno University.

**Rizwana Mobin** is presently working as Assistant Professor in the Department of Industrial Chemistry, Government College for Women, Cluster University, Srinagar. She has received her Bachelor's, Masters' and Ph. D. degree from Aligarh Muslim University. She has been gold medallist at Master's level. Her research expertise includes thin-layer chromatographic analysis of surfactants, food dyes and their densitogram. She has also worked in the area of "Development of new methodologies involving green solvent system for the analysis of surfactants and food dyes". Several green chromatographic systems have been proposed for the identification of target analytes in various commercial and pharmaceutical products. She has published her work in number of peer review journals. She has presented her research at several national and international conferences. She was conferred P.D. Sethi Award 2014.

**Altaf Ahmad Najar** is presently working as Assistant Professor in the Department of Industrial Chemistry, Govt. College for Women, M.A.Road, Srinagar.He has completed Hon's in Industrial Chemistry from A.M.U. Aligarh in 2004. He received his Master's in Industrial Chemistry in 2006. He has worked as Teaching Assistant from 2007-2012.

**Tauseef Ahmad Rangreez** is working as SERB-Post Doctoral Fellow at National Institute of Technology, Srinagar. He has received his Master's degree from Jamia Hamdard University, New Delhi. He did his Ph. D from Aligarh Muslim University. His research expertise includes synthesis of organic-inorganic composite ion-exchangers, ion-exchange kinetics, dc conductivity, development of membrane sensors for heavy metals, proton conductivity, cyclic and linear sweep voltammetry and synthesis, characterization and application (loading of drugs) of biocompatible surfactant based hydrogels. He has published his work in several international reputed journals and attended and presented his work at both international and national conferences, seminars.

**Victor Shikuku** is an Assistant Lecturer in Department of Chemistry, Kaimosi Friends University College. Research interests includes application of microporous and mesoporous materials in water and wastewater treatment regimes; fabrication, Characterization and application.

# Index

## I



## L



## M



## N



## O



## P



## R



## S

## T



## U



## W

www.ingramcontent.com/pod-product-compliance
Lightning Source LLC
LaVergne TN
LVHW060424300926
849162LV00006B/1516
* 9 7 8 1 5 2 2 5 5 7 5 4 8 *